Trade-offs in Conservation

Conservation Science and Practice Series

Published in association with the Zoological Society of London

Wiley-Blackwell and the Zoological Society of London are proud to present our *Conservation Science and Practice* series. Each book in the series reviews a key issue in conservation today. We are particularly keen to publish books that address the multidisciplinary aspects of conservation, looking at how biological scientists and ecologists are interacting with social scientists to effect long-term, sustainable conservation measures.

Books in the series can be single or multi-authored and proposals should be sent to:
Ward Cooper, Senior Commissioning Editor, Wiley-Blackwell, John Wiley & Sons,
9600 Garsington Road, Oxford OX4 2DQ, UK
Email: ward.cooper@wiley.com

Each book proposal will be assessed by independent academic referees, as well as our Series Editorial Panel. Members of the Panel include:
Richard Cowling, Nelson Mandela Metropolitan University, Port Elizabeth, South Africa
John Gittleman, Institute of Ecology, University of Georgia, USA
Andrew Knight, University of Stellenbosch, South Africa
Georgina Mace, Imperial College London, Silwood Park, UK
Daniel Pauly, University of British Columbia, Canada
Stuart Pimm, Duke University, USA
Hugh Possingham, University of Queensland, Australia
Peter Raven, Missouri Botanical Gardens, USA
Helen Regan, University of California, Riverside, USA
Alex Rogers, Institute of Zoology, London, UK
Michael Samways, University of Stellenbosch, South Africa
Nigel Stork, University of Melbourne, Australia

Previously published

Urban Biodiversity and Design
Edited by Norbert Müller, Peter Werner and John G. Kelcey
ISBN: 978-1-4443-3267-4 Paperback; ISBN 978-1-4443-3266-7 Hardcover; 640 pages; April 2010

Wild Rangelands: Conserving Wildlife While Maintaining Livestock in Semi-Arid Ecosystems
Edited by Johan T. du Toit, Richard Kock and James C. Deutsch
ISBN: 978-1-4051-7785-6 Paperback; ISBN 978-1-4051-9488-4 Hardcover; 424 pages; January 2010

Reintroduction of Top-Order Predators
Edited by Matt W. Hayward and Michael J. Somers
ISBN: 978-1-4051-7680-4 Paperback; ISBN: 978-1-4051-9273-6 Hardcover; 480 pages; April 2009

Recreational Hunting, Conservation and Rural Livelihoods: Science and Practice
Edited by Barney Dickson, Jonathan Hutton and Bill Adams

ISBN: 978-1-4051-6785-7 Paperback; ISBN: 978-1-4051-9142-5 Hardcover; 384 pages; March 2009

Participatory Research in Conservation and Rural Livelihoods: Doing Science Together
Edited by Louise Fortmann
ISBN: 978-1-4051-7679-8 Paperback; 316 pages; October 2008

Bushmeat and Livelihoods: Wildlife Management and Poverty Reduction
Edited by Glyn Davies and David Brown
ISBN: 978-1-4051-6779-6 Paperback; 288 pages; December 2007

Managing and Designing Landscapes for Conservation: Moving from Perspectives to Principles
Edited by David Lindenmayer and Richard Hobbs
ISBN: 978-1-4051-5914-2 Paperback; 608 pages; December 2007

Conservation Science and Practice Series

Trade-offs in Conservation: Deciding What to Save

Edited by

Nigel Leader-Williams, William M. Adams
and Robert J. Smith

A John Wiley & Sons, Inc., Publication

This edition first published 2010, © 2010 by Blackwell Publishing Ltd

Blackwell Publishing was acquired by John Wiley & Sons in February 2007. Blackwell's publishing program has been merged with Wiley's global Scientific, Technical and Medical business to form Wiley-Blackwell.

Registered office: John Wiley & Sons Ltd, The Atrium, Southern Gate, Chichester, West Sussex, PO19 8SQ, UK

Editorial offices: 9600 Garsington Road, Oxford, OX4 2DQ, UK
The Atrium, Southern Gate, Chichester, West Sussex, PO19 8SQ, UK
111 River Street, Hoboken, NJ 07030-5774, USA

For details of our global editorial offices, for customer services and for information about how to apply for permission to reuse the copyright material in this book please see our website at www.wiley.com/wiley-blackwell

Library of Congress Cataloguing-in-Publication Data
Trade-offs in conservation : deciding what to save / edited by Nigel Leader-Williams, William M. Adams, and Robert J. Smith.
    p. cm. – (Conservation science and practice series)
  Based on presentations at a meeting held in London in November 2007.
  "Published in association with the Zoological Society of London."
  Includes bibliographical references and index.
  ISBN: 978-1-4051-9383-2 (pbk.: alk. paper) – ISBN 978-1-4051-9384-9 (hardcover: alk. paper)
  1. Wildlife conservation – Decision making – Congresses. 2. Wildlife conservation – Social aspects – Congresses. I. Leader-Williams, N. II. Adams, W. M. (William Mark), 1955– III. Smith, Robert J. (Robert James), 1971–
  QL81.5.T73 2010
  333.95′16 – dc22
                                                                2010016467

A catalogue record for this book is available from the British Library.

Set in 10.5/12.5 pt Minion by Laserwords Private Limited, Chennai, India
Printed and bound in Malaysia by Vivar Printing Sdn Bhd

1   2010

Contents

Contributors

William M. (Bill) Adams is Moran Professor of Conservation and Development at the University of Cambridge, where he has taught in the Department of Geography since 1984. His research focuses on social dimensions of conservation and the evolution of conservation policy and strategy, with an emphasis on conservation in the UK and in Africa. He is a Trustee of Fauna & Flora International. Department of Geography, University of Cambridge, Downing Place, Cambridge CB2 3EN, UK; wa12@cam.ac.uk.

Helen Anthem is Programme Manager for the Conservation Livelihoods and Governance Programme at Fauna & Flora International. Helen's expertise bridges the development and environment sectors and her research interests include the impact of conflict on livelihood options for women, as well as issues of human rights in relation to the environment. Her recent work has included analysis of environmental partnerships in post-conflict and post-disaster situations in Eurasia, Asia-Pacific, Africa and the Americas. Fauna & Flora International, Jupiter House, Station Road, Cambridge CB1 2JD, UK; helen.anthem@fauna-flora.org.

Rosalind Aveling is Deputy Chief Executive Officer at Fauna & Flora International. A primatologist, Rosalind has set up conservation initiatives for Great Apes in both Indonesia and Central Africa. Her focus is on conservation within landscapes that include human use and, as former Director of Program Design with the African Wildlife Foundation, she has concentrated on issues of environmental governance and sustaining conservation finance. Within FFI, she has established the division of Conservation Partnerships, which explores the economic, social, ecological and institutional determinants of effective conservation. Her current interests include maintaining a focus on biodiversity within climate adaptation and mitigation strategies. Fauna & Flora International, Jupiter House, Station Road, Cambridge CB1 2JD, UK; rosalind.aveling@fauna-flora.org.

Andrew Balmford is Professor of Conservation Science in the University of Cambridge. His main research interests are the costs and benefits of effective conservation, quantifying the changing state of nature, conservation planning, evaluating the success of conservation interventions, and exploring how best to reconcile conservation and farming, especially in developing countries. He tries to tackle these problems through collaborations across disciplines and with conservation practitioners. He helped found the Cambridge Conservation Forum and the Cambridge Conservation Initiative, which is building carefully targeted collaborations among 10 organizations, as well as the Student Conference on Conservation Science, which each year attracts ~200 young scientists from ~60 countries. Conservation Science Group, Department of Zoology, University of Cambridge, Downing Street, Cambridge CB2 3EJ, Cambridge, UK; a.balmford@zoo.cam.ac.uk.

Michael Bode is Research Fellow at the Applied Environmental Decision Analysis (AEDA) Research Hub, University of Melbourne. He is primarily interested in quantitative conservation decision making. Applied Environmental Decision Analysis Research Hub, Department of Botany, University of Melbourne, Melbourne, Victoria 3010, Australia; bodem@unimelb.edu.au.

J. Peter Brosius is Professor of Anthropology at the University of Georgia, and Director of the Center for Integrative Conservation Research. His research focuses on political ecology and the cultural politics of conservation at both local and global scales. He previously worked on international environmental politics in Sarawak, with an early focus on Penan, and more recently with the Kelabit community. With the ACSC initiative, Pete focuses on global conservation and the politics of scale, with a particular focus on ecoregional planning and conservation finance. He was past president of the Anthropology and Environment Section, American Anthropological Association, and is a member of the IUCN Commission on Economic, Environmental and Social Policy (CEESP) Co-Management Working Group and the World Commission on Protected Areas/CEESP Theme on Indigenous and Local Communities, Equity and Protected Areas (TILCEPA). He was awarded the Lourdes Arizpe Award in Anthropology and Environment in 2005. Center for Integrative Conservation Research, Department of Anthropology, University of Georgia, 250A Baldwin Hall, Jackson Street, Athens, GA 30602-1619, USA; pbrosius@uga.edu.

Aaron Bruner is Director of Economic Incentives and Protected Area Finance at Conservation International. Aaron's work focuses on improving the effectiveness of protected area management and on developing the use of conservation agreements as a means to engage communities in conservation. Past work has covered a range of scales and issues, including analyses of protected area effectiveness and the cost of moving towards effective protected area management, economic assessments of a range of forest products, and analysis of strategies for increasing the contribution of the tourism industry to biodiversity conservation. He holds a Bachelor's degree in Economics from Wesleyan University. Conservation International, 2011 Crystal Drive, Suite 500, Arlington, VA 22202, USA; a.bruner@conservation.org.

Philip Bubb is Senior Programme Officer at the United Nations Environment Programme World Conservation Monitoring Centre (UNEP-WCMC). Philip has 20 years' experience in project management, design and training for international biodiversity conservation. At UNEP-WCMC, he works with assessments and indicators, including in the UK National Ecosystem Assessment. He also supports countries and organizations to develop and use biodiversity indicators, building on experience as manager of the Biodiversity Indicators in National Use project from 2002 to 2005. He is particularly involved in capacity building, including training and the production of guidelines in biodiversity assessment and monitoring for protected areas. Prior to UNEP-WCMC, Philip worked for 7 years in Mexico on forest conservation and rural development projects. UNEP World Conservation Monitoring Centre, 219c Huntingdon Road, Cambridge CB3 0DL, UK; philip.bubb@unep-wcmc.org.

Peter Carey is Director of Bodsey Ecology Limited. Pete is an independent academic researcher specializing in biogeography, climate change and the evaluation of agri-environment schemes. He is an Affiliated Lecturer in the Department of Plant Sciences at the University of Cambridge. He spent 18 years at ITE/CEH Monks Wood, where he worked on The Countryside Survey from 2004 to 2009 and was the lead author of the National Reports. Bodsey Ecology Limited, 4 Bodsey Cottages, Ramsey, Cambridgeshire PE26 2XH, UK; bodsey.ecology@btinternet.com.

Richard M. Cowling is Professor of Botany at Nelson Mandela Metropolitan University. Richard has published extensively – in the scientific and popular literature – on the ecology and conservation of the South Africa's fynbos, succulent karoo and subtropical thicket biomes. He is widely acclaimed for his contribution to the theory and application of conservations science. He was rated a world leader in conservation science by the National Research Foundation, and has been awarded a Pew Fellowship, a Distinguished Service Award by the Society for Conservation Biology, a Gold Award for Innovating Conservation by the Cape Action for People and the Environment, and a Flora Conservation Award by the

Botanical Society of South Africa. He was elected Foreign Associate of the National Academy of Sciences USA. Department of Botany, Nelson Mandela Metropolitan University, PO Box 77000, Port Elizabeth 6031, South Africa; rmc@kingsley.co.za.

Gretchen C. Daily is Bing Professor of Environmental Science and Director of the Center for Conservation Biology at Stanford University, and Senior Fellow in the Woods Institute for the Environment. She is also Chair of The Natural Capital Project, a partnership among The Nature Conservancy, World Wildlife Fund and Stanford University, whose goal is to align economic forces with conservation. An ecologist by training, Gretchen's work spans scientific research, teaching, public education and working with leaders to advance practical approaches to environmental challenges. Department of Biological Sciences, Stanford University, 371 Serra Mall, Gilbert Hall, Stanford, CA 94305-5020, USA; gdaily@stanford.edu.

Juan Luis Dammert is Assistant Professor at the Pontificia Universidad Católica del Peru (PUCP). He holds a BA in Sociology from PUCP. His research interests include decentralization, political representation, local conflicts and citizen participation. He is currently developing projects for the PUCP on 'Political representation in Latin America', and the Instituto de Estudios Peruanos on 'Gobiernos locales y conflictividad social en la Región Puno'. He has published articles in academic journals such as *Debate Agrario* on 'Participación, concertación y confrontación en espacios locales: el Caso de la Mesa de Concertación para la Lucha Contra la Pobreza del Departamento de Puno', co-authored with Aldo Panfichi (2006), and in *Palestra* on 'Análisis de la Representación Congresal por Regiones' in 2006. Department of Social Sciences, Catholic University of Peru, Avenida Universitaria 1801, San Miguel, Lima 32, Peru; jdammert@spda.org.pe.

Abigail Entwistle is Director of Science at Flora & Fauna International, where she has worked for over 13 years, after completing a PhD on bats. During this time she has held a range of positions, and has undertaken direct conservation management across a wide range of countries and environments. She has a particular interest in conservation planning and impact monitoring. Fauna & Flora International, Jupiter House, Station Road, Cambridge CB1 2JD, UK; abigail.entwistle@fauna-flora.org.

Rebecca L. Goldman is Senior Scientist in the Central Science Division at The Nature Conservancy. Rebecca joined TNC in 2008, after completing her PhD at Stanford University in Environment and Resources. Her main responsibilities include working with scientists on the Board of Directors, and on ecosystem services topics and projects. Most recently her focus has been on water fund projects in Latin America. Specifically, she is helping design technical monitoring plans for these ecosystem services-based projects to ensure that conservation investments are effective and is analyzing the potential for replicating the water fund approach globally. The Nature Conservancy, 4245 N Fairfax Drive, Suite 100, Arlington, VA 22203, USA; rgoldman@tnc.org.

Hedley Grantham is Research Fellow at the University of Queensland. He is a conservation scientist with research interests in conservation decision making in marine and terrestrial ecosystems. He is currently working on marine spatial planning projects in Indonesia and South Africa. Ecology Centre, University of Queensland, Brisbane, Queensland 4072, Australia; h.grantham@uq.edu.au.

Stuart R. Harrop is Professor of Wildlife Management Law at the University of Kent, and since 2009 Director of the Durrell Institute of Conservation and Ecology. His research interests encompass the full spectrum of the regulation of human relationships with wildlife. The context of his research is predominantly the international policy and regulatory matrix with occasional excursions into the regional and national. Durrell Institute of Conservation and Ecology, School of Anthropology and Conservation, Marlowe Building, University of Kent, Canterbury CT2 7NR, UK; s.r.harrop@kent.ac.uk.

Jon Hoekstra is Managing Director of the Global Climate Change Program at The Nature Conservancy. Jonathan provides strategic and scientific leadership for policy and field-based efforts to reduce greenhouse gas emissions by protecting and restoring forests, and to help people and nature adapt to unavoidable climate change impacts. Jonathan collaborates with experts from around the world to develop innovative, practical solutions to climate change problems based on top-notch science and real-world experience. Jonathan previously directed TNC's Emerging Strategies Unit and led the Conservancy's Global Habitat Assessment Team. He earned BS and MS degrees from Stanford University, and a PhD from the University of Washington. The Nature Conservancy, 1917 First Avenue, Seattle, WA 98101, USA; jhoekstra@tnc.org.

David G. Hole was in the School of Biological and Biomedical Sciences at Durham University, but is now Climate Change Researcher at Conservation International. His research focuses principally on the nexus between biodiversity, ecosystem services and human well-being, and seeks to build understanding across these still disparate fields, focusing particularly on the consequences of climate-driven shifts in biodiversity on the mitigation and adaptation potential of conservation resources (e.g. protected areas) and ecosystems more widely. Understanding how to increase the resilience of both ecosystems and human communities to climate change, and other global change drivers, is a principal goal of his research. Conservation International, 2011 Crystal Drive, Suite 500, Arlington, VA 22202, USA; d.hole@conservation.org.

Katherine M. Homewood is Professor in Anthropology at University College London. She works on the interplay between conservation and development, with a focus on African pastoralist people and savanna environments. Her Human Ecology Research Group integrates natural and social sciences approaches to analyze implications of developing country rural resource use and land use change for environment and wildlife on the one hand, and of environmental conservation policy and management for human welfare and livelihoods on the other. Recent publications include *Ecology of African Pastoralist Societies* published by James Currey and Ohio University Press in 2008, and *Staying Maasai: livelihoods, conservation and development in East African rangelands* published by Springer in 2009. Department of Anthropology, University College London, Gower Street, London WC1E 6BT, UK; k.homewood@ucl.ac.uk.

John Hopkins is Principal Specialist at Natural England. John has worked in the statutory conservation sector for nearly 30 years, including roles as Grassland Specialist, Head of Biotopes Team and Scientific Advisor. In his current role, John is responsible for developing an ecosystems approach to the work of Natural England, the government's statutory advisor on biodiversity and landscape conservation in England. Natural England, Northminster House, Peterborough, Cambridgeshire PE1 1UA, UK; john.hopkins@naturalengland.org.uk.

Brian Huntley is Professor in the School of Biological and Biomedical Sciences at Durham University. His research investigates the relationships between environmental change and changes in the distributions of organisms, as well as in the composition, structure and dynamics of ecosystems. He also studies the interactions between the land surface and the atmosphere, especially biospheric feedbacks to the climate system. His research uses a combination of palaeoecological, ecological and biogeographic methods, including techniques like pollen analysis, palaeoclimate reconstruction, modelling of ecological processes and of species distributions, geographic information systems and Earth observation. In 2007, Brian acted as a consultant to the Council of Europe Bern Convention Group of Experts on Biodiversity and Climate Change and is now an observer on this group. In 2008, Brian joined the Steering Group for the North-East Regional Climate Change and Biodiversity Study. Centre for Ecosystem Sciences, School of Biological and Biomedical Sciences, Durham University, South Road, Durham DH1 3LE, UK; brian.huntley@durham.ac.uk.

Valerie Kapos is Senior Programme Officer at the United Nations Environment Programme World Conservation Monitoring Centre (UNEP-WCMC). Valerie trained as a tropical forest ecologist and conducted field research in Latin America and the Caribbean for 15 years, including studies of the ecological effects of forest fragmentation in Amazonia and remote sensing of forest cover change. She has since focused on applying ecological knowledge to support policy and practice in conservation and natural resource management. Her recent work has examined development and use of biodiversity indicators to support policy and decision making at international and national scales, and developing measures of the effectiveness of conservation action. UNEP World Conservation Monitoring Centre, 219c Huntingdon Road, Cambridge CB3 0DL, UK; val.kapos@unep-wcmc.org.

Peter Kareiva is Chief Scientist at The Nature Conservancy. Peter moved to TNC after working as a university professor and at NOAA. His past publications and research have concerned such diverse fields as mathematical biology, fisheries science, insect ecology, risk analysis, population viability analysis, landscape ecology and global climate change. Peter maintains connections with several universities, and still advises students, as well as teaching courses on occasion. His responsibilities at TNC include reporting to the Board of Directors on science at TNC, mentoring TNC scientists, identifying opportunities and shortcomings that warrant science attention, advising leadership on emerging conservation challenges, and serving as one of several external spokespeople for TNC Science. Central Science, The Nature Conservancy, 4722 Latona Avenue NE, Seattle, WA 98105, USA; pkareiva@tnc.org.

Andrew T. Knight is Senior Lecturer at the University of Stellenbosch. During 7 years as a conservation planner with the New South Wales National Parks and Wildlife Service, Australia, Andrew assisted planning expansion of the state's reserve network, and co-developed conservation plans for priority bioregions. He completed a PhD examining implementation of conservation plans at the Nelson Mandela Metropolitan University in

Port Elizabeth, South Africa, which included a 2-year post as Implementation Specialist for a conservation plan. His research interests include spatial prioritizations fusing social and ecological information, social learning institutions promoting adaptive management, and the gap between research and implementation. Department of Conservation Ecology and Entomology, University of Stellenbosch, Matieland 7602, Stellenbosch, Western Cape, South Africa; tawnyfrogmouth@gmail.com.

Annette Lanjouw heads the grant making, convening and strategy formulation of the Arcus Foundation's Great Apes Fund, the world's largest private funder of Great Ape conservation. Annette has worked with chimpanzees, bonobos and gorillas in the wild. She joined the Arcus Foundation after gaining years of previous experience in the areas of conservation strategy, programme implementation, research and fieldwork. Most recently she served as International Program Officer to the Howard G. Buffett Foundation. She previously served for 10 years as Director of the International Gorilla Conservation Programme (IGCP) – a three-way collaboration of WWF, African Wildlife Foundation and Fauna & Flora International, which successfully worked to secure safety for nearly 800 mountain gorillas that range across the three countries of Rwanda, Democratic Republic of Congo (DRC) and Uganda – and as Project Manager/Field Director for the Frankfurt Zoological Society's Chimpanzee Conservation Project in eastern DRC. Arcus Foundation, Wellington House, East Road, Cambridge CB1 1BH, UK; annette@arcusfoundation.org.

Nigel Leader-Williams was Professor of Biodiversity Management at the University of Kent, and Director of the Durrell Institute of Conservation and Ecology until 2009. He is currently Director of Conservation Leadership in the Department of Geography at the University of Cambridge. His research focuses on issues related to sustainable use and human–wildlife conflict. Department of Geography, University of Cambridge, Downing Place, Cambridge CB2 3EN, UK; nl293@cam.ac.uk.

Georgina M. Mace is Professor of Conservation Science at Imperial College London. Her research interests are in the assessment of biodiversity and the implications of its loss. Centre for Population Biology, Imperial College London, Silwood Park, Ascot, Berkshire SL5 7PY, UK; g.mace@imperial.ac.uk.

Douglas C. MacMillan is Professor in Conservation and Applied Resource Economics, and Head of the School of Anthropology and Conservation, at the University of Kent. His research interests lie in the economics of wildlife conservation, ecological modelling, and collaboration in land and wildlife management and forest resources. Durrell Institute of Conservation and Ecology, School of Anthropology and Conservation, Marlowe Building, University of Kent, Canterbury CT2 7NR, UK; d.c.macmillan@kent.ac.uk.

Andrea Manica is Senior Lecturer in Population Biology at the University of Cambridge. As well as working on biological problems, he is interested in questions at the boundary between biology and the social sciences. Current topics of research are which factors predict sustainable exploitation of wild resources, methods of assessing the success of conservation initiatives, and the quantification of factors that affect public attitudes towards environmental issues, such as biodiversity, conservation and climate change. Department of Zoology, University of Cambridge, Downing Street, Cambridge CB2 3EJ, UK; am315@cam.ac.uk.

Bruno Monteferri is a lawyer from the Peruvian Society for Environmental Law (SPDA) and co-ordinator of SPDA's decentralized office at Iquitos in Loreto. Bruno graduated from the Faculty of Law of the Pontificia Universidad Católica del Peru (PUCP) with a spell at Barcelona University, Spain. He holds a Diploma in Integral Management of Coastal Marine Areas from Guadalajara University, Mexico. He specializes in legal environmental issues and especially on governance, and public and private conservation mechanisms. He is a member of the Initiative for Private Conservation Team, of the Private Conservation Network and the Management Committee of Pacaya Samiria National Reserve. He has published several articles on the analysis of legal conservation opportunities and participated in two books *Management Committees: constructing governance for the natural protected areas from Peru* and *Essay of the National Context: the coast and its people.* Peruvian Society for Environmental Law, Prolongación Arenales 438, Lima 27, Peru; bmonteferri@spda.org.pe.

Teresa Mulliken is Co-ordinator of Programme Development and Evaluation at TRAFFIC International. Teresa has a degree in biology from the University of Maryland and an MSc in Environment and Development from the University of East Anglia. She has worked for TRAFFIC since 1989 on research and advocacy on a wide range of wildlife trade issues, including the wild bird trade, trade in medicinal and aromatic plants, the development implications of wildlife trade, and implementation of the Convention on International Trade in Endangered Species of Wild Fauna and Flora (CITES). She is specifically interested in increasing understanding of the factors driving unsustainable and illegal use of wild resources and the effectiveness of different approaches to bringing harvest and trade within sustainable levels. TRAFFIC International, 219a Huntingdon Road, Cambridge CB3 0DL, UK; teresa.mulliken@traffic.org.

William Murdoch is Professor at the University of California at Santa Barbara. Bill gained his BSc at Glasgow University and his DPhil at Oxford University. His research has involved mainly theory and experiments in population regulation and predator–prey dynamics. He became interested in efficient decision making in conservation when serving on the Board of Directors of The Nature Conservancy. Department of Ecology, Evolution and Marine Biology, University of California, Santa Barbara, CA 93106, USA; murdoch@lifesci.ucsb.edu.

Eduard T. Niesten is Senior Director of Economics and Planning at Conservation International. Eduard received his PhD in Applied Economics from Stanford University, with a focus on natural resource economics as well as agricultural and development economics. Before joining Conservation International, he was an Associate with Hardner and Gullison, where he conducted feasibility studies and cost assessments for conservation incentive agreements throughout the tropics. He currently concentrates on incentive-based conservation approaches, and promoting awareness and understanding of such approaches. He also supplies economics expertise to projects implemented by other CI programmes and partners, such as estimating costs of protected area management in Liberia and sustainable development planning in Suriname. Conservation International, 2011 Crystal Drive, Suite 500, Arlington, VA 22202, USA; e.niesten@conservation.org.

Steve Polasky is Fesler-Lampert Professor of Ecological/Environmental Economics at the University of Minnesota. Stephen received a PhD in Economics from the University of Michigan. His research interests include ecosystem services, natural capital, biodiversity conservation, endangered species policy, integrating ecological and economic analysis, common property resources and environmental regulation. He was Senior Staff Economist for Environment and Resources for the President's Council of Economic Advisers. He is a Fellow of the American Academy of Arts and Sciences, and of the American Association for the Advancement of Science. Department of Applied Economics and Department of Ecology, Evolution and Behavior, University of Minnesota, 1994 Buford Avenue, St Paul, MN 55112, USA; polasky@umn.edu.

Hugh P. Possingham is Director of the Ecology Centre and ARC Federation Fellow in Maths and Biology at the University of Queensland. Hugh accidentally completed a BSc in Applied Mathematics in 1984 with top honours and gained a DPhil as a Rhodes Scholar at Oxford University in 1987. He became Professor and Chair at Adelaide University in 1995, and Director of the Ecology Centre at the University of Queensland in 2000. He was elected Fellow of the Australian Academy of Sciences in 2005. He has co-authored 300 papers generating 5000 citations. His research helped stop land clearing in Queensland and NSW, thereby securing at least 1 billion tonnes of carbon dioxide. He used Marxan to rezone the Great Barrier Reef. This software is now used in 100 countries, and is changing the face of the planet. Hugh suffers from obsessive bird watching. He campaigns on biodiversity issues, directly informs policy, deals with media and advises government every day. Ecology Centre, University of Queensland, Brisbane, Queensland 4072, Australia; h.possingham@uq.edu.au.

Manuel Pulgar-Vidal has been the executive director of the Peruvian Society for Environmental Law (SPDA) since 1994. SPDA is arguably one of the most important and influential environmental law organizations in Latin America. Manuel's areas of work and expertise include environmental policy, with an emphasis on promoting dialogue between the public and private for-profit and non-for-profit sectors, and pollution prevention in productive sectors, especially mining and fisheries. He works frequently as a consultant for national and international organizations on environmental policy in Peru and throughout Latin America. He has been a conference speaker at numerous national and international fora. Manuel was a visiting professor in the North South Center at the University of Miami from 1999 to 2000, and is currently Professor of Environmental Law at the Universidad Católica del Perú and Universidad Peruana de Ciencias Aplicadas. Manuel is currently a Board Member of FONDEBOSQUE – the national fund for sustainable management of forests – and was on the Board of PROFONANPE – the trust fund for protected areas in Peru – until 2000. He also chaired the Board of the Permanent Seminar for Agrarian Research (SEPIA) from 2003 to 2005. Peruvian Society for Environmental Law, Prolongación Arenales 438, Lima 27, Peru; mpulgar-vidal@spda.org.pe.

Richard E. Rice is Chief Conservation Officer at Save Your World. Richard has more than 20 years' experience in natural resource and public policy analysis, most recently as Chief Economist at Conservation International. While at CI, he conducted extensive research on the costs and effectiveness of different approaches to tropical biodiversity conservation. He has supervised research projects and protected area development throughout the tropics. He has published widely on the economics of sustainable forest management and was instrumental in pioneering CI's first conservation concession – an approach to habitat protection that involves annual payments to resource owners in exchange for long-term commitments to conservation. Save Your World, LLC, 7404 Cedar Avenue, Takoma Park, MD 20912, USA; Richarderice@gmail.com>

Dilys Roe is Senior Researcher in the Natural Resources Group at the International Institute of Environment and Development. Dilys Roe has worked for IIED – a sustainable development policy research institute based in London – since 1992. Her work focuses on exploring the links between biodiversity conservation and poor people's livelihoods and informing policy in this field. Specific activities have included co-ordinating an international 'Learning Group' on poverty–conservation linkages; working with international conservation organizations to explore the development of a human rights charter to guide practice in developing countries; reviewing the impacts and achievements of community-based wildlife management in Africa; and providing technical support on biodiversity to the UK Department for International Development. International Institute for Environment and Development, 3 Endsleigh Street, London WC1H 0DD, UK; dilys.roe@iied.org.

Alison M. Rosser was until recently Lecturer in Biodiversity Conservation at the University of Kent. She has worked in the NGO sector for TRAFFIC in Tanzania and for the IUCN Species Programme responsible for much of IUCN's input to the Convention on International Trade in Endangered Species of Wild Fauna and Flora (CITES). Her interests include investigating means to ensure that the use of wild species will be sustainable and developing pragmatic methods to support conservation decision making. She is now Senior Programme Officer in the Biodiversity, Biomass and Food Security Programme at the United Nations Environment Programme World Conservation Monitoring Centre (UNEP-WCMC). UNEP World Conservation Monitoring Centre, 219c Huntingdon Road, Cambridge CB3 0DL, UK; alison.rosser@unep-wcmc.org.

Roger Safford is Senior Programme Manager at BirdLife International. Roger Safford has worked on tropical conservation programmes and projects since 1988. He spent 5 years in the field in Mauritius, Madagascar and other Western Indian Ocean islands with special attention to the endemic birds of Mauritius; this region and its biodiversity has remained a focus ever since. Roger's other main areas of work have been on wetlands and rainforests and their ecology and conservation, and building the capacity of national conservation institutions in developing countries, particularly at BirdLife where he has worked since 2000. BirdLife International, Wellbrook Court, Girton Road, Cambridge CB3 0NA, UK; roger.safford@birdlife.org.

Michael J. Samways is Professor and Chair of the Department of Conservation Ecology and Entomology, University of Stellenbosch. Michael completed his PhD at London University. His research is mainly aimed at designing landscapes for the future. Michael is very involved internationally, especially representing invertebrates. He has published eight books, three special issues, 43 book chapters and peer-reviewed scientific papers. He is a Fellow of Royal Society of South Africa, and Member of the Academy of Science of South Africa. He is a John Herschel Medallist of the Royal Society of South Africa, Senior Captain Scott Medallist of the South African Academy of Sciences and Arts, and Gold Medallist of the Academy of Science of South Africa. Department of Conservation Ecology and Entomology, University of Stellenbosch, Matieland 7602, Stellenbosch, South Africa; samways@sun.ac.za.

Robert J. (Bob) Smith is Research Fellow at the University of Kent and a Senior Fellow at the United Nations Environment Programme World Conservation Centre (UNEP-WCMC). Much of his work focuses on designing conservation landscapes and protected area networks, especially as part of long-term projects in southeast Africa and the English Channel. He has worked on projects in 14 countries in Africa, Asia and Europe and this has given him a broad interest in the factors that affect conservation policy and practice. In particular, he has published work on the impacts of corruption, measuring project effectiveness, the role of positive incentives and how marketing influences the conservation agenda. Durrell Institute of Conservation and Ecology, School of Anthropology and Conservation, Marlowe Building, University of Kent, Canterbury CT2 7NR, UK; r.j.smith@kent.ac.uk.

Alison Stattersfield is Head of Science, Policy and Information at BirdLife International. Ali has had a life-long interest in wildlife and conservation. After completing a BA in Zoology at Cambridge University in 1978, she worked as a teacher before joining the International Council for Bird Preservation (BirdLife International's precursor) as a research assistant in 1986. She was involved in the first comprehensive evaluation of the Red List status of the world's birds, and in pioneering work on broad-scale priority setting using Endemic Bird Areas. She is now Head of Science at

BirdLife's Secretariat with an interest in using scientific analyses to guide advocacy and policy decisions. BirdLife International, Wellbrook Court, Girton Road, Cambridge CB3 0NA, UK; ali.stattersfield@birdlife.org.

Diogo Veríssimo is a PhD student at the Durrell Institute of Conservation and Ecology (DICE) in the University of Kent. His first job in conservation was as an educator and guide at the Lisbon Zoological Park in Portugal. He completed a BSc in Environmental Biology at Lisbon University and an MSc in Conservation Biology at DICE. His research interests lie at the interface between human and wildlife communities but also in applied terrestrial and marine ecology. Diogo is currently undertaking research on conservation marketing, focusing on optimizing the use of flagship species in different contexts. Durrell Institute of Conservation and Ecology, School of Anthropology and Conservation, Marlowe Building, University of Kent, Canterbury CT2 7NR, UK; dv38@kent.ac.uk.

Matthew J. Walpole is Head of the Ecosystem Assessment Programme at the United Nations Environment Programme World Conservation Monitoring Centre (UNEP-WCMC). Matt is a conservation biologist with almost two decades of multidisciplinary research, consultancy, teaching and project/programme management experience at the interface between biodiversity conservation and rural livelihoods. He has worked and travelled in over 50 countries, with a particular focus on Africa and Asia. After leading the development of a global programme on biodiversity and human needs for Fauna & Flora International, Matt joined UNEP-WCMC where he heads a team developing biodiversity and ecosystem service indicators and assessment tools and products for national and intergovernmental policy makers. UNEP World Conservation Monitoring Centre, 219c Huntingdon Road, Cambridge CB3 0DL, UK; matt.walpole@unep-wcmc.org.

Stephen G. Willis is Lecturer at Durham University. His recent research examines the role of climate and habitat in determining species distributions, with a major emphasis on examining the effects of environmental change, particularly global climate change, on ecosystems and the mechanisms by which environmental change acts upon species. Stephen also examines how environmental change impacts upon factors such as biodiversity and causes range shifts in both native and invasive species. Most of his research in this area involves ecological modelling using spatially explicit models and GIS, often incorporating remote-sensed data, although he also undertakes experimental manipulations in the field. Centre for

Ecosystem Sciences, School of Biological and Biomedical Sciences, Durham University, South Road, Durham DH1 3LE, UK; s.g.willis@durham.ac.uk.

Kerrie A. Wilson is Senior Lecturer at the University of Queensland. She holds a BSc in Environmental Science from University of Queensland (University Medallist in 1999) and obtained a DPhil in Ecology from the University of Melbourne in 2004. Her research into the socioeconomic aspects of conservation involves collaborations with national and international government and non-government organizations. She has authored ~50 scientific publications, including in *Science* and *Nature*, and edited one book. She was awarded an Australian Leadership Award and European Erasmus Mundus Fellowship in 2009. Kerrie has previously held leadership positions with NGOs, including as Director of Conservation for The Nature Conservancy, Australia. University of Queensland, School of Biological Sciences, Brisbane, Queensland 4072, Australia; k.wilson2@uq.edu.au.

Preface and Acknowledgments

Conservation practitioners aim to address one of the greatest challenges facing human kind during the 21st century, that of saving global biodiversity and all its associated services. In order to conserve biodiversity, conservationists make trade-offs on a regular basis during the course of their work. Yet such trade-offs are only recognized infrequently, nor debated explicitly. This book is based on presentations made at a 2-day meeting entitled 'Trade-offs in conservation: deciding what to save', held in London in November 2007, that sought to address the important issue of trade-offs in conservation. This meeting was a Zoological Society of London (ZSL) Symposium in Conservation Biology that was organized by the editors of this book. In particular, the editors would like to thank the presenters of talks and authors of chapters for the diligence and panache that they have brought to talking and writing about their allotted topics.

We would like to thank ZSL very warmly for hosting the meeting that led to this book. Joy Hayward and Linda DaVolls of ZSL worked tirelessly and with great good grace to pull off a faultlessly organized 2 days. They, the organizers and the speakers were rewarded with a full house and lots of interesting discussion on this important topic. We are very grateful to Dan Brockington, Tim Coulson, E-J. Milner-Gulland and Mark Wright for chairing sessions that ran perfectly to time. Cathy Dean, Toby Gardner, Bruno Monteferri and Ana Rodrigues made thought-provoking summary presentations at the end of the meeting to crystallize thoughts on the previous 2 days' proceedings. Rosalind Aveling of Fauna & Flora International and Sheila O'Connor of Advancing Conservation in a Social Context provided generous sponsorship for the meeting and much moral support to the organizers.

We are most grateful to Ward Cooper, Delia Sandford, Kelvin Matthews and Rosie Hayden at Wiley-Blackwell for their support while producing this book. We also thank Jane Andrew for her careful editing of the book. Philip Stickler and David Watson at the Department of Geography in the University of Cambridge drew the final maps and diagrams with great skill. Finally, we thank our own institutions for their support in running the meeting and producing the book: the Durrell Institute of Conservation and Ecology at the University of Kent, and the Department of Geography at the University of Cambridge.

Nigel Leader-Williams, Bill Adams and Bob Smith
February 2010

Introduction

(**1**)

Deciding What to Save: Trade-offs in Conservation

Nigel Leader-Williams[1], William M. Adams[2]
and Robert J. Smith[1]

[1]Durrell Institute of Conservation and Ecology, University of Kent,
Canterbury, UK
[2]Department of Geography, University of Cambridge, Cambridge,
UK

Introduction

Conservation action inevitably involves choices, between the populations of different species and the states of various ecosystems, between preservation and transformation by economic forces, between the needs of people and those of other species, between the interests of some people over others. However, these choices are rarely explicitly recognized or debated by practising conservationists (Mace *et al.*, 2007). To encourage further improvements in their professional practices, conservationists need to be more explicit about what hidden choices are made in their conservation policies. Furthermore, conservationists need to carefully weigh up the trade-offs that they make every day in deciding what to save.

Setting goals and policies for conservation is increasingly seen as a scientific activity, where outcomes should follow rational, indeed evidence-based, choices (Sutherland *et al.*, 2004). However, different kinds and pieces of evidence may suggest different strategies, and conservation planners rarely

Trade-offs in Conservation: Deciding What to Save, 1st edition.   Edited by N. Leader-Williams,
W.M. Adams and R.J. Smith.   © 2010 Blackwell Publishing Ltd.

have full information. In practice, different factors or philosophical positions may constrain or influence the choices made. Very often trade-offs are made, whether consciously or unconsciously, for example by selecting some species and ecosystems for conservation action, while abandoning others, or by taking one approach to conservation while ignoring others.

Such trade-offs are surprisingly common, indeed probably universal, in conservation. However, practitioners may be slow to recognize them, and often reluctant to draw attention to them as would be necessary to understand them better. But does this matter? The answer to that question depends on what further biodiversity is being lost because of the trade-offs conservationists make. Conservationists certainly need to understand how and why trade-offs are made, and need to think very hard about what, if anything, to do about them. These are the questions that this book seeks to address. More specifically, the book explores how to manage conservation responsibly in a world of trade-offs. In particular, we wish to ask:

- Are choices in conservation explicitly recognized or debated?
- What choices remain hidden in conservation policy?
- How can the trade-offs that are made daily be carefully weighed up?

The evidence of trade-offs is everywhere once you look. By way of an everyday example, a visitor display board has been erected on a boardwalk at Heron's Carr, a not greatly inviting woodland edge of a not particularly well-known area of wetland in Britain's Norfolk Broads. This board (Figure 1.1) notes that '*London Zoo has about 690 species of animal and that's including everything!*' The board goes on to say '*The scientists who studied this wood before the boardwalk was built reckon there are more than 1500 different sorts of invertebrates – that's insects, spiders, beetles, flies, snails, worms and so on – jumping and creeping and flying and crawling and swimming around. This means Heron's Carr is a real wildlife treasure house.*' In seeking to further emphasize the importance of Heron's Carr, the board also notes that only 210 different breeding birds and 1400 flowering plants are recorded in the whole of the British Isles! This display board unfolds a whole series of subtle trade-offs to Heron Carr's advantage, those between *in situ* and *ex situ* conservation, those between the attention devoted to more charismatic species than to creepy crawlies, and to those between local and national conservation objectives.

This chapter outlines the thinking behind this book, and the original symposium upon which it is based, held in November 2007 at the Zoological

Figure 1.1   **A signboard at Heron's Carr in the Norfolk Broads, UK, that outlines a series of subtle trade-offs to the advantage of Heron's Carr. (Photograph by Nigel Leader-Williams.)**

Society of London. Here, we seek to ensure that the title of the volume and the issues it addresses are understood. This first section of the book explains some of the terms that appear throughout this volume. The second section seeks to understand some of the current approaches and toolkits used in conservation. The third section examines the influence of different value systems in setting conservation objectives. The fourth section examines issues related to economics and governance. The fifth section addresses some key institutional constraints. The final section examines emerging drivers of biodiversity loss.

Understanding terms

The first key term to understand is that of *conservation*. A simple definition suggests that '*conservation comprises actions that directly enhance the chances of habitats and species persisting in the wild*'. However, this definition does not help explain: *why* should we conserve; *what* should we conserve; *how* should we conserve; and, *how* much should we pay to conserve? The Convention on Biological Diversity (CBD), agreed at the World Summit on Sustainable Development in 1992, provides some aspirational guidance for international efforts to conserve biological diversity worldwide. For example, Article 1 notes that the overall goals of conservation are to:

- maintain biological diversity;
- allow sustainable use of its components; and
- promote equitable sharing of its benefits.

Subsequent articles elaborate some of the measures that conservationists might take to achieve each of these goals. Article 8a stresses the importance of '*establishing a system of protected areas or areas where special measures need to be taken to conserve biological diversity*' to *in situ* efforts to conserve biological diversity. Meanwhile, Article 8j notes that '. . . *subject to its national legislation, respect, preserve and maintain knowledge, innovations and practices of indigenous and local communities embodying traditional lifestyles relevant for the conservation and sustainable use of biological diversity* . . .', while Article 11 notes that: '. . . *as far as possible and appropriate adopt economically and socially sound measures that act as incentives for conservation and sustainable use* . . .'.

Professional conservationists, however, often adopt polarized positions over how to implement these sometimes apparently opposing aspirations. The one goal to which most committed conservationists aspire is that of conserving as much biodiversity as feasibly possible. Nevertheless, all conservation entails some form of cost for someone, somewhere. Meanwhile, resources to offset these costs are limited, and so choices are socially determined, often with little or no consensus over understanding what this goal means. By way of another example, a tiger *Panthera tigris* (Figure 1.2) might conjure up the image of a flagship species for armchair conservationists in developed countries, or threat to life and livelihood for farmers and agriculturalists living cheek by jowl with tigers (Leader-Williams & Dublin, 2000).

Some have likened the choices that conservationists and wider society now face to those that faced doctors manning the trenches in the First World War, where the French coined the term *triage* to help guide choices over which wounded soldiers should be treated. Currently, *Webster's Dictionary* contains two definitions for *triage*. The first reflects the origin of the word in the trenches, where '*triage comprises the sorting and allocation of treatment to*

Figure 1.2   **A Sumatran tiger *Panthera tigris* caught in a camera-trap in Kerinci Seblat National Park in Sumatra. Tigers can be a flagship species for conservationists, but a source of conflict for neighbours of protected areas in Asia. (Photograph by kind permission of Matthew Linkie.)**

patients to maximise numbers of survivors'. A second generalizes the original and highly specific definition of triage to one with which conservationists, seeking to answer some of the questions outlined above, might well identify. Thus triage can also comprise *'assigning priority order to projects on basis of where funds and other resources can best be used, are most needed, or are most likely to achieve success'*. The practice of triage is likely to become increasingly important for conservationists (Bottrill *et al.*, 2008). However, we have instead adopted the term *trade-off* for the title of this volume, which *Webster's* defines as *'a balancing of factors all of which are not attainable at the same time'*. In seeking to make explicit some of the many trade-offs that conservationists face, we hope this volume will contribute to the process of conservation getting itself increasingly into position to practice more effective triage. We now outline our logic behind why we asked our authors to write different chapters for different sections of the book.

Current approaches and toolkits

In Chapter 2, Wilson *et al.* discuss how decision theory may allow the more explicit formulation of conservation problems and the optimizing of trade-offs in conservation. These authors recognize that conservation is not just about protected areas, and that there is more to conservation than the underlying patterns of biodiversity. Furthermore, shortage of data is often not the problem that many conservationists claim it is, when suggesting that further research is needed before a particular problem is addressed. Instead, Wilson *et al.* stress the critical importance of clearly defining problems, and ensuring decisions are explicit, based on the data that are available.

At its 5th Conference of the Parties in 2000, the CBD adopted an ecosystem approach to conservation. Many international non-govermental organizations (NGOs) had already identified global conservation priority regions, including Conservation International's Biodiversity Hotspots (Myers *et al.*, 2000) and WWF's 200 Ecoregions (Olson *et al.*, 2001), to better guide the considerable investments they made in global conservation. These global priority regions were basically designed to protect the maximum biodiversity per dollar spent. However, returns on investment in biodiversity represent a trade-off between costs and conservation achieved, as Murdoch *et al.* discuss in Chapter 3.

Furthermore, the concept of biodiversity is poorly understood, and of limited public appeal, as a goal for conservation. In contrast, a functioning biosphere is vital to human welfare, in terms of providing clean air and water, and for recycling nutrients. Therefore, many have argued that conservation should increasingly focus on the importance of maintaining ecosystem services to broaden support for conservation objectives. However, as Goldman *et al.* discuss in Chapter 4, this in turn can represent a trade-off between different goals in conservation, because areas rich in biodiversity and in ecosystem services do not always overlap.

That said, whatever goals are set for conservation, it is necessary to have tools that can measure the successes and, as importantly, failures of conservation, given that much conservation activity remains unaccounted for in terms of its impact. Many conservation projects can account for their outputs, such as the numbers of vehicles bought and the number of person hours spent on different activities. However, the extent to which impacts, measured in terms of *the chances given to habitats and species persisting in the wild*, are less often considered. In order to optimize trade-offs between different conservation goals, we need to be able to clearly define problems, to ensure that decisions are explicit, *and* that their conservation impacts can be assessed, as Kapos *et al.* discuss in Chapter 5. Nevertheless, setting goals in conservation can involve socially important value judgments, which the next section addresses.

Influence of value systems

Debates over the conservation of surrogate species continue, and there is now a good understanding of the role and limitations of this approach (Karieva & Levin, 2003). However, many conservationists still continue to focus on the conservation of charismatic or *flagship* species, whose plight might appeal to the general public, and who may not understand fuzzy concepts like biodiversity and ecosystem services. In turn, this can lead to many ignoring the small and uncharismatic species that the visitor board at Heron's Carr sought to highlight. Nevertheless, as Samways discusses in Chapter 6, the conservation of such species is often vital, given that invertebrates make up the majority of the 13 million species estimated to occur on Earth today, and that the ecosystem services they generate are so valuable that the adage 'bugs drive the world' is probably not far from the truth.

Furthermore, the ongoing focus on saving charismatic species can lead to much polarized debate over whether conservation seeks to ensure the welfare of individuals or protect viable populations of species. Hence, many charismatic species attract direct use values as a key part of their total economic value. Indeed, the objectives of the CBD seek to promote sustainable use of components of biological diversity, *and* equitable sharing of its benefits with poor people living among biological diversity. But as Harrop questions in Chapter 7, how does this square with trade-offs between consumptive use and the welfare of the individuals that are the subject of that use? Likewise, Rosser and Leader-Williams discuss in Chapter 8 how to manage trade-offs between conserving charismatic species *and* promoting traditional and well-tried practices of consumptive use.

Given, however, that the CBD supports the principle of sustainable use, Roe and Walpole consider in Chapter 9 how this has been linked to poverty reduction in recent policy debates, and outline possible trade-offs between these two once separate areas of policy. They question how the components of biodiversity, and the services they provide, might contribute to poverty reduction. They also explore situations where local priorities for poverty reduction and international priorities for conserving biodiversity are very different, and how any resulting trade-offs might be resolved. Likewise, the CBD also supports the importance of tradition in using and conserving biodiversity. Therefore, in Chapter 10, Homewood examines how different local traditions operate within the conservation arena, and suggests that conventional wisdom often overlooks conservation benefits of local land use practices. Equally, conservation initiatives that seek to build on local conservation traditions often face long histories of mistrust through previous misappropriation of resources, stewardship rights and benefits. In turn, this may have resulted in distortions of grassroots democratic processes, making it difficult to implement approaches that might provide local incentives to conserve biodiversity without recourse to the recurrent public funding that is necessary to ensure enforcement of more formal measures to protect biodiversity. In turn, this raises the issue of funding shortages and better selling conservation, which the next section addresses.

Economics and governance

Most businesses spend 10% of the value of their capital assets on maintaining those assets (Mace *et al.*, 2007). Even though based on very crude estimates of the total economic value of the world's ecosystem services and the money spent formally on protecting biodiversity (Costanza *et al.*, 1997; James *et al.*, 1999), conservation only spends $\sim0.02\%$ of the total value of its capital assets on protecting the biodiversity that provides those assets. In turn, this raises two inter-related trade-offs: how to circumvent funding shortfalls in a world of trade-offs and how to better 'market' conservation so that it rises higher up political and funding agendas. As Bruner *et al.* discuss in Chapter 11, even though we heavily under-invest in conservation, the funds that are actually spent on conservation are not spent very effectively. Second, as Smith *et al.* address in Chapter 12, there is a clear need to ensure the public and their elected political decision makers realize that conservation is a public good that is vital for their long-term welfare, so conservation becomes better understood and better funded. In other words, conservation needs to be better marketed even though this also creates trade-offs.

Equally, many biodiversity-rich areas are underlain by non-renewable natural resources, where economically valuable extractive industries are, or could be, practiced, often in countries with poor governance. In turn, this can make it exceedingly difficult to achieve effective conservation in the face of such development pressures. Therefore, Pulgar-Vidal *et al.* discuss in Chapter 13 how trade-offs can be achieved between the development of extractive industries and conservation. Likewise, many biodiversity-rich areas occur in conflict and post-conflict situations, where the immediate priorities of life and death take on much greater precedence than conserving biodiversity and its associated ecosystem services. In situations of conflict, institutions can collapse and protected area staff can face extreme pressures that prevent them from undertaking their normal duties, while local people facing humanitarian disaster can take refuge in areas from which they are normally excluded. Therefore, Aveling *et al.* discuss how trade-offs should be confronted for conservation to succeed in the wake of conflict and disaster, in Chapter 14.

Social and institutional constraints

We have earlier noted that setting goals and policies for conservation is increasingly seen by some as a scientific activity, where outcomes should follow rational choices. However, as we progressed through the papers at the symposium and put together the chapters of this book, it has become increasingly evident that conservation should largely be a social process that engages science, rather than a scientific process that engages society. Nevertheless, as Knight and Cowling discuss in Chapter 15, conservation biologists commonly place great emphasis on collecting more and more biological data at the fine scale, while ignoring rudimentary social data that would allow for more successful implementation of conservation projects. Furthermore, as Adams discusses in Chapter 16, conservationists often lock into particular policy approaches that persist as dominant narratives, whether or not those approaches are successful in conserving biodiversity. Finally in this section, Brosius in Chapter 17 discusses the importance of the politics of knowledge in determining how to frame the concept of trade-offs, which different disciplines frame in different ways. He argues that understanding the politics of knowledge in conservation is key to understanding the processes by which trade-offs are identified, calculated, analyzed and negotiated. He further suggests that effective conservation decision making and calculation of trade-offs requires more than scientific information and must be premised within the broader concept of credibility that recognizes the different contexts in which academics, practitioners, state authorities, community members and other actors interact.

Future challenges

Extinction is a natural process, but the current rates of extinction are hundreds or thousands of times higher than background rates such that Earth now faces the sixth, and possibly greatest, extinction spasm in its history (Lawton & May, 1995). Indeed, the world is changing faster than recently thought and the 'evil quartet' of factors that once caused most known extinctions (Diamond, 1989) has now been joined by the additional threat of climate change. As Willis *et al.* describe in Chapter 18, this threat was largely unforeseen when most protected area networks were established, so requiring trade-offs to ensure that protected areas conserve representative suites of biodiversity over the long term. Given that the nature of threats to biodiversity, and the interactions between those threats, are now changing rapidly, Mace asks how the relative importance of drivers of species and biodiversity loss will change over time. Given that we have previously defined *conservation* as comprising *actions that directly enhance the chances of habitats and species persisting in the wild*, Chapter 19 takes a look into the future, to ask whether and how species might increase their resilience and adaptability to each major driver?

Finally, in Chapter 20, Smith *et al.* take a retrospective look over the issues raised in the chapters of this book and a look forward at how conservationists may improve the ways in which they address policy issues in future, in the light of the trade-offs that they have always made in their professional lives.

References

Bottrill, M., Joseph, L.N., Carwardine, J. *et al.* (2008) Is conservation triage just smart decision making? *Trends in Ecology and Evolution*, 23, 649–654.

Costanza, R., d'Arge, R., De Groot, R. *et al.* (1997) The value of the world's ecosystem services and natural capital. *Nature*, 387, 253–260.

Diamond, J.M. (1989) Overview of recent extinctions. In *Conservation for the Twenty-first Century*, eds D. Western & M. Pearl, pp. 37–41. Oxford University Press, New York.

James, A.N., Gaston, K.J. & Balmford, A. (1999) Balancing the Earth's accounts. *Nature*, 401, 323–324.

Karieva, P. & Levin, S.A. (eds) (2003) *The Importance of Species: perspectives on expendability and triage*. Princeton University Press, Princeton, NJ.

Lawton, J.H. & May, R.M. (eds) (1995) *Extinction Rates*. Oxford University Press, Oxford.

Leader-Williams, N. & Dublin, H.T. (2000) Charismatic megafauna as 'flagship species'. In *Priorities for the Conservation of Mammalian Diversity: has the panda had its day?* eds A. Entwistle & N. Dunstone, pp. 53–81. Cambridge University Press, Cambridge.

Mace, G.M., Possingham, H.P. & Leader-Williams, N. (2007) Prioritizing choices in conservation. In *Key Topics in Conservation Biology*, eds D.M. Macdonald & K. Service, pp. 17–34. Blackwell Publications, Oxford.

Myers, N., Mittermeier, R.A., Mittermeier, C.G., da Fonseca, G.A.B. & Kent, J. (2000) Biodiversity hotspots for conservation priorities. *Nature*, 403, 853–858.

Olson, D.M., Dinerstein, E., Wickramanayake, E.D. et al. (2001) Terrestrial ecoregions of the world: a new map of life on Earth. *Bioscience*, 51, 933–938.

Sutherland, W.J., Pullin, A.S., Dolman, P.M. & Knight, T.M. (2004) The need for evidence-based conservation. *Trends in Ecology and Evolution*, 19, 305–308.

Part I
Current Approaches and Toolkits

Prioritizing Trade-offs in Conservation

Kerrie A. Wilson[1], Michael Bode[2], Hedley Grantham[3] and Hugh P. Possingham[3]

[1]University of Queensland, School of Biological Sciences, Brisbane, Queensland, Australia
[2]Applied Environmental Decision Analysis Research Hub, Department of Botany, University of Melbourne, Melbourne, Victoria, Australia
[3]Ecology Centre, University of Queensland, Brisbane, Queensland, Australia

Introduction

Trade-offs are an unavoidable part of decision making when multiple considerations or choices are not perfectly aligned, and where resources are limited. Despite their ubiquity, trade-offs have not played a central role in conservation because practitioners and theorists have traditionally been unwilling to acknowledge that they cannot achieve all of their objectives and so satisfy all stakeholders, or to consider constraints. Put another way, conservationists find it hard to accept that they cannot save everything.

In this chapter, we show how trade-offs can be systematically explored and optimized, and the opportunity costs of decisions evaluated. First, we outline the key knowledge required to solve any problem, given that understanding the structure of a problem is essential for exploring trade-offs. Second, we deal with trade-offs in one of two ways: either by making one consideration a

Trade-offs in Conservation: Deciding What to Save, 1st edition.   Edited by N. Leader-Williams, W.M. Adams and R.J. Smith.   © 2010 Blackwell Publishing Ltd.

constraint, and the other something that needs to be maximized or minimized, or by weighting different objectives and showing a variety of solutions that deliver different levels of satisfaction with respect to different objectives. Third, we work through examples of both of these alternative problem formulations. Our overall aim is to describe, through example, methods to optimize and evaluate a broad range of trade-offs in conservation.

Background

The process of making any decision requires consideration of a number of issues. Take, for example, the decision process associated with buying food. When shopping for food, a broad *goal* or *problem* might be to buy a week's food for the least cost, sufficient to satisfy and sustain a family. Hence, a specific *objective* is to minimize the cost and that is subject to several *constraints*: for example, each meal has to be representative of the major food groups and the family will have to enjoy eating it. The shopper's *knowledge* of the system and its dynamics might include some idea of food prices, which fluctuate somewhat predictably with seasons but that also show an element of unpredictability. And finally what the shopper *controls* is whether or not, and how much, they purchase of each item. All resource allocation problems have a similar structure, with objective(s), constraints and our knowledge of the system and its control variables (Wilson *et al.*, 2006).

As with buying food, financial considerations will probably be fundamental in any conservation investment problem. However, many conservation assessments have ignored the cost of the conservation action (Balmford *et al.*, 2000; Odling-Smee, 2005; Brooks *et al.*, 2006). Equally, when costs have been accounted for, they are often considered secondary to biological factors or evaluated *post hoc*, following an analysis based on other considerations such as biological data. The *post-hoc* consideration of any factor that we actually aim to optimize will always lead to suboptimal outcomes. Conservationists have most likely been hesitant to explicitly account for the costs of conservation, because accurate, high-resolution data are generally unavailable, and because biologically focused conservationists are hesitant to allow non-biological factors to influence decisions about where funds are allocated. Nevertheless, without explicit consideration of the costs of conservation, useful and accurate guidance cannot be provided on how and when funds should be distributed between different regions, activities or commodities. Fortunately, the number of studies that explicitly account for costs have grown, and datasets on the costs of conservation are becoming increasingly available (reviewed in Wilson *et al.*, 2009). But perhaps more worrying is that conservation problems are generally poorly defined, and rarely follow the generic problem-solving framework outlined above. Consequently, the explicit consideration of trade-offs is often impossible.

Allocation of funds to achieve conservation goals

A common objective in conservation is to achieve certain conservation goals at minimal cost. Most conservation planning software is designed to provide good solutions to this problem. The aim might be to minimize the total cost of land or sea reserved. It is, of course, entirely possible that reservation may not be the only possible land use under consideration, in which case conservationists might want to consider the potential contribution of a range of land uses to their conservation outcomes (Box 2.1).

Box 2.1 **The potential contribution of alternative land uses to biodiversity conservation: the case of tropical rainforests in Borneo (Wilson *et al.*, 2010)**

Different land uses make different contributions to the conservation of biodiversity. For example, in tropical regions some land uses provide habitat throughout all levels of forest strata, along with a diversity of food sources for fauna species occupying the forest. Other land uses are more restricted in their provision of food and habitat, and their floristic and faunal diversity reflects these differences.

   Meijaard *et al.* (2006) have contributed to our understanding of the value of different land uses to biodiversity conservation by analyzing the sensitivity of Bornean vertebrates to the direct and indirect effects of land use modification and timber harvesting. They found that the most sensitive mammals tend to have narrow ecological niches and many have strictly frugivorous, carnivorous or insectivorous feeding habits. These sensitive mammals live in particular forest strata, especially ground or upper canopy levels, rather than ranging throughout all levels

---

(Meijaard & Sheil, 2008). On the other hand, less sensitive mammals tend to be herbivorous or omnivorous, and live in the lower vegetation strata, although some are found at all levels. Since selective logging in Borneo primarily targets dipterocarps, vertebrates depending on these and other commercial timber species will be affected most by land uses involving timber harvesting. Hunting, an indirect effect of timber harvesting, particularly affects those species important for human food or animal trade such as bearded pigs *Sus barbatus*, long-tailed porcupine *Trichys fasciculate* and muntjac *Muntiacus* species, but also monkeys and deer.

Along with the relative sensitivity of species to different land uses, the degree of protection offered by different land uses also varies. In some regions protection status is not necessarily synonymous with a high contribution to biodiversity conservation, and there is much evidence that in some cases production forests can contribute a great deal (Jepson *et al.*, 2001; Curran *et al.*, 2004; Meijaard *et al.*, 2006; Nakagawa *et al.*, 2006; Meijaard & Sheil, 2007; Wells *et al.*, 2007).

In terms of problem definition and the analysis of trade-offs, this situation presents a variety of challenges. First, the suite of possible land uses is complex and the control variable is not simply to reserve a site or not. Second, there may be constraints, upper and lower, on how much land can be put into forestry, reserves, oil palm plantations and other land uses.

From the perspective of problem formulation, therefore, it is necessary to move beyond the consideration of the landscape (or seascape) as a binary variable involving only protected areas and an unprotected matrix, to prioritizing investments in a diversity of land uses and exploring the trade-offs involved. Thus, while traditional conservation planning has considered only protected areas nested within an unprotected matrix, the basic problem definition can be extended to consider multiple land use zones, which have multiple costs and varying contributions to biodiversity conservation. It can also open up new opportunities for implementation, due to constraints on the places where protected areas can be established. Multiple zones imply different trade-offs.

For example, conservation outside protected areas may make less contribution to biodiversity, but cost less to implement and manage. Therefore, the fundamental trade-offs associated with multiple zones is between their costs, benefits, security and whether there are constraints on particular outcomes that can be delivered.

In order to undertake multiple zone conservation planning, a substantial amount of data are required, some of which might be difficult to obtain. First of all, a map is required of the current land (or sea) uses, which is relatively straightforward to obtain in most regions. Second, an understanding is required of the relative contribution of each land use zone to achieving predefined biodiversity targets (see Box 2.1). This value might vary between 0 and 1: if the contribution of a zone is 1, then each hectare of investment will contribute 1 ha towards the target for this species in this zone. If the zone contribution was 0.5, then each hectare of habitat would contribute 0.5 ha towards the targets for the species in this zone. These zone contribution values could be either species-specific or constant across all species.

Another key piece of information is the possible zones in which that parcel might occur. This might be guided by region-specific planning guidelines, but could also reflect the constraints of the problem under consideration. For example, planning units that are already cleared or converted may not be able to change zones, as would be the case if restorative activities were not being accounted for. In addition, protected planning units may not be able to change zones, and this would be determined by the security of protected tenure in the region under consideration. It would also be necessary to know the cost of changing land use zones. For example, to convert a production forest to a strict protected area might require consideration of start-up costs, management costs and opportunity costs (Naidoo et al., 2006). However, if certain extractive activities are allowed in the region, for example selective logging, but not agricultural production, these opportunity costs will be lower.

Regardless of their potentially important contribution to biodiversity conservation, unprotected zones are unlikely to result in maximal local conservation outcomes. Nevertheless, given a limited budget they may represent part of the most cost-effective option for the conservation region: so their relative benefits and costs must be understood. With such information, it is possible to identify a conservation landscape that optimizes the trade-off between different land uses and prioritizes investments across land uses in order to achieve conservation goals in a cost-effective manner.

Allocation of funds through time subject to a fixed budget

Another type of conservation problem might be to maximize biodiversity persistence. To ensure the persistence of biodiversity, it will probably be necessary to invest in a range of conservation actions through time, not just reservation at a single time. Therefore, the conservation problem can be redefined to a multi-action, multi-time problem with a fixed budget constraint. Then the problem is to maximize biodiversity conserved through annual investment in a suite of conservation actions (one objective) given a fixed annual budget (the other objective is set as a constraint). In order to solve this problem it is necessary to know the suite of available conservation actions. The management decision is how much of the available budget to allocate to each conservation action each year, and the resulting trade-off is between how much to allocate to the different conservation actions.

In order to optimize this trade-off it is necessary to know several things about each conservation action, including the area of land currently receiving and requiring the action, its cost per unit area, its benefit to biodiversity and the likelihood of success (Box 2.2). With such information, dynamic investment schedules can be generated that reflect shifts in the allocation of funds as the return from investing in each conservation action changes. The investment schedule is determined by: (i) the relationship between the additional area invested in each conservation action and the biodiversity benefit; (ii) the cost of this investment; and (iii) the existing level of investment. Actions can then be prioritized that result in the greatest biodiversity benefit per dollar invested. Alternatively, if the rate of habitat loss is known, strategies can be prioritized that will minimize the loss of biodiversity (Murdoch et al., 2007). This type of approach can help to make explicit the trade-offs between different conservation actions and has been shown to outperform more traditional approaches to conservation funding allocation focusing solely on protected area establishment (Box 2.2).

Box 2.2 **Maximizing the conservation of biodiversity through investment in a range of conservation actions through time: a case from Mediterranean ecoregions**

Wilson *et al.* (2007) developed a framework for guiding the allocation of funds among alternative conservation actions that address specific threats, and applied it to 17 of the world's 39 Mediterranean climate ecoregions. In total, they evaluated 51 ecoregion–conservation action combinations that were termed 'ecoactions'. For example, for the action of invasive predator control in an Australian ecoregion, they determined that 30% of the remaining natural habitat requires this action, 8% already receives it, the biodiversity benefit is estimated to be around 100 vertebrate species, and the cost to apply this action over $1\,km^2$ would be around US$7000. They assumed that the marginal biodiversity benefit decreases for each new unit area receiving investment, and, therefore, the biodiversity benefit diminishes with cumulative investment.

Only 24 of the 51 possible ecoactions were prioritized to receive investment during the first 5 years. In particular a conservation action comprising a mix of land protection and off-reserve management in South Africa was prioritized, along with invasive plant control in Chile, California and South Africa. These conservation actions will yield the greatest marginal return on investment over 5 years because the biodiversity benefits are high and the costs are comparatively low. However, these conservation actions are not necessarily the cheapest actions, nor are the ecoregions prioritized necessarily the most species-rich. Instead, this represents a trade-off between the key characteristics of each action. The action-specific framework was then compared to a simplified model of conservation that focused only on land acquisition. The decision steps in the resource allocation process were identical regardless of investment approach, with the exception that the land acquisition-only approach considered only the single conservation action of land acquisition. Over 5 years almost four times as many (2780 versus 703) species can be protected using the ecoaction approach.

Therefore it is possible to find good solutions to well-defined problems that incorporate complex trade-offs between several considerations: an annual budget, the region in which conservation action is being undertaken, the type of action, different sorts of biodiversity, and different threats. Scoring systems or threshold approaches to prioritization (Murdoch *et al.*, this volume, Chapter 3) may appear to be the best way to deal with this complexity. However, there are rational and irrational approaches to finding good solutions to dealing with trade-offs: proper problem formulation of questions is key to rational approaches.

There is more to conservation than biodiversity

Conservation planners and practitioners have historically focused on con-serving elements of biodiversity, such as species or habitat types. Increasingly there is also interest in incorporating ecosystem services into conservation plans. While the benefits that people derive from nature have long been acknowledged (Helliwell, 1969; Costanza *et al.*, 1997), interest in their dual protection stems largely from two initiatives: the creation of schemes to allow for payments for ecosystem service delivery, for example the Katoomba Initia-tive,[1] and the Millennium Ecosystem Assessment, which helped to clarify the importance of ecosystem services to human well being (Millennium Ecosys-tem Assessment, 2005; Goldman *et al.*, this volume, Chapter 4). However, incorporating ecosystem services into conservation plans is still in its infancy. There are myriad technical challenges, including the process of mapping and valuing ecosystem services and, perhaps, most fundamentally in challenges associated with making judgments about the relative importance of conserv-ing biodiversity versus ecosystem services. In order to inform discussions on the latter, the spatial and temporal congruence of ecosystem services and biodiversity features can be assessed and the trade-offs that exist between biodiversity conservation and the protection and maintenance of ecosystem services can be quantified.

If an analysis of congruence shows that areas of high biodiversity and ecosystem service value coincide, then conservation interventions will simul-taneously benefit the conservation and maintenance of both biodiversity and ecosystem services. However, there are several reasons why congruence may

---

[1] www.Katoomba.org.

not arise and a number of analyses have found variable overlap between ecosystem services delivery and biodiversity conservation (Chan *et al.*, 2006; Turner *et al.*, 2007). In such cases a properly posed problem will need to be explicit about how to trade-off biodiversity against one or more ecosystem services, and ecosystem services against each other. In essence, a three-way trade-off presents itself, between biodiversity, ecosystem services and money. Such trade-offs cannot be avoided by setting one objective as a constraint and minimizing or maximizing the other objective, because there are at least two other objectives. The trade-off needs to be faced head on and an explicit weighting be made that values the different objectives relative to each other.

Using decision analysis methods, conservation investments can then be prioritized, in order to maximize a weighted sum of biodiversity and ecosystem services, across the entire landscape. The weighted sum defines the relative importance of biodiversity conservation and ecosystem service delivery, and is either defined quantitatively or implicitly. When the relative importance of protecting biodiversity is varied against the protection of ecosystem services a 'pareto-frontier' (Nalle *et al.*, 2004) is identified (Figure 2.1). The highest and lowest points in this frontier correspond to landscapes planned exclusively for ecosystem services and biodiversity respectively, with the points in between being defined by different weights, and representing compromises across both objectives.

How much time and resources should be spent on learning?

A trade-off is also associated with collecting more information for increased certainty in decision making versus taking more immediate conservation action. So the question might be asked 'how much investment is required to learn about ecological, socioeconomic, and political systems before we make an investment decision?' What is the relative return on investment associated with waiting for an improved information base, given the opportunity costs of ongoing habitat loss, and a limited budget available for conservation endeavours? In essence, knowledge becomes another objective that must be traded off against other objectives. Interestingly, in this situation knowledge itself has no value, but knowledge does help to make better decisions in the future and so influences efficiency. In the most extreme case, with no knowledge no decision can be made.

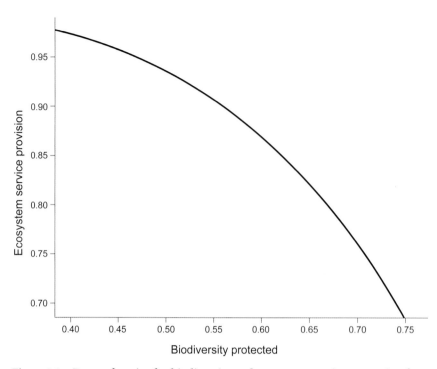

Figure 2.1 **Pareto-frontier for biodiversity and ecosystem service protection for a hypothetical landscape.**

For example, a common constraint in conservation planning assessments is the availability of data on the distribution of biodiversity (Possingham *et al.*, 2007; Rodrigues & Brooks, 2007). Despite this acknowledged constraint, very limited research has been undertaken on how different levels of investment in biodiversity data improve the quality of decisions made (Balmford & Gaston, 1999; Gardner *et al.*, 2008). However, the return on investment can be explicitly analyzed for different amounts of biodiversity, and for other types of data, and the trade-off between waiting for improved data, and acting immediately based on currently available data, can be determined. Such analyses can be done retrospectively (Box 2.3).

Box 2.3 **Case from the Fynbos region of South Africa (Grantham
et al., 2008, 2009)**

The Proteas are a plant group characteristic of the Fynbos region of
South Africa (Figure 2.2). Many Proteas have very narrow distributional
ranges and the Fynbos region is globally recognized for its conservation
importance (Myers *et al.*, 2000). In this region, the *Protea Atlas* has been
developed over 10 years, involving volunteer surveys of the distribution

Figure 2.2   **The silver-edge pincushion *Leucospermum patersonii* occurs on
limestone soils in the Agulhas region at the southern tip of Africa. It is one of
381 taxa of Proteaceae, which occur in the species-rich Fynbos biome of the
Cape Floral Kingdom. They were the focus of a 10-year atlas project, available
at www.sanbi.org/protea. Parts of the Fynbos biome are under threat from
habitat clearing due to agriculture. (Photograph by kind permission of Justine
Grantham.)**

of *Protea* species and subspecies across 220 000 records resulting in one of the most comprehensive species locality datasets available (Grand *et al.*, 2007). Using this database, the return on investment in different types of biodiversity data was investigated. The return on investment in species survey data was found to be large initially, although this benefit quickly diminishes (Figure 2.3). Similar, if not more drastic, results are associated with the trade-off between data collation and implementation (Figure 2.4). Here, conservation outcomes are measured by the retention of species in the landscape. Under this scenario, the analysis shows it is

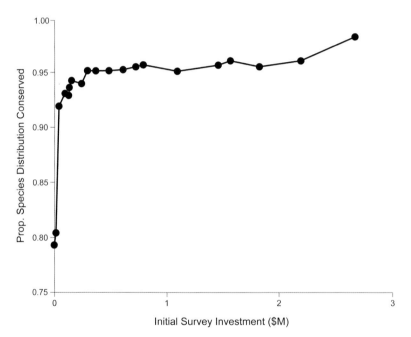

Figure 2.3 **The return on investment in terms of the coverage of *Protea* species in new and existing protected areas after different levels of initial investment in survey data ($US). The patterns remain the same regardless of the level of background habitat destruction. Zero investment reflects the use of a habitat map and no species data. (From Grantham *et al.*, 2008 and reproduced by kind permission of ©2008 Wiley Periodicals, Inc.)**

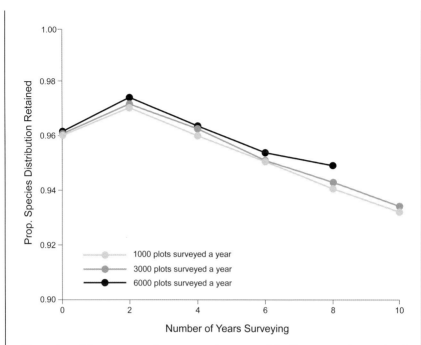

Figure 2.4 **The return on investment in terms of the *Protea* species retained in the landscape after different levels of initial investment in survey data. The patterns remain the same regardless of the level of survey intensity. Zero investment reflects the use of a habitat map and no species data. (From Grantham *et al.*, 2009 and reproduced by kind permission of ©2009 Blackwell Publishing Ltd/CNRS.)**

better not to use any species data and just use a habitat map, than to wait 10 years for comprehensive data, during which time species may be overtaken by threats. In this case, waiting for improved data will result in a worse outcome despite having gained the capacity to make better decisions.

Initial case studies such as that presented in Box 2.3 have illustrated that the return on investment associated with collecting more data is large initially, but then quickly diminishes (Grantham *et al.*, 2008, 2009). In the case of collecting species locality data, any data are better than no species data, although trade-offs do exist, particularly when seeking to maximize species retained in the landscape in the context of ongoing habitat loss (Box 2.3). In such cases, planners are forced to trade-off the improvement in conservation outcomes that result from better information against the opportunity costs of waiting for information to be gathered. Again, there are rational and irrational ways of performing such a trade-off. There is no doubt that scoring the value of knowledge in units of biodiversity would be inappropriate (Murdoch *et al.*, this volume, Chapter 3). Performing the types of analyses presented in Box 2.3 is useful as funds spent on improving our knowledge base leaves less time and resources available for on-the-ground conservation action. It is important to be aware of such temporal trade-offs and analyze the return on investment in learning versus doing.

Trading off biodiversity benefit and investment security

Risk pervades conservation decision making. In the previous section we showed how we can gain knowledge to reduce uncertainty, however some risk is irreducible. For example we can reduce our risk of heart disease but not make it zero. Optimizing algorithms, such as stochastic dynamic programming, can be used to deliver optimal solutions to conservation problems in the context of risk. However, since these are limited to small problems, with a small number of sites or states, heuristics have been sought to approximate optimal solutions (Meir *et al.*, 2004).

Two heuristics that we have tested, and that are related to the common objectives outlined earlier, are to maximize short-term gains and to minimize short-term loss. Figure 2.5 illustrates that the *minimize short-term loss* heuristic most closely approximates the optimal decision, calculated using stochastic dynamic programming (SDP). However, the *maximize short-term gain* heuristic results in the greatest number of endemic species reserved at early time steps.

There are multiple sources of risk that may need to be considered such as the probability of project failure due to political instability and corruption, the impact of natural catastrophes, and a lack of budget continuity. These sources of risk might be regarded as transaction risk (investment ceases) and performance risk (investment fails). McBride *et al.* (2007) have illustrated that when there is transaction risk it is best to be precautionary and opportunistic, and it is therefore best to maximize gains. When performance risk is incorporated, the optimal solution becomes a complex trade-off between the immediate biodiversity benefits of acting in a region, and the perceived

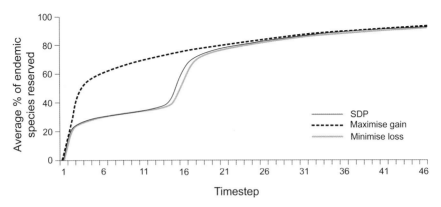

Figure 2.5   **Performance of two heuristics compared to an optimal solution. (From Wilson *et al.*, 2006 and reproduced by kind permission of ©2006 Nature Publishing Group.)**

longevity of the investment. Again, decision analysis tools provide mathematically credible approaches to dealing with risk, something that *ad hoc* scoring methods cannot provide where the trade-offs are so complicated.

Conclusions

The examples outlined in this chapter illustrate just a few of the trade-offs faced by conservation decision makers. However, by using a decision theory framework and appropriate decision analysis tools, we have also demonstrated that trade-offs can be rationally explored. We have shown that superficially daunting trade-offs can be evaluated – first by considering the simple trade-offs between biodiversity and money, then by increasingly complex problems that consider multiple land uses, actions and benefits, and further by accommodating uncertainty reduction and risk management into the decision-making framework. Too often, conservation biologists and practitioners seek algorithms to solve a problem – like scoring systems or threshold methods – before they have defined the problem. Once a problem is properly posed, economic and mathematical tools abound that can deliver optimal or near-optimal solutions. Where disparate objectives cannot be reconciled into a single currency, trade-off curves provide decision makers with a clear suite of options. The secret to success is proper problem formulation and the collation of data often overlooked in favour of more information on biodiversity. Bode *et al.* (2008) found global funding allocation schedules to be less sensitive to variation in taxon assessed than to variation in cost and threat data. Conservation outcomes were also found to be most sensitive to uncertainty in the cost data, with errors potentially leading to inefficient funding strategies that could result in 20 times as many extinctions as errors of the same proportional size in the biodiversity data. To ensure that trade-offs are analyzed in a comprehensive and accurate way, the types of information routinely collated for conservation assessment will probably require broadening.

References

Balmford, A. & Gaston, K.J. (1999) Why biodiversity surveys are good value. *Nature*, 398, 204–205.

Balmford, A., Gaston, K.J., Rodrigues, A.S.L. & James, A. (2000) Integrating costs of conservation into international priority setting. *Conservation Biology*, 14, 597–605.

Bode, M., Wilson, K.A., Brooks, T.M. *et al.* (2008) Cost-effective global conservation spending is robust to taxonomic group. *Proceedings of the National Academy of Sciences of the USA*, 105, 6498–6501.

Brooks, T.M., Mittermeier, R.A., da Fonseca, G.A.B. *et al.* (2006) Global biodiversity conservation priorities. *Science*, 313, 58–61.

Chan, K.M.A., Shaw, M.R., Cameron, D.R., Underwood, E.C. & Daily, G.C. (2006) Conservation planning for ecosystem services. *PLoS Biology*, 4, 2138–2152.

Costanza, R., d'Arge, R., de Groot, R. *et al.* (1997) The value of the world's ecosystem services and natural capital. *Nature*, 387, 253–260.

Curran, L.M., Trigg, S.N., McDonald, A.K. *et al.* (2004) Lowland forest loss in protected areas in Indonesian Borneo. *Science*, 303, 1000–1003.

Gardner, T.A., Barlow, J., Araujo, I.S. *et al.* (2008) The cost-effectiveness of biodiversity surveys in tropical forests. *Ecology Letters*, 11, 139–150.

Grand, J., Cummings, M.P., Rebelo, T.G., Ricketts, T.H. & Neel, M.C. (2007) Biased data reduce efficiency and effectiveness of conservation reserve networks. *Ecology Letters*, 10, 364–374.

Grantham, H.S., Moilanen, A., Wilson, K.A., Pressey, R.L., Rebelo, T.G. & Possingham, H.P. (2008) Diminishing return on investment for biodiversity data in conservation planning. *Conservation Letters*, 1, 190–198.

Grantham, H.S., Wilson, K.A., Moilanen, A., Rebelo, T. & Possingham, H.P. (2009) Delaying conservation actions for improved knowledge: how long should we wait? *Ecology Letters*, 12, 293–301.

Helliwell, D.R. (1969) Valuation of wildlife resources. *Regional Studies*, 3, 41–49.

Jepson, P., Jarvie, J.K., Mackinnon, K. & Monk, K.A. (2001) Decentralization and illegal logging spell the end of Indonesia's lowland forests? *Science*, 292, 859–861.

McBride, M. J, Wilson, K.A., Bode, M. & Possingham, H.P. (2007) Incorporating the effects of socioeconomic uncertainty into priority setting for conservation investment. *Conservation Biology*, 21, 1463–1474.

Meijaard, E. & Sheil, D. (2007) A logged forest in Borneo is better than none at all. *Nature*, 446, 974–974.

Meijaard, E. & Sheil, D. (2008) The persistence and conservation of Borneo's mammals in lowland forests managed for timber: observations, overviews and opportunities. *Ecological Research*, 23, 21–34.

Meijaard, E., Sheil, D., Nasi, R. & Stanley, S.A. (2006) Wildlife conservation in Bornean timber concessions. *Ecology and Society*, 11, 47.

Meir, E., Andelman, S. & Possingham, H.P. (2004) Does conservation planning matter in a dynamic and uncertain world? *Ecology Letters*, 7, 615–622.

Millennium Ecosystem Assessment (2005) *Ecosystems and Human Well-being: Synthesis*. Island Press, Washington, DC.

Murdoch, W., Polasky, S., Wilson, K.A., Possingham, H.P., Kareiva, P. & Shaw, R. (2007) Maximising return on investment in conservation. *Biological Conservation*, 139, 375–388.

Myers, N., Mittermeier, R.A., Mittermeier, C.G., da Fonseca, G.A.B. & Kent, J. (2000) Biodiversity hotspots for conservation priorities. *Nature*, 403, 853–858.

Naidoo, R., Balmford, A., Ferraro, P.J., Polasky, S., Ricketts, T.H. & Rouget, M. (2006) Integrating economic costs into conservation planning. *Trends in Ecology and Evolution*, 21, 681–687.

Nakagawa, M., Miguchi, H. & Nakashizuka, T. (2006) The effects of various forest uses on small mammal communities in Sarawak, Malaysia. *Forest Ecology and Management*, 231, 55–62.

Nalle, D.J., Montgomery, C.A., Arthur, J.L., Schumaker, N.H. & Polasky, S. (2004) Modeling joint production of wildlife and timber. *Journal of Environmental Economics and Management*, 48, 997–1017.

Odling-Smee, L. (2005) Conservation: dollars and sense. *Nature*, 437, 614–616.

Possingham, H.P., Grantham, H. & Rondinini, C. (2007) How can you conserve species that haven't been found? *Journal of Biogeography*, 34, 758–759.

Rodrigues, A.S.L. & Brooks, T.M. (2007) Shortcuts for biodiversity conservation planning: the effectiveness of surrogates. *Annual Review of Ecology, Evolution and Systematics*, 38, 717–737.

Turner, W.R., Brandon, K., Brooks, T.M. *et al.* (2007) Global conservation of biodiversity and ecosystem services. *Bioscience*, 57, 868–873.

Wells, K., Kalko, E.K.V., Lakim, M.B. & Pfeiffer, M. (2007) Effects of rain forest logging on species richness and assemblage composition of small mammals in Southeast Asia. *Journal of Biogeography*, 34, 1087–1099.

Wilson, K.A., Carwardine, J. & Possingham, H.P. (2009) Setting conservation priorities. *Annals of the New York Academy of Sciences*, 1162, 237–264.

Wilson, K.A., McBride, M.F., Bode, M. & Possingham, H.P. (2006) Prioritising global conservation efforts. *Nature*, 440, 337–340.

Wilson, K.A., Meijaard, E., Drummond, S. *et al.* (2010) Conserving biodiversity in production landscapes. *Ecological Applications*, 20, in press.

Wilson, K.A., Underwood, E.C., Morrison, S.A. *et al.* (2007) Conserving biodiversity efficiently: what to do, where and when. *PLoS Biology*, 5, 1850–1861.

$$\textbf{3}$$

Trade-offs in Identifying Global Conservation Priority Areas

William Murdoch[1], Michael Bode[2], Jon Hoekstra[3], Peter Kareiva[4], Steve Polasky[5], Hugh P. Possingham[6] and Kerrie A. Wilson[7]

[1]Department of Ecology, Evolution and Marine Biology, University of California, Santa Barbara, California, USA
[2]Applied Environmental Decision Analysis Research Hub, Department of Botany, University of Melbourne, Melbourne, Victoria, Australia
[3]The Nature Conservancy, Seattle, Washington, USA
[4]The Nature Conservancy, Arlington, Virginia, USA
[5]Departments of Applied Economics and of Ecology, Evolution and Behavior, University of Minnesota, St Paul, Minnesota,USA
[6]Ecology Centre, University of Queensland, Brisbane, Queensland, Australia
[7]University of Queensland, School of Biological Sciences, Brisbane, Queensland, Australia

Introduction

Global conservation priority areas claim to tackle the fundamental trade-off of biodiversity conservation: 'how to support the most species, at the least cost' (Myers *et al.*, 2000). These schemes attempt to satisfy both of these contrasting

Trade-offs in Conservation: Deciding What to Save, 1st edition.   Edited by N. Leader-Williams, W.M. Adams and R.J. Smith.   © 2010 Blackwell Publishing Ltd.

demands at a global scale, and have rapidly become ubiquitous conservation tools. However, these priority schemes have not been without criticism by a number of scientists.

In this chapter, we examine some of the rationale for, and the shortcomings of, different global conservation priority areas. First, we review the two major protocols, comprising scoring systems and thresholds, used over the last decade to define these priority sets. Second, we discuss them in the context of the two alternative forms of conservation prioritization, comprising proactive and reactive types. Third, we explore briefly an alternative approach: analysis of conservation return on investment (ROI).

Background to global conservation priority schemes

Conservation non-governmental organizations (NGOs) share a common mission: to conserve Earth's rapidly disappearing biodiversity. The logistics of large-scale conservation, and the need for ecosystem-based management, has increasingly pushed conservation attention away from individual species towards the construction of a global protected area network (James *et al.*, 1999a). At present, this network covers ~11.5% of the Earth's land surface (IUCN, 2003). However, despite this significant investment in land for conservation, it is estimated that an adequate global protected area system will cost an additional US$17 billion annually, over the next 30 years (James *et al.*, 1999b). This is far in excess of current expenditure, estimated at around US$6 billion per year (Balmford *et al.*, 2003). Therefore, these scarce financial resources must be prioritized to protect the regions with the greatest need.

This realization has spurred conservation organizations to designate 'global conservation priority areas', those small subsets of the Earth's surface that demand the most urgent conservation attention. For example, Conservation International (CI) has focused its flexible global conservation resources on so-called Biodiversity Hotspots (Myers *et al.*, 2000; Mittermeier *et al.*, 2003). These hotspots cover only 1.4% of the world's terrestrial surface area, and if they were entirely enclosed within the global protected areas network, then ~35% of terrestrial vertebrate species and 44% of vascular plant species would be protected. The Biodiversity Hotspots were the first proposed set of global conservation priority areas (Myers, 1988). However, other conservation organizations quickly followed with their own schemes. Table 3.1 lists the major priority-setting schemes of the last decade, and their sponsoring conservation organizations. The different sets were all defined by one of two protocols:

Table 3.1 Summary of methods used by major NGOs to select global priority regions for conservation, see text for description of goals.

Scheme, and sponsoring conservation NGO	Goals	Prioritization method	Method detail	Key references
WWF **Global 200 Ecoregions**	Reactive + representative	Scoring system	The final score used to rank sites is a function of: number of species (from 7 taxonomic groups) number of endemic species unusual phenomena rarity of biome intactness of biome	Olson & Dinerstein (1998, 2002)
Conservation International **Biodiversity Hotspots**	Reactive	Threshold	≥1500 endemic vascular plant species *and* ≥ 70% habitat lost	Myers (1988), Myers *et al.* (2000), Mittermeier *et al.* (2004)
Birdlife International and others **Key Biodiversity Areas**	Reactive	Threshold	Significant population of a globally threatened species *or* ≥5% of world population of ≥1 restricted-range species *or* ≥1% of world population of a congregatory species *or* Significant fraction of species are biome-restricted *or* Source of ≥1% of global population (marine)	Eken *et al.* (2004)
The Nature Conservancy **Crisis Ecoregions**	Reactive + representative	Threshold	>20% habitat lost *and* Habitat converted/habitat protected > 2	Hoekstra *et al.* (2005)
Alliance for Zero Extinction	Reactive	Threshold	Contains entire (or almost) world population of ≥1 highly threatened species	Ricketts *et al.* (2005)
Conservation International **High Biodiversity Wilderness Areas**	Proactive	Threshold	≥70% of original habitat remains intact *and* ≥1 million hectares in size *and* Human density ≤5 persons per km^2	Mittermeier *et al.* (2003)

scoring systems or thresholds. Consequently, we organize our discussion of these schemes under these two headings.

Methods in relation to goals

Whether they are described by scoring systems or thresholds, priority areas aim to cost-effectively reduce the loss of Earth's biodiversity, and they may also be aimed at unique habitat types, charismatic megafauna and ecosystem services. The different prioritization schemes operationalize this common mission using two contrasting approaches. The first is to maximize the amount of biodiversity contained in conservation reserves: 'proactive conservation'. The second is to minimize or prevent likely extinctions, generally by focusing on immediately threatened species or habitats: 'reactive conservation'. Although both approaches attempt to minimize biodiversity loss, the decisions that each enact are driven by different time scales. Proactive conservation aims to protect species long before they become threatened, whereas reactive conservation is focused on species or areas that are threatened now. Global conservation priority areas follow one of these goals, and one of the two protocols described below.

Thresholds for identifying priority regions

Conservation prioritization, at its most basic, is a decision about which areas should receive resources and which should not (Bottrill et al., 2008). Threshold protocols define a criterion for each feature of conservation interest. If a particular region does not satisfy any or each of these criteria, it is not a priority. The most famous example of a threshold-based global conservation prioritization scheme is that of Biodiversity Hotspots (Box 3.1).

Box 3.1 **Global conservation prioritization areas defined by thresholds**

Biodiversity Hotspots

Conservation International (CI) used thresholds to select 34 'Biodiversity Hotspots' (extended from an original, tropical forest-based set of 10 (Myers, 1988)) from among the world's terrestrial ecoregions. CI argued these were the most important reactive conservation regions on the planet and, if protected, would represent a 'silver bullet' that would 'stem the mass extinction of species' (Myers *et al.*, 2000). To be included, an ecoregion must satisfy two criteria: (i) it must have at least 1500 endemic vascular plant species; and (ii) at least 70% of the original vegetation/habitat in that ecoregion must be already degraded. Endemic vascular plant species were used to measure biodiversity to avoid the problems inherent in a compound biodiversity currency, and because plants have the best distribution data.

Crisis Ecoregions

Hoekstra *et al.* (2005) used a conservation risk index (CRI), the ratio of percent habitat lost to percent protected, to help define 'Crisis Ecoregions'. For example, 64 ecoregions are critically endangered, having >50% of their original habitat lost and a CRI > 25. These thresholds were chosen to correspond to the IUCN, the International Union for the Conservation of Nature's Red List thresholds for threatened species (IUCN, 2009), although it is not known how appropriate this analogy is. No analysis has yet tried to determine how well these existing loss thresholds relate to the likely rate of future ecoregional loss.

Key Biodiversity Areas

Birdlife International use thresholds to define Key Biodiversity Areas (KBAs) (Eken *et al.*, 2004). KBAs are reactive conservation tools designed to '*tackle the main causes of extinctions*' and to identify '*sites where . . . biodiversity must be conserved in the short term.*' KBAs are designed at a much smaller resolution and are also many times more numerous than other priority areas. Also, an area is selected if it passes *any one* of the five thresholds in Table 3.1. The first threshold (vulnerability) is clearly appropriate for reactive conservation. The remaining thresholds, however, are based on 'irreplaceability' (Margules & Pressey, 2000) and seem aimed at protecting 'important' sites (Langhammer *et al.*, 2007) rather than those facing immediate threats. KBAs have a high international profile, but have attracted additional criticism for their omission of landscape processes, and their inflexibility (Knight *et al.*, 2007).

The first problem with threshold prioritization is the particular set of features chosen as criteria. Any choice will inevitably focus on a particular subset of conservation features and ignore the rest. However, because a failure to satisfy any criterion is often sufficient to exclude regions from the prioritization set, a poorly chosen criterion will not just disadvantage valuable regions, but it will exclude them entirely. For example, to qualify as a Biodiversity Hotspot, an ecoregion must have already lost more than 70% of its natural habitat (Table 3.1; Box 3.1). Threats to biodiversity are certainly a key concern for reactive conservation, but the 70% loss threshold may be a poor estimator of ongoing and future extinction risk. Bode *et al.* (2008) calculated expected habitat loss rates based on probabilities of extinction of IUCN-defined at-risk terrestrial vertebrates in the hotspots. However, these predicted loss rates do not correlate with the percentage of habitat lost (Figure 3.1). The reason may be that historical habitat loss does not indicate ongoing degradation, or 30% of a huge habitat, particularly if intact, may still provide adequate protection for many species. Nevertheless, it is clear that this choice of criterion could exclude important areas from the set of priorities.

The second problem with thresholds is their critical value. Thresholds are somewhat arbitrary points on a continuum, and this can easily lead to peculiar decisions, and most typically to contentious omissions. For example, the Queensland wet tropics in Australia is not a Biodiversity Hotspot, despite containing 1200 endemic vascular plant species in only 11 000 km². While

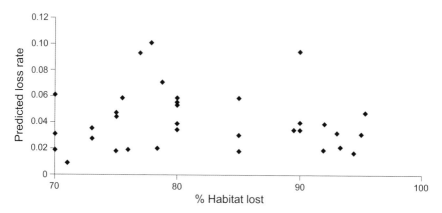

Figure 3.1　**Predicted habitat loss rate versus current fraction of habitat lost in the 34 Biodiversity Hotspots.**

the region falls short of the 1500 species criterion, the density of endemic plants at 10.9 species/km^2 is an order of magnitude higher than in some hotspots, for example central Chile, which contains only 1.1 species/km^2. This would seem to make the Queensland wet tropics a more suitable candidate for '[supporting] *the most species at the least cost*' (Myers *et al.*, 2000). Threshold prioritization will also reject an area that greatly exceeds several criteria but only marginally fails one, while including an area that barely meets all criteria.

The analytical simplicity of threshold prioritization schemes has made them popular tools for designing global priority sets. Other well-known examples include: the Key Biodiversity Areas of Birdlife International and The Nature Conservancy's Crisis Ecoregions (see Box 3.1). The Alliance for Zero Extinction mainly avoids the above pitfalls by prioritizing only sites that contain threatened species (see Table 3.1).

Scoring systems for identifying priority regions

Scoring systems assume that candidate regions can be ranked unambiguously according to a single metric of 'conservation value'. Prioritization is then simply a matter of identifying the highest valued regions. However, conservation value is made up of many different components: the essence of a scoring system is a formula that amalgamates the multiple important facets of conservation value into a single metric.

The single metric offered by scoring systems is necessary because studies frequently show that the different components of biodiversity, for example taxonomic groups or endemic and threatened species, are not reliably congruent in space. Even within endemic terrestrial vertebrates, the number of species of any one class, whether birds, mammals, amphibians or reptiles, in an ecoregion predicts only 24–37% of the variation in the number of species in the other three classes combined (Lamoreux *et al.*, 2006). Therefore, the areas most important for endemic mammals, for example, are unlikely to also be the most important for endemic reptiles. The scoring system approach tries to resolve this ambiguity.

Merging the multiple facets of conservation value is not a simple matter, however, and depends on difficult societal value judgments – for example, how many plant species are of equivalent value to a single primate species? Since no single formula is scientifically 'correct', it is critical that the often contentious value judgments are both explicit and transparent. Unfortunately, the amalgamation that is essential to the process of scoring is antithetical to transparency. We illustrate this problem by discussing the 'Global 200 Ecoregions', a global priority region set introduced by WWF-US, and based on a scoring system described in Box 3.2.

Box 3.2 **A global conservation prioritization area defined by a scoring system**

WWF Fund's Global 200 Priority Ecoregions

WWF's Global 200 Priority Ecoregions constitute 142 terrestrial, 53 freshwater and 43 marine ecoregions (Olson & Dinerstein, 2002), prioritized to protect '*the areas richest in biodiversity*' and '*ecosystems harboring globally important biodiversity and ecological processes*' (Olson & Dinerstein, 1998, 2002). These are proactive objectives, as they do not consider relative urgency.

To rank candidate ecoregions, Olson and Dinerstein (2002) first summed the logarithm of species richness of different taxonomic groups (for North America, mammals, birds, amphibians, reptiles, land snails, butterflies and vascular plants) to get a single 'richness score'. For all ecoregions in each biome (because the scheme seeks to be representative), these scores were divided into four groups by visual inspection and were assigned a value of either 100 (globally outstanding), 15 (high), 10 (medium) or 5 (low). This categorization was then repeated for endemic species, with values of 100, 25, 15 and 5. The authors also used expert opinion to rate the global rarity of each biome, and to score ecoregions within each biome for 'unusual ecological or evolutionary phenomena', assigning scores of 100, 5 or 0. The final score for an ecoregion was the sum of these four scores, with ties resolved in favor of the more intact ecoregion. A final score of ≥ 40 was needed for the ecoregion to be considered a priority, and be placed in the Global 200.

The Global 200 was the first set of global conservation priority areas to cover every habitat type, and to be based on biogeographic principles. However, despite its groundbreaking nature, there are considerable problems with the methods used. By necessity, the formula contains value judgments about the relative importance of its constituent parts. Unfortunately, its opacity makes it difficult to understand why certain regions were prioritized, and in fact hides several variable and arbitrary decisions. The most obvious example is the double counting of endemic species, and possibly triple counting if those species were used to designate a biome as 'rare' (Box 3.2). It is unlikely that conservationists would agree with such a weighting, but the opacity of the scoring formula means they may not realize this decision is being made. The conversion of continuously varying diversity scores into four categories, and then their re-conversion into arbitrarily chosen continuous values (e.g. 100, 15, 10 and 5) is poorly motivated. The logarithmic transformations are also problematic: species counts were transformed so that taxonomically diverse groups did not overwhelm the score. For example, plants can be 100 times more speciose than vertebrates, and the logarithmic transform achieves this. However, it confers a decreasing value to later-counted species, in a manner that is not supported by ecological considerations, and the relative value of protecting, for example, an additional plant versus an additional vertebrate species changes as more species are protected. Each of these arbitrary assignments of relative conservation value is hidden by the single final score for each ecoregion, and it is not clear conservationists would knowingly make them. The complexity and opacity increase when the components are amalgamated: why is an unusual evolutionary phenomenon worth 6.67 times more than a 'high' value of species richness?

Two recent suggestions would make the assignment of value more transparent. First, to increase the focus on endemic species, which removes the need to account for complementarity (Possingham *et al.*, 2006). These endemic species should only be counted once when compiling the metric. Second, the Global Environment Facility (GEF) suggests transforming the number of species in each taxon in an ecoregion into the fraction they form of the total number of species in the taxon (GEF, 2005). Therefore, equal value is assigned to every, say, 10% of plants or vertebrates. This last is, of course, a value judgment, with which there may be disagreement. However, at least it is simple, clear and explicit.

A separate problem is that scoring systems lend themselves to adding incommensurables (Wolman, 2005). The GEF biodiversity score for each country adds components (weighted in various ways) for number of species, number of threatened species, number of ecoregions and number of threatened ecoregions (GEF, 2005; Knight *et al.*, 2007). These scores decide how funds will best 'maximize the impact on global environmental improvements'. However, it is likely that different conservation approaches are needed to maximize the protection of threatened species, the protection of biodiversity in general, or the preservation of ecosystem services. Simply summing these factors results in a metric with less useful information than the individual scores, and obscures the relative importance of each component.

General issues relating to existing approaches

Despite these weaknesses, there is a large overlap in the areas selected by different protocols when goals are similar (Brooks *et al.*, 2006). For example, the 25 Biodiversity Hotspots originally defined by CI overlap by 68%, 82% and 92% the Endemic Bird Areas, Centres of Plant Diversity and the most critical and endangered Global 200 Ecoregions (Myers *et al.*, 2000), respectively. On the other hand, a total of 79% of the Earth's surface has been prioritized by at least one scheme (Brooks *et al.*, 2006).

Most fundamentally, there are two critical omissions from every one of these prioritization schemes. First, conservationists have long understood the importance of costs (Ando *et al.*, 1998; Wilson *et al.*, 2006; Naidoo *et al.*, 2007). Here, needs far exceed resources, and conservation costs vary enormously across the globe, by over seven orders of magnitude (Balmford *et al.*, 2003). However, none of these schemes considers global variation in cost, even though every introduction cites cost-effectiveness as a motivation, For example, 'How can we protect the most species per dollar invested?' (Myers *et al.*, 2000). In recent years, global conservation cost datasets have become freely available, although they do have limitations. Second, current approaches are static, and provide a list of priorities. However, conservation is a dynamic problem because the landscape is dynamic, as land is continually being converted or protected.

Although these shortcomings could be addressed within the threshold/scoring system framework of the existing protocols, their opacity would only be exacerbated. Furthermore, the lack of clear connection between their objective, to conserve biodiversity, and the methods used by the different protocols would make incorporation difficult. *Return on investment* is a new method that can incorporate additional conservation considerations such as costs and dynamics into global conservation prioritization, in a rational and transparent manner.

Return on investment

Return on investment (ROI) analysis is a method for allocating scarce resources among alternative places or activities (or both). In so doing, ROI provides an analytical framework for including complexities such as costs, threats and dynamics into priority setting (Ando *et al.*, 1998; Wilson *et al.*, 2006; Murdoch *et al.*, 2007). Here, we provide a brief primer on ROI.

The mathematical nature of ROI demands an unambiguous definition of the conservation objective. This might be, for example, number of endemic plants protected, the total area of habitat reserved/revegetated, or the monetary value of ecosystem services provided within a system. The explicit nature of the ROI objective facilitates debate about what conservation is designed to achieve, and also provides a clear benchmark for a *post hoc* analysis of how successful planned actions were (Ferraro & Pattanayak, 2006).

The ROI approach assumes a finite conservation resource, typically money. However, it could equally be conservation expertise or available habitat. ROI demands specific estimates of how much each potential conservation action will cost which, although currently scarce in developing countries, are likely to be much easier to obtain than biodiversity data. ROI also demands a method for translating amount invested into conservation benefit obtained. Finally, as in all conservation activities, ROI will usually want information on threats, for example the likelihood of losing a species or habitat if no conservation action is taken.

Two ROI approaches correspond to proactive and reactive conservation (Wilson *et al.*, 2006; this volume, Chapter 2). The 'maximize gain' approach corresponds to the first, seeking to maximize the number of species, or other benefit, protected per dollar invested. The 'minimize loss' approach corresponds to the second, aiming to minimize the number of otherwise expected extinctions (or amount of habitat lost). Consider endemic plant species as an example of conservation benefit. One way to translate budgetary investment into expected conservation benefit is to convert the well-known species–area curve (Figure 3.2a) into a species–investment curve (Figure 3.2b), in this case by assuming the cost of land is the same in both hypothetical ecoregions ($1 million buys 1 km^2). Figure 3.2b now defines how the number of species protected increases with dollars invested. If the objective is to maximize the number of species protected by a reserve, and if there is $10 million to invest

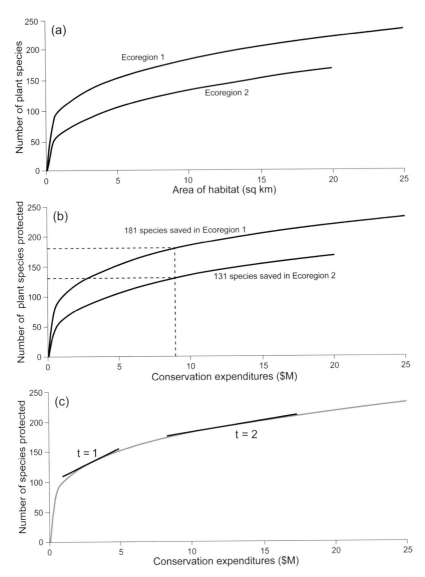

Figure 3.2 (a) Hypothetical species–area curves for two ecoregions. Ecoregion 1 has more species and more remaining habitat. (b) Species–investment curve where the *x*-axis of (a) has been converted to dollars invested. (c) Marginal ROI in ecoregion 1 at two times. At $t = 1$, $3 million has already been invested, and ROI is approximately 11 species/$1 million; at $t = 2$, $13 million has been invested and marginal ROI is ∼4 species/$1 million.

in one of the ecoregions, we can use what we will call a 'simple ROI' analysis (Figure 3.2b). Here the ROI values (namely, species protected per $1 million) are 18.1 and 13.1 in ecoregion 1 and 2, respectively, so we should invest all $10 million in ecoregion 1.

This result assumes that nothing was already protected in either of the ecoregions and that ROI is constant. Frequently, though, some areas will have been protected and we will be interested in the *marginal ROI*, that is, the incremental number of species protected for the next, say, $1 million invested. For simplicity consider just ecoregion 1 and assume that, at time

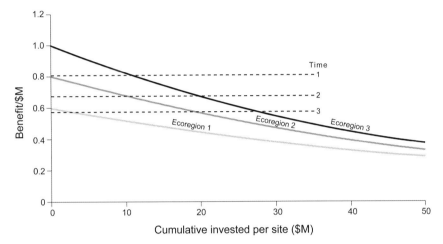

Figure 3.3   **Maximizing marginal ROI across ecoregions, assuming a budget of $50 million over 3 years ($10 million in year 1, $20 million in years 2 and 3): the marginal ROI for three hypothetical ecoregions is shown plotted against resources invested. Each point is the tangent to a species–investment curve (see Figure 3.2c) at the appropriate x value. With no investment, $1 million invested in ecoregion 3 protects one species, but the first $1 million invested in ecoregion 2 will protect only 0.8 species. At $t = 1$, $10 million has been invested. Moving along the x-axis to $10 million, we see that all of it will be allocated to ecoregion 3 because the marginal ROI there (0.805 species per $1 million) exceeds that in the other two ecoregions. At $t = 2$, another $20 million has been invested, this time comprising $10 million in each of ecoregions 3 and 2, by which time the ROI in these two ecoregions is 0.67. At $t = 3$, another $20 million has been invested, this time spending $8 million in ecoregion 3, $9 million in ecoregion 2, and $3 million in ecoregion 1. By $t = 3$, $28 million, $19 million and $3 million has been invested in ecoregions 3, 2 and 1, respectively.**

$t = 1$, \$3 million has already been invested, protecting approximately 140 species (Figure 3.2c). The marginal ROI is the slope of the tangent to the curve at $x = $ \$3 million: approximately 11 additional species per \$1 million. If the existing investment was \$13 million at $t = 2$, the next \$1 million will protect an additional four species (Figure 3.2c). Thus, there are diminishing returns as more is invested in conservation.

The optimal maximize gain rule for investing across multiple areas is to allocate so that the marginal ROI is equal for all regions with positive investment, and to make no investment in areas with lower marginal ROI. This is illustrated for the maximum gain approach in Figure 3.3, which plots the marginal ROI as a function of amount invested for three hypothetical ecoregions, each with its own species – investment curve. A total of \$50 million is invested over 3 years, at the end of which time the ROI has been equalized in all three ecoregions. In addition to a more transparent, optimal method for identifying global priority regions, a ROI approach may frequently solve the problem that different components of conservation value are not well correlated in space, as described in Box 3.3.

Because conservation benefit is counted as all species accumulated from the origin outwards, the maximize gain ROI approach assumes that all species not protected will eventually be lost. If a fraction of the total species in the ecoregion were in fact certain to persist outside the protected area, we would be overestimating the contribution of our protected area to the overall number of species that would persist into the future. The second, minimum loss, approach would instead specifically target species (or areas) that are predicted to be lost in the absence of conservation action. Box 3.4 illustrates an alternative approach that incorporates threat into the maximize gain approach.

Box 3.3 **Return on investment and low feature congruence**

Murdoch *et al.* (2007) demonstrated that a ROI approach to the 21 North American Temperate Forest Ecoregions circumvents the problem of low correlation among different features in space. Vertebrate species richness can predict only about 40% of the variation in plant richness among ecoregions (Figure 3.4a). Next, using plant and vertebrate species richness as separate objectives for conservation resource allocation, we calculate the optimal maximize gain ROI prioritization. In Figure 3.4b, each point is now the simple ROI, i.e. the number of species saved in an ecoregion per \$1 million spent on vertebrates (*x*-axis) or plants (*y*-axis).

Vertebrate ROI now predicts 86% of the variation in plant ROI: we achieve essentially the same allocation whether plants or vertebrates are used. The reason for the huge increase in congruence is that costs vary among ecoregions more (>10-fold) than does species richness (<3-fold). Finally, there is essentially perfect agreement between the optimal allocations using plants or vertebrates when we include the amount of existing protection in each region using a marginal ROI calculation (Figure 3.4c).

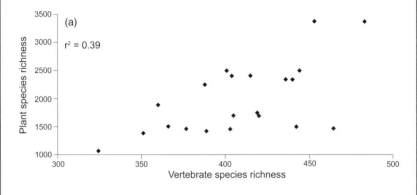

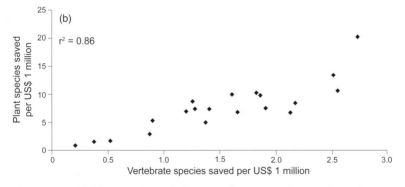

Figure 3.4 (a) **The number of plant species versus the number of vertebrate species in 21 North American temperate deciduous forests. Vertebrate diversity explains 39% of the variation in plant diversity. (b) The number of species saved per $1 million invested, using vertebrates to calculate the simple ROI (see text), plotted against the simple ROI result using plants for the 21 ecoregions. Vertebrate ROI explains 86% of variation in plant ROI. This calculation takes costs into account.** *Continued over page.*

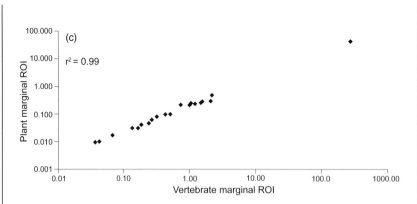

Figure 3.4   (c) The number of additional species saved per additional $1 million invested (marginal ROI), calculated using vertebrates, is plotted against the marginal ROI using plants, for the 21 ecoregions. Vertebrate marginal ROI now explains 99% of the variation in plant marginal ROI (note the logarithmic scale); $r^2 = 0.93$ when the outlying ecoregion is omitted. This analysis adds the effect of area already protected in each ecoregion. (Note that two data points are superimposed in (b) and (c).)

Box 3.4 Incorporating threats into return on investment decisions

An important strategic question is: 'to preserve global biodiversity, is it better to do proactive or reactive conservation?' If resources are to be allocated in a sequence of planning periods stretching into the future, each period covering perhaps 10 or 20 years, it seems optimal to direct funds to reactive conservation – to projects aimed at minimizing, for example, the number of species that will go extinct in the planning period. Indeed, it seems self-evident that a continual process of minimizing the loss of species over each interval must come close to maximizing the number of species that persist. Against this, it might be argued that it is better to protect species in currently safe areas, while they are cheap, rather than later when they may be threatened and more expensive. But protecting unthreatened, inexpensive areas cannot be optimal unless the real cost of preserving them is expected to increase faster (and how much faster would need to be determined) than in currently threatened areas. Alternatively, it might be argued it is foolish to waste money on actions that are doomed to failure in highly threatened areas. However, probability of success can also be incorporated into the ROI calculation.

Threats can be readily incorporated by modifying the conservation benefit to reflect the probability of loss. Suppose, for example, that by spending C dollars in an area we aim to protect an additional S endemic species. But suppose also that without conservation action, the probability of losing these species in the next 10 years is P_L. The conservation benefit, B, derived from the action would then be $B = SP_L$.

We might also weight expected benefit by the probability, P_S, that our action will successfully protect the species (McBride et al., 2007); this is a rational way to include factors such as habitat intactness, level of governance or predicted effect of climate change, rather than adding them to a biodiversity score. We can include these additional factors in a new ROI formulation:

$$\mathrm{ROI} = \frac{B}{C} = \frac{SP_LP_S}{C}.$$

Here no benefit is derived from protecting species that are not threatened during the planning period (i.e. $P_L = 0$), so no funds would be invested in their protection. Despite this, apparently proactive conservation may still be pursued by targeting species with relatively low threats. Consider, for example, two areas, each with 100 species, where the ROI units represent the species saved per $1 million:

Approach	S	P_L	P_S	$C(\$)$	ROI
High threat/low success/high cost	100	0.9	0.2	1×10^7	1.8
Low threat/high success/low cost	100	0.01	0.9	2×10^5	4.5

This analysis suggests we should invest preferentially in the low threat area (even though there is only a 1% chance of losing the species over the next 10 years) because it costs much more to protect the high threat area and our chances of success are low. This implies triage: to maximize protected biodiversity, given limited resources, we may have to let some species or habitats go extinct.

This modified formulation of the maximize gain approach also shows that the ROI approach provides an analytical framework for making trade-offs in conservation – the subject of this book.

Conclusions

Despite their short history, global conservation priority schemes have proven effective tools for focusing conservation actions into a subset of the globe, and for raising conservation awareness and funds. Indeed, Biodiversity Hotspots had helped raise more than US$750 million dollars for conservation by 2003 (Myers, 2003). However, new theoretical insights cast doubt on their optimality through, for example, the omission of cost. They are also static plans, fixed in the dynamic environment of conservation actions, theory and data. Since the last re-definition of the Biodiversity Hotspots (Mittermeier et al., 2004), global datasets of mammal (Carwardine et al., 2008) and amphibian (NatureServe, 2007) species' distributions have been compiled. New socioeconomic datasets are also being continually released on the rates of habitat loss (Hansen et al., 2008) and on the provision of ecosystem services (Turner et al., 2007) as well as on cost.

In spite of their weaknesses, it is unlikely that the priority areas championed by the different NGOs will be abandoned any time soon. But the well-nigh universal goal of protecting the greatest possible biodiversity, given the limited resources available, mandates a robust discussion of whether and how best priority-setting schemes can achieve that goal.

Acknowledgments

We thank David Bael for help with the North American Temperate ROI calculation.

References

Ando, A., Camm, J., Polasky, S. & Solow, A. (1998) Species distributions, land values, and efficient conservation. *Science*, 279, 2126–2128.

Balmford, A., Gaston, K.J., Blyth, S., James, A. & Kapos, V. (2003) Global variation in terrestrial conservation costs, conservation benefits, and unmet conservation needs. *Proceedings of the National Academy of Sciences of the USA*, 100, 1046–1050.

Bode, M., Wilson, K.A., Brooks, T.M. *et al.* (2008) Cost-effective global conservation spending is robust to taxonomic group. *Proceedings of the National Academy of Sciences of the USA*, 105, 6498–6501.

Bottrill, M., Joseph, L., Carwardine, J. *et al.* (2008) Is conservation triage just smart decision-making? *Trends in Ecology and Evolution*, 23, 649–654.

Brooks, T.M., Mittermeier, R.A., da Fonseca, G.A.B. *et al.* (2006) Global biodiversity conservation priorities. *Science*, 313, 58–61.

Carwardine, J., Wilson, K.A., Ceballos, G. *et al.* (2008) Cost-effective priorities for global mammal conservation. *Proceedings of the National Academy of Sciences of the USA*, 105, 11446–11450.

Eken, G., Bennun, L., Brooks, T.M. *et al.* (2004) Key biodiversity areas. *BioScience*, 54, 1110–1118.

Ferraro, P.J. & Pattanayak, S.K. (2006) Money for nothing? A call for empirical evaluation of biodiversity conservation investments. *PLoS Biology*, 4, 482–488.

GEF (Global Environment Facility) (2005) *Implementing the GEF Resource Allocation Framework*. Report GEF/C.27/5/Rev.1. Global Environment Facility, Washington, DC.

Hansen, M.C., Stehman, S.V., Potapov, P.V. *et al.* (2008) Humid tropical forest clearing from 2000 to 2005 quantified by using multitemporal and multiresolution remotely sensed data. *Proceedings of the National Academy of Sciences of the USA*, 105, 9439–9444.

Hoekstra, J.M., Boucher, T.M., Ricketts, T.H. & Roberts, C. (2005) Confronting a biome crisis: global disparities of habitat loss and protection. *Ecology Letters*, 8, 23–29.

IUCN (International Union for the Conservation of Nature) (2003) *United Nations List of Protected Areas*. IUCN, Gland, Switzerland.

IUCN (International Union for the Conservation of Nature) (2009) *The IUCN Red List of Threatened Species*. Available at: http//www.iucnredlist.org.

James, A.N., Gaston, K.J. & Balmford, A. (1999a) Balancing the Earth's accounts. *Nature*, 401, 323–324.

James, A.N., Green, M.J.B. & Paine, J.R. (1999b) *Global Review of Protected Areas Budgets and Staff*. World Conservation Monitoring Centre, Cambridge.

Knight, A.T., Smith, R.J., Cowling, R.M. *et al.* (2007) Improving the key biodiversity areas approach for effective conservation planning. *BioScience*, 57, 256–261.

Lamoreux, J.F., Morrison, J.C., Ricketts, T.H. *et al.* (2006) Global biodiversity conservation priorities. *Nature*, 440, 212–214.

Langhammer, P.F., Bakarr, M.I., Bennun, L.A. *et al.* (2007) *Identification and Gap Analysis of Key Biodiversity Areas: targets for comprehensive protected area systems.* IUCN World Commission on Protected Areas, Gland, Switzerland.

Margules, C.R. & Pressey, R.L. (2000) Systematic conservation planning. *Nature*, 405, 243–253.

McBride, M.F., Wilson, K.A., Bode, M. & Possingham, H.P. (2007) Incorporating the effects of socioeconomic uncertainty into priority setting for conservation investment. *Conservation Biology*, 21, 1463–1474.

Mittermeier, R.A., Mittermeier, C.G., Brooks, T.M. *et al.* (2003) Wilderness and biodiversity conservation. *Proceedings of the National Academy of Sciences of the USA*, 100, 10309–10313.

Mittermeier, R.A., Robles-Gil, P., Hoffmann, M. *et al.* (2004) *Hotspots Revisited: Earth's biologically richest and most endangered terrestrial ecoregions.* CEMEX, Mexico City.

Murdoch, W.W., Polasky, S., Wilson, K.A. *et al.* (2007) Maximizing return on investment in conservation. *Biological Conservation*, 139, 375–388.

Myers, N. (1988) Threatened biotas: hotspots in tropical forests. *The Environmentalist*, 8, 1–20.

Myers, N. (2003) Biodiversity hotspots revisited. *Bioscience*, 53, 916–917.

Myers, N., Mittermeier, R.A., Mittermeier, C.G., da Fonseca, G.A.B. & Kent, J. (2000) Biodiversity hotspots for conservation priorities. *Nature*, 403, 853–858.

Naidoo, R., Balmford, A., Ferraro, P.J. *et al.* (2007) Integrating economic costs into conservation planning. *Trends in Ecology and Evolution*, 21, 681–687.

NatureServe (2007) *InfoNatura: animals and ecosystems of Latin America* [web application]. Version 5.0. NatureServe, Arlington, VA.

Olson, D.M. & Dinerstein, E. (1998) The Global 200: a representative approach to conserving the Earth's most biologically valuable ecoregions. *Conservation Biology*, 12, 502–515.

Olson, D.M. & Dinerstein, E. (2002) The Global 200: priority ecoregions for global conservation. *Annals of the Missouri Botanical Garden*, 89, 199–224.

Possingham, H.P., Wilson, K.A., Andelman, S. & Vynne, C.H. (2006) Protected areas: goals, limitations, and design. In *Principles of Conservation Biology*, eds G.K. Meffe, M.J. Groom & C.R. Carroll, pp. 509–533. Sinauer Associates, Sunderland, MA.

Ricketts, T.H., Dinerstein, E., Boucher, T. *et al.* (2005) Pinpointing and preventing imminent extinctions. *Proceedings of the National Academy of Sciences of the USA*, 102, 18497–18501.

Turner, W.R., Brandon, K., Brooks T.M. *et al.* (2007) Global conservation of biodiversity and ecosystem services. *BioScience*, 57, 868–873.

Wilson, K.A., McBride, M.J., Bode, M. & Possingham, H.P. (2006) Prioritising global conservation efforts. *Nature*, 440, 337–340.

Wolman, A.G. (2005) Measurement and meaningfulness in conservation science. *Conservation Biology*, 20, 1626–1634.

$$4$$

Trade-offs in Making Ecosystem Services and Human Well-being Conservation Priorities

Rebecca L. Goldman[1,2], Gretchen C. Daily[1]
and Peter Kareiva[2]

[1]Department of Biological Sciences, Stanford University, Stanford, California, USA
[2]The Nature Conservancy, Seattle, Washington, USA

Introduction

The Millennium Ecosystem Assessment (MA) reframed conservation as an activity that could benefit people as much as iconic species or biodiversity writ large (MA, 2005). The keys to this hypothesis are two-fold: the obvious fact that nature provides many societal benefits, and the supposition that protecting biodiversity will to help ensure the protection of natural assets to the advantage of people. This is a stark contrast to debates about 'endangered species versus people' or 'conservation refuges'. The MA also emphasized that many ecosystem services are declining at a global scale and issued a call for major initiatives aimed at protecting these services (MA, 2005). This call is not just for biodiversity, but also because economic prosperity and other dimensions of human well-being would suffer if conservation did not advance. The results of the MA (2005) have been impressive. Many

Trade-offs in Conservation: Deciding What to Save, 1st edition.　Edited by N. Leader-Williams, W.M. Adams and R.J. Smith.　© 2010 Blackwell Publishing Ltd.

countries and international institutions have begun to pay attention to their ecosystem services, and conservation non-government organizations (NGOs) now routinely pay attention to resource use and human well-being, as well as to biodiversity.

The emergence of ecosystem services as a new bandwagon in conservation is unarguable (Daily *et al.*, 2009). But it is too early to tell whether conservation aimed at protecting ecosystem services will be successful also at enhancing the conservation of biodiversity. There have been numerous studies demonstrating both the interconnectedness of different ecosystem services and the possibility of trade-offs among different services. Trade-offs are especially important, because they might imply that the conservation of one service is done at the expense or degradation of another (e.g. Heal *et al.*, 2001; Pereira *et al.*, 2005; Rodríguez *et al.*, 2005, 2006). Of particular importance is the possibility that the new-found emphasis on ecosystem services could actually detract from biodiversity conservation. However, other trade-offs may also be important. For example, if priority sites for conservation are selected on the basis of high carbon sequestration or flood control value, will those sites also possess high biodiversity value? Or if projects are designed and managed with special attention to how people benefit, will the attention paid to people take attention away from biodiversity (Kareiva *et al.*, 2008)? In this chapter, we examine the empirical evidence for three critical potential trade-offs involving ecosystem services. First, we examine the lack of spatial congruence in priority areas for ecosystem services and biodiversity. Second, we investigate the lack of congruence in project financing and design that may require favouring either ecosystem services or biodiversity conservation goals. Third, we examine trade-offs in project outcomes when biodiversity is an objective of World Bank development projects. These empirical analyses suggest that biodiversity conservation and ecosystem services or human well-being are not automatically linked, but neither are there any automatic trade-offs. This conclusion points us to the final section, in which we present a modelling, mapping and valuation tool that seeks to integrate multiple ecosystem services so that decision makers can identify win–win solutions as much as possible, and avoid conservation actions that fail to pay sufficient attention to human needs.

Congruence between important sites for biodiversity and ecosystem service provision

A quick answer to the question of whether sites of high biodiversity importance are congruent with sites of high ecosystem service provision (Figure 4.1) is 'not necessarily'. However, this answer indicates that the devil lies in the detail.

Figure 4.1   **The Natural Capital Project is seeking to mainstream biodiversity and ecosystem services into major resource decisions. This project, a partnership among Stanford University, The Nature Conservancy and WWF, is working in a suite of countries globally, including China. Over 18% of China's land area has been designated as Ecological Function Conservation Areas, priority areas to be managed for protecting vital ecosystem services and biodiversity. Here, in the Laojun Mountains of the Upper Yangtze River basin, conservationists work with the Chinese government to ensure provisioning of critical hydrological services and protection of habitat for the endemic Yunnan golden monkey, *Rhinopithecus bieti*. (Photograph by kind permission of Christine Tam.)**

It would be wrong to assume there is always overlap. Equally, it would also be wrong to assume the complete lack of win–win geographic priorities.

Setting geographic priorities for conservation is a well-developed science, supported by research and conservation practice internationally. For many decades, conservation practitioners have focused on preserving species, habitats and ecological communities. To this end, there is a vast literature on 'best' methods for identifying priority areas to achieve a particular conservation goal. More often than not, this goal is to conserve biodiversity and often involves particular attributes of landscapes, such as size, connectivity and uniqueness (e.g. Pressey *et al.*, 1994; Margules & Pressey, 2000; Williams, 2001). Site-selection approaches typically rely on optimization techniques that seek to preserve a maximum amount of biodiversity at minimum cost (Murdoch *et al.*, this volume, Chapter 3). Data on presence and absence of each species for each site are often needed, but algorithms can be employed that use the expected number of species represented in protected areas where these data are lacking (Polasky *et al.*, 2000). More recently, these algorithms have been enhanced to include spatial and temporal dynamics of populations (Cabeza & Moilanen, 2003), information on threats (O'Connor *et al.*, 2003; Costello & Polasky, 2004; Meir *et al.*, 2004), costs (Murdoch *et al.*, 2007) and feasibility (O'Connor *et al.*, 2003). Ecosystem services remain a notable omission from this list. As yet, we know of no implemented conservation plan or protected area network designed with ecosystem services in mind.

In light of the growing attention paid to ecosystem services, both globally and locally (MA, 2005), it is essential to ask whether conservation plans and priorities might change if ecosystem services were to be elevated in importance when selecting conservation areas and strategies. To date, two studies have explicitly sought to quantify the overlap between conservation priorities based on traditional biodiversity dimensions, and priorities based on the delivery of ecosystem services (Chan *et al.*, 2006; Naidoo *et al.*, 2008). Both consider the spatial flow of a particular set of services as compared to areas rich in biodiversity (Figure 4.1), but do so at dramatically different scales.

Chan *et al.* (2006) estimated carbon storage, crop pollination, flood control, forage production, outdoor recreation and water provision for the Central Coast ecoregion of California. The mapping for this analysis was done at the scale of a planning unit measuring 500 ha, over a total of *c.* 12 000 planning units. The service inputs for the mapping had a variety of different grid sizes and effectiveness of conservation metrics associated with them. However, each was scaled to match the 500 ha hexagonal planning units. Naidoo *et al.* (2008) estimated carbon sequestration, carbon storage, production of livestock on grasslands and water provision, on a global scale. They evaluated biodiversity and ecosystem service concentration and overlap along a continuum of scales moving from local, at just over 0 km^2, to global at 100 million km^2. In addition, they analyzed the ability of existing priority areas, comprising Biodiversity Hotspots, High-biodiversity Wilderness Areas and Global 200 Ecoregions (Murdoch *et al.*, this volume, Chapter 3) to conserve ecosystem services. These areas span a range of spatial scales. For example, hotspots can range in area from under $20\,000 \text{ km}^2$ to just over 2 million km^2, as a result of which some scale peculiarities are created (Murdoch *et al.*, this volume, Chapter 3). Naidoo *et al.*'s (2008) study did not account for the effects of actual land use on ecosystem service provision but rather analyzed potential ecosystem service flows from the landscape.

The most striking result from both the global-scale analysis and the regional California analysis is that there was no spatial correlation between metrics of biodiversity and ecosystem service delivery. For the global-scale analysis, see Figure 4.2. Likewise, in the smaller-scale California analyses, similar scatterplots were obtained. In particular, the correlation coefficients between biodiversity and the ecosystems services examined ranged from -0.04 to $+0.12$ (Chan *et al.*, 2006).

Clearly, conservation organizations would be in error making the assumption that regions of high biodiversity importance always deliver high ecosystem services (see Figure 4.1). Similarly, as ecosystem services gain momentum as a focus for conservation, spatial data will make it clear that areas prioritized for delivering high ecosystem services cannot necessarily be expected to also deliver high biodiversity value (Figure 4.3). Importantly, the absence of any spatial correlation between biodiversity and ecosystem services reported by Chan *et al.* (2006) and Naidoo *et al.* (2008) needs to be more widely investigated in other ecosystems and at other spatial scales. For now, the operating assumption should be there is no such correlation.

But all is not lost. Despite the absence of large-scale, significant spatial correlations, there remain opportunities to conserve areas both ripe for biodiversity protection and ecosystem service provision. These win–win situations appear to be affected by scale and by consideration of side benefits not often included in initial planning phases. The areas in quadrant 1 of Figure 4.2 represent areas of high biodiversity and high ecosystem service benefits. However, on closer examination these cases support the importance

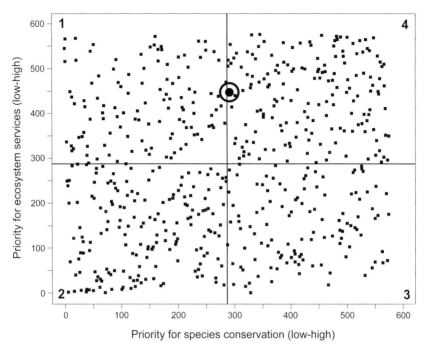

Figure 4.2   **Relationship between ecosystem service importance and species con-servation importance. Each axis is the rank order (low to high) of 574 terrestrial ecoregions. The y-axis shows the per-area carbon storage; the x-axis shows the area-corrected number of endemic species, calculated as (number of endemic species/area$^{0.25}$). Lines indicate median values for each variable. The circle in quadrant 4 represents the California Central Coast ecoregion. Examination of the pairwise correlation coefficients between particular services such as carbon sequestration, carbon storage, grassland production and water provision, shows no correlation >0.20. Correlation coefficients with biodiversity overlap ranged from -0.12 to $+0.58$, again with no significant correlations. (From Naidoo et al., 2008 and reproduced by kind permission of ©2008 The National Academy of Sciences of the USA.)**

of scale for conservation win–wins. While changing the scale does not necessarily change the correlation pattern, it does allow opportunities to find specific places where high ecosystem service values align with high biodiversity values. These regional win–wins indicate essential geographic areas to look for locally advantageous conservation priority sites.

Figure 4.3   **Provision of reliable and abundant water for irrigation is a key ecosystem service, vital to the productivity of wheat grown on the terraces surrounding Stone City, Shitoucheng, in northwest Yunnan Province, China. (Photograph by kind permission of Christine Tam.)**

Optimal site selection for a particular goal does not necessarily lead to optimal conservation design. Important side benefits can be missed if seeking to conserve only areas of 'optimal' overlap. At times, networks of conservation areas can maximize the full suite of benefits, including both biodiversity and ecosystems services, by conserving a set of 'suboptimal' but valuable sites that together coincide to actually provide more benefits overall for both goals. Most effective network design, from the perspective of maximizing ecosystem service and biodiversity values, will be configured to maximize all benefits even if this means conserving areas that are neither the best for conserving ecosystem services nor biodiversity (Chan *et al.*, 2006).

A very appealing notion in conservation is the dream of win–win projects, those places or projects where both biodiversity and human well-being benefit. If there were a positive correlation between biodiversity and ecosystem services, these win–win outcomes might be easy to come by, almost as an expected ancillary benefit of working to protect biodiversity. In the absence of such a correlation, win–wins remain possible, and perhaps even relatively easy with appropriate models and data. The idea here is that even the most ambitious conservation plans anticipate protecting no more that 30–40% of the Earth's land surface. Given that goal, can those places be selected to identify regions that deliver both biodiversity and ecosystem services? The answer from Chan *et al.* (2006) and Naidoo *et al.* (2008) is in the affirmative. However, it is messy and overlap cannot be assumed, although the possibility should not be discarded either.

Lastly, even if there is not perfect spatial overlap, ecosystem service goals may entice new stakeholders, whether funders or partners, to join conservation efforts. On the other hand, there is the risk of distracting attention and resources away from biodiversity conservation by including ecosystem service provision. It is to these trade-offs that we turn next.

Financial support for conserving ecosystem services and biodiversity

A major motivation for broadening the ambit of conservation to include ecosystems services (see Figures 4.1, 4.3) is the potential to broaden public support, both financial and otherwise. The risk is that trying to both conserve biodiversity and provide ecosystem services may mean failing at both. Everyone has personal experience of trying to do too many different things that impairs an ability to do any one thing well. So the hypothesis of a financial gain due to including ecosystems services in conservation objectives needs to be tested, along with the risk of reduced attention to biodiversity conservation. To date, only one study has examined this issue, by asking two questions: (i) do ecosystem service projects attract new revenue sources into conservation or do they draw down on currently available, limited resources; and (ii) if new revenue sources are generated for ecosystem service projects, do such projects still pay attention to, or veer away from, biodiversity conservation objectives?

A study of 60 conservation projects conducted by The Nature Conservancy and its partners revealed that ecosystem service projects appear to attract more new money for conservation (Goldman *et al.*, 2008). To reach this conclusion, the researchers used interviews to understand the project implementation of 34 ecosystem service projects and 26 biodiversity projects. They defined biodiversity projects as those that only have species and habitat goals, while ecosystem service projects have at least one explicit goal of a service in addition to biodiversity goals. The interviews allowed them to populate a database of over 500 project attributes (Goldman *et al.*, 2008; Goldman & Tallis, 2009; Tallis *et al.*, 2009).

In terms of revenue sources, Goldman *et al.* (2008) found two significant results for trade-offs associated with biodiversity conservation and ecosystem service provision approaches. Ecosystem service projects tend to draw in more revenue overall, and significantly more overall and on a per project basis, from private and corporate sources. Expanding conservation goals beyond species and habitats attracts a new set of conservation funders, including major energy companies, bottling companies, palm oil firms, soft drink companies and others. Each of these funders had a direct stake in a particular ecosystem service of concern to their business.

There were significantly more corporate funding sources for ecosystem service projects than for biodiversity projects. In fact, a subset of the ecosystem service and biodiversity projects showed that about 44% of all funding for ecosystem service projects came from these sources. In contrast, only about 9% of the revenue supporting biodiversity conservation originated from these sources. Additionally, in terms of overall percentage of funder types per project, corporate funders were significantly more involved in ecosystem service projects than in biodiversity projects. All other funder types per project were about the same on average, and these included: federal government, state government, local government, non-profit sources, individual private landowners and various university funders.

This still begs the question of whether this new revenue is being completely diverted to conservation actions that do *not* follow traditional approaches for species and habitat conservation, or if the additional funding is being spread between actions that target both biodiversity conservation and/or ecosystem service provision. Encouragingly, Goldman *et al.* (2008) found that there is more money available for ecosystem services without neglecting biodiversity. To measure this, they defined a set of traditional conservation actions and activities, namely: supporting current, and creating new, protected areas; and hiring guards to ensure compliance with protected area management goals. The frequencies with which both ecosystem service and biodiversity projects encouraged these activities were very similar. Thus, it would appear that ecosystem service projects encourage extra revenue to expand protected areas on the landscape, thereby helping conserve species and habitats using more traditional strategies, while still providing ecosystem services.

Beyond these financial trade-offs, Goldman *et al.* (2008) discovered ecosystem service projects are far more likely to include private lands in conservation. In other words, the key conservation strategies should not include purchasing the land or changing land ownership. Instead, new people, often-times poor landowners, should be engaged in conservation efforts. Inevitably, this will lead to more explicit development questions and the ability of conservation efforts to meaningfully include development issues. On the flip side, some development agencies include conservation goals in their mission, such as the World Bank. Therefore, we next ask what trade-offs arise from these associations and how explicit development goals fare in the face of efforts to conserve biodiversity.

Do World Bank projects achieve conservation and development goals?

Given population growth and the current economic crisis, it is essential to determine whether there is a trade-off between economic development and conservation. In theory, ecosystem services should be connected to human welfare when populations depend on nature for food and water, as is often the case in developing economies. But are development goals truly compatible with conservation? A recent study used World Bank projects to address this question (see Kareiva *et al.*, 2008 for further details). Drawing randomly from a much larger sample of projects, Kareiva *et al.* (2008) selected 97 projects with conservation goals and 97 projects without conservation goals, matched by country and all implemented between July 1998 and August 2006. They found that when ranked on overall performance, development-only as opposed to development + environment rankings were similar. Environment projects were defined fairly broadly to include irrigation and water quality improvements.

Looking more specifically at biodiversity trade-offs, there was again no evidence to indicate that the addition of biodiversity goals led to any negative changes in project success as measured by development progress, even on a 'per objective' basis. In other words, biodiversity projects did not under-perform compared to development-only projects with regards to financial, poverty alleviation or gender objectives. However, there were a couple negative trade-offs associated with biodiversity projects. First, if biodiversity was not included as a goal (but was instead subsumed under an 'environment'-only goal), then environmental outcomes suffered. This seems to indicate that unless the conservation-related goal is made explicit, it is neglected. Second, within biodiversity projects, the more money that was spent on achieving biodiversity aims, the less likely was the project to succeed with regards to poverty reduction aims. Thus, it is possible that social goals suffered due to inclusion of biodiversity conservation goals (Kareiva *et al.*, 2008).

Interestingly, Kareiva *et al.* (2008) found that the only outcomes associated with successful biodiversity projects were sustainable financing and market mechanisms. This finding implies that market development and new finance sources are integral to conservation success. More broadly, less than 20% of all projects, whether development, development + environment, development + environment + biodiversity, were rated highly satisfactory. The main con-clusion from Kareiva *et al.* (2008) is that, while combining development goals and biodiversity goals does not necessarily lead to a win–win conservation scenario, it would be wrong to make the *a priori* assumption that such a combination is perilous. Moreover, because there is evidence that the envi-ronment suffers if conservation goals are not present, then it would seem that all development projects should explicitly include conservation goals.

There is a rich literature of case studies of conservation and development projects that do not take as strongly statistical an approach as did Kareiva *et al.* (2008), but that are nevertheless informative. What the case studies can reveal that the large statistical analyses cannot, is the nature of the specific mechanisms by which conservation and development might represent a trade-off. For example, protected area strategies can exacerbate poverty as former land users are evicted from their lands and separated from access to critical resources (e.g. Geisler & de Sousa, 2001; Colchester, 2002; Fortwangler, 2003). In fact, most of the costs of biodiversity conservation are suffered by the relatively poor and impoverished (Roe & Elliot, 2004; Roe & Walpole, this volume, Chapter 9). This ongoing debate and trade-off assessment (see Adams *et al.*, 2004) led to an international agreement at the World Summit on Sustainable Development in Johannesburg in 2002 that biodiversity and resource conservation need to be integrated into economic development, to avoid having to make choices between humans and conservation objectives (LWAG, 2002). Despite this global decree and intellectual understanding of the interconnectedness between human health and the environment (MA, 2005), very few projects demonstrate win–win scenarios for conservation and development. Viewed in the light of Kareiva *et al.*'s (2008) statistical analyses, this poor track record for pro-development conservation projects might simply arise because conservation by itself is difficult, development by itself is difficult, and when two difficult challenges are combined the chances of succeeding at both are very low. Note that this argument does not require any assumptions of trade-offs – only the realization that succeeding at two difficult tasks is the product of succeeding at each task by itself, assuming independence, which is a statistical way of saying there is no trade-off.

Integrating multiple ecosystem services in a mapping tool

The clear message from empirical examinations of the link between ecosystem services and biodiversity is that there are no guarantees, one way or another (Table 4.1). Spatially, protecting areas rich in biodiversity does not necessarily mean that an abundance of ecosystem services will also be captured. Conservation that successfully protects biodiversity and advances human well-being is very hard to accomplish.

If ecosystem services are to become a priority for conservation, conservation planners need to be really smart in their approach. To that end, it is essential to generate explicit maps that depict the geographic variation of multiple ecosystem services and allow assessment of trade-offs among ecosystem services, including biodiversity, that arise as a result of different land uses and land management strategies. One attempt to produce maps of ecosystem services was led by Costanza *et al.* (1997), who produced what amounted to 'look-up tables' of the total dollar value of different habitat types (Table 4.2). For example, wetlands and marshes were assigned a value of US\$14 785/ha, from which it was relatively easy to produce maps of the value of ecosystem services. However, the approach taken by Costanza *et al.* (1997) was criticized by economists for various reasons (for a discussion of these and other criticisms, see Costanza *et al.*, 1998). Paramount among the reasons was that, although simple, this 'benefits transfer' approach incorrectly assumed that every hectare of a given habitat type was of equal value. In addition,

Table 4.1 **Summary of trade-offs between ecosystem services and biodiversity, where '+' means that there can be positive correlations between the conservation goal and the project feature (e.g. biodiversity only can be conserved successfully at a local scale), '–' indicates there are mostly negative correlations, '+ and –' means that both can be found, and '+/–' means the relationship is unclear (i.e. there are no clear patterns and conservation of biodiversity does not necessarily mean conservation of ecosystem services). '++' indicates the positive effect is amplified.**

Conservation goal	Spatial design for conservation priorities			Financial	Outcomes/ success
	Local	Regional	Global		
Biodiversity only	+	+	+	+	?
Ecosystem services and biodiversity	+	+ and –	-	++	+/–
Biodiversity plus ecosystem services and/or development	+ and –	+/–	+/–	++	+ and –

information on the impacts of land use management on ecosystem service provision is necessary to design policies or payment programmes that provide a desired level of ecosystem services.

An alternative approach, therefore, is to explicitly model the production of specific ecosystem services and to then use the mapped production of multiple services to evaluate the effects on flows of different land use scenarios (Nelson *et al.*, 2009). This approach has been used by the Natural Capital Project, which has developed a suite of models that convert land use and land cover maps into rates of production for the following ecosystem services: agricultural production, forest production, carbon sequestration, flood control, water quality, non-timber forest products, recreation and sediment retention. These models have been applied to the Willamette Valley in Oregon across three different future scenarios for land use: (i) development, defined as a loosening of current regulations to allow freer reign to market forces; (ii) planned trend, defined as a future where existing regulations and policies stay as they are and current trends continue; and (iii) conservation, defined as placing greater emphasis on ecosystem service and biodiversity protection. The trade-off

Table 4.2 **Estimated economic value per year for a sample of ecosystem types.** (**Values and table adapted from Costanza *et al.*, 1997, with permission.**)

Ecosystem	Services provided	Value per ha per year	Total global value per year
Coastal estuaries	Disturbance regulation; nutrient cycling; biological control; habitat; food production; raw materials; recreation; cultural	$22 832	$4.11 billion
Tropical forest	Climate regulation; disturbance regulation; water regulation; water supply; erosion control; soil formation; nutrient cycling; waste treatment; food production; raw materials; genetic resources; recreation; cultural	$2007	$3.813 billion
Wetlands	Gas regulation; disturbance regulation; water regulation; water supply; waste treatment; biological control; food production; raw materials; recreation; cultural	$14 785	$4.879 billion
Lakes and rivers	Water regulation; water supply; waste treatment; food production; recreation	$8498	$1.7 billion

question is addressed by asking whether the different ecosystem services change in concert or in opposing directions across the different land use scenarios.

In this specific example, Nelson *et al.* (2009) found little evidence of any trade-offs among the different ecosystem services. For example, relative to development or existing planned trend, the conservation scenario resulted in improvements in all of the ecosystem services. This analysis was done at a scale of 30 × 30 m grid cells. However, the movement towards conservation represented by the conservation scenario would result in lower market yields from agriculture and forestry. In other words, there was a trade-off between conservation and economic growth. Interestingly, under the assumption of a carbon market value of $43/megaton, the so-called social value of carbon sequestration, then economic growth and conservation are aligned. Thus the conservation scenario improves biodiversity status, all ecosystem services and market returns.

The results reported by Nelson *et al.* (2009) are unlikely to be generally replicable because of the unusually pro-environment leanings of Oregon. In particular, none of the three scenarios considered represented severe land conversion or resource extraction. If the scenarios were more disparate, and the range of land uses more extreme, trade-offs among ecosystem services could well be detected. The important lesson is the clarity that can be achieved by a formal quantitative analysis of ecosystem services and economic activity. The hope is that the new tools being developed by the Natural Capital Project and others will help decision makers identify land use scenarios that effectively merge human development concerns with the protection and preservation of our natural resources.

Conclusions

The importance of understanding the nature of trade-offs among ecosystem services, or trade-offs between people and conservation, cannot be overestimated. Here, we have evaluated several recent studies that have examined such trade-offs (summarized in Table 4.1). To date, there is no compelling evidence of striking trade-offs between ecosystem service provision and biodiversity conservation, or between biodiversity conservation and economic development. More certain, however, is the lack of evidence that pursuing biodiversity conservation leads automatically to enhanced delivery of ecosystem services to people. Given the absence of simple trade-offs or clear correlations, efforts that combine biodiversity conservation and ecosystem service provision will need to invest heavily in adaptive management. Unfortunately, ecosystem services projects, like most conservation projects, do a poor job of monitoring and evaluating effectiveness (Ferraro & Pattanayak, 2006; Goldman *et al.*, 2008; Kapos *et al.*, this volume, Chapter 5). Given that ecosystem service projects will have to make claims about direct links between conservation efforts and human welfare, conservationists need to make sure those claims are backed up by solid data.

Acknowledgments

We thank our collaborators on the Natural Capital Project, including WWF, The Nature Conservancy and Stanford University.

References

Adams, W.M., Aveling, R., Brockington, D. *et al.* (2004) Biodiversity conservation and the eradication of poverty. *Science* 306, 1146–1149.

Cabeza, M. & Moilanen, A. (2003) Site-selection algorithms and habitat loss. *Conservation Biology*, 17, 1402–1413.

Chan, K.M.A., Shaw, R., Cameron, D.R., Underwood, E.C. & Daily G.C. (2006) Conservation planning for ecosystem services. *PLoS Biology*, 4, 2138–2152.

Colchester, M. (2002) *Salvaging Nature: indigenous peoples, protected areas and biodiversity conservation.* World Rainforest Movement, Montevideo, Uruguay.

Costanza, R., d'Arge, R., de Groot, R. *et al.* (1997) The value of the world's ecosystem services and natural capital. *Nature*, 387, 253–260.

Costanza, R., d'Arge, R., de Groot, R. *et al.* (1998) The value of ecosystem services: putting the issues in perspective. *Ecological Economics*, 25, 67–72.

Costello, C. & Polasky, S. (2004) Dynamic reserve site selection. *Resource and Energy Economics*, 26, 157–174.

Daily, G.C., Polasky, S.M., Goldstein, J.H. *et al.* (2009) Ecosystem services in decision-making: time to deliver. *Frontiers in Ecology and the Environment*, 7, 21–28.

Ferraro, P.J. & Pattanayak, S.K. (2006) Money for nothing? A call for empirical evaluation of biodiversity conservation investments. *PLoS Biology*, 4, 482–488.

Fortwangler, C.L. (2003) The winding road: incorporating social justice and human rights into protected area policies. In *Contested Nature: promoting international biodiversity with social justice in the twenty-first century*, eds S.R. Brechin, P.R. Wilshusen, C.L. Fortwangler & P.C. West, pp. 25–40. State University of New York Press, Albany, NY.

Geisler, C. & de Sousa, R. (2001) From refuge to refugee: the African case. *Public Administration and Development*, 21, 159–170.

Goldman, R.L. & Tallis, H. (2009) A critical analysis of ecosystem services as a tool in conservation projects: the possible perils, the promises, and the partnerships. In *Annals of the New York Academy of Sciences: The Year in Ecology and Conservation Biology*, eds R.S. Ostfeld & W.H. Schlesinger, pp. 63–79. Wiley-Blackwell, Boston, MA.

Goldman, R.L., Tallis, H., Kareiva, P. & Daily, G.C. (2008) Field evidence that ecosystem service projects support biodiversity and diversify options. *Proceedings of the National Academy of Sciences of the USA*, 105, 9445–9448.

Heal, G., Daily, G.C., Ehrlich, P.R. *et al.* (2001) Protecting natural capital through ecosystem service districts. *Stanford Environmental Law Journal*, 20, 333–364.

Kareiva, P., Chang, A. & Marvier, M. (2008) Development and conservation goals in World Bank projects. *Science* 321, 1638–1639.

LWAG (Livestock and Wildlife Advisory Group) (2002) *Wildlife and Poverty Study*. Department for International Development, London.

Margules, C.R. & Pressey, R.L. (2000) Systematic conservation planning. *Nature*, 405, 243–253.

MA (Millennium Ecosystem Assessment) (2005) *Ecosystems and Human Well-being: synthesis*. Island Press, Washington, DC.

Meir, E., Andelman, S. & Possingham, H.P. (2004) Does conservation planning matter in a dynamic and uncertain world? *Ecology Letters*, 7, 615–622.

Murdoch, W.W., Polasky, S., Wilson, K.A. *et al.* (2007) Maximizing return on investment in conservation. *Biological Conservation*, 139, 375–388.

Naidoo, R., Balmford, A., Costanza, R. *et al.* (2008) Global mapping of ecosystem services and conservation priorities. *Proceedings of the National Academy of Sciences of the USA*, 105, 9495–9500.

Nelson, E., Mendoza, G., Regetz, J. *et al.* (2009) Modeling multiple ecosystem services, biodiversity conservation, commodity production, and tradeoffs at landscape scales. *Frontiers in Ecology and the Environment*, 7, 1–11.

O'Connor, C., Marvier, M. & Kareiva, P. (2003) Biological vs. social, economic and political priority-setting in conservation. *Ecology Letters*, 6, 706–711.

Pereira, H.M., Reyers, B., Watanabe, M. *et al.* (2005) Condition and trends of ecosystem services and biodiversity. In *Millennium Ecosystem Assessment: sub-global assessments and working group*, Vol. 4, eds C. Samper, D. Capistrano, C. Raudsepp-Hearne & M.J. Lee, pp. 171–203. Island Press, Washington, DC.

Polasky, S., Camm, J.D., Solow, A.R. *et al.* (2000) Choosing reserve networks with incomplete species information. *Biological Conservation*, 94, 1–10.

Pressey, R.L., Johnson, I.R. & Wilson, P.D. (1994) Shades of irreplaceability: towards a measure of the contribution of sites to a reservation goal. *Biodiversity and Conservation*, 3, 242–262.

Rodríguez, J.P., Beard, T.D. Jr., Agard, J. *et al.* (2005) Interactions among ecosystem services. In *Millennium Ecosystem Assessment: ecosystems and human-well being, scenarios*, Vol. 2, eds S.R. Carpenter, P.L. Pingali, E.M. Bennett & M.B. Zurek, pp. 431–448. Island Press, Washington, DC.

Rodríguez, J.P., Beard, T.D. Jr., Bennett, E.M. *et al.* (2006) Trade-offs across space, time, and ecosystem services. *Ecology and Society*, 11, 28.

Roe, D. & Elliott, J. (2004) Poverty reduction and biodiversity conservation: rebuilding the bridges. *Oryx*, 38, 137–139.

Tallis, H., Goldman, R., Uhl, M. & Brosi, B. (2009) Integrating conservation and development in the field: implementing ecosystem service projects. *Frontiers in Ecology and the Environment*, 7, 12–20.

Williams, P.H. (2001) Complementarity. In *Encyclopedia of Biodiversity*, ed. S.A. Levin, pp. 813–829. Academic Press, London.

5

Defining and Measuring Success in Conservation

Valerie Kapos[1,2], Andrea Manica[3], Rosalind Aveling[4], Philip Bubb[2], Peter Carey[5], Abigail Entwistle[4], John Hopkins[6], Teresa Mulliken[7], Roger Safford[8], Alison Stattersfield[8], Matthew J. Walpole[4], and Andrew Balmford[3]

[1]Cambridge Conservation Forum, c/o Department of Zoology, University of Cambridge, Cambridge, UK
[2]United Nations Environment Programme World Conservation Monitoring Centre, Cambridge, UK
[3]Department of Zoology, University of Cambridge, Cambridge, UK
[4]Fauna & Flora International, Cambridge, UK
[5]Centre for Ecology and Hydrology, Monks Wood, Huntingdon, Cambridgeshire, UK
[6]Natural England, Peterborough, Cambridgeshire, UK
[7]TRAFFIC International, Cambridge, UK
[8]BirdLife International, Cambridge, UK

Introduction

It is well recognized that conservation action cannot take place everywhere and that trade-offs have to be made in choosing priority locations for conservation efforts. Similarly, limitations on the resources available for conservation mean that choices must be made about what conservation actions to take and what

Trade-offs in Conservation: Deciding What to Save, 1st edition.   Edited by N. Leader-Williams, W.M. Adams and R.J. Smith.   © 2010 Blackwell Publishing Ltd.

approaches to use. Therefore, it is increasingly important to assess conservation achievements and to identify approaches that use scarce resources efficiently and effectively (Salafsky *et al.*, 2002; Christensen, 2003; Sutherland *et al.*, 2004; Ferraro & Pattanayak, 2006).

While there is a great deal of accumulated experience in conservation, which should be a rich source of information on achievements and factors affecting them, that experience is often poorly assessed and documented, and not available in any systematic form. The need for information to identify successful examples and approaches creates a further tension in the use of scarce conservation resources. Consequently, there is a trade-off between using resources for action and using them for monitoring and evaluation and for sharing experience amongst practitioners. There is strong pressure to use all available resources for action, but if the effectiveness of that action is not assessed there is no basis for adaptive management or for learning systematically from a range of conservation experiences. The need for rigour in such assessments creates a further dilemma. Experimental or quasi-experimental approaches to programme design and evaluation can help to provide rigour (Ferraro & Pattanayak, 2006), but they are frequently politically, logistically and/or scientifically impractical and place further demands on capacity and resources.

In this chapter we present experience gained in developing tools to help conservation practitioners address these trade-offs. The tools help practitioners in assessing the achievements of conservation projects, identifying the approaches that work best and the factors that help to determine success, and demonstrating the cumulative impacts of conservation efforts. The users and audiences for such tools and their application include: conservation professionals and organizations, who need to design new projects and programmes and prioritize their investments; donors, who wish to ensure that their investments in conservation are as effective as possible; decision makers in governments and elsewhere, who must prioritize actions and allocate resources to support them and are accountable for progress and achievements; and the public, to whom decision makers and conservation organizations are accountable and who wish to understand whether and how progress is being made in conserving biodiversity.

Background to assessing success in conservation

In the past, the most commonly reported measures of project success were those addressing *inputs* (such as money and time spent) and *implementation* (activities completed). More recently, there has been increasing emphasis on reporting project *outputs*, concrete and countable products such as numbers of publications produced, numbers of hits on websites and numbers of people participating in events or courses. However, assessing success and identifying effective approaches in conservation require moving beyond such commonly employed indicators to measures that assess an intervention's *outcomes* (how it affects a problem) and their *conservation effects* (project-scale impacts on target ecosystems, habitats, species or populations).

While growing efforts in biological monitoring are providing the basis for assessing changes in conservation status, such changes are often apparent only over the long term and may be influenced by complex interactions among many different factors and interventions. Some recent studies have shown how rigorous analyses can help to elucidate the longer term effectiveness of specific conservation interventions (e.g. Gusset *et al.*, 2007; Andam *et al.*, 2008; Linkie *et al.*, 2008), but such studies require significant investment of time and resources that is beyond the reach of many conservation projects and programmes. Moreover, by the time biological change has happened (or failed to happen), it may be too late to address its causes. Distinguishing between more and less successful approaches over shorter time scales, and managing interventions adaptively, require tools for measuring threat reduction (Salafsky & Margoluis, 1999) and other intermediate outcomes (Salafsky *et al.*, 2002; CMP, 2004; Stem *et al.*, 2005).

Recent efforts to provide such tools have mostly fallen into three broad categories. Firstly, planning-based approaches such as logical frameworks and results chains (Salafsky *et al.*, 2001) help practitioners to articulate how interventions are expected to lead to conservation outcomes and, if properly applied, should help to identify key parameters for monitoring. A second approach to improving conservation practice and associated monitoring has been through the development and application of standards of good practice, which often include the use of planning-based tools. Examples include the Open Standards for Conservation Practice formulated by the Conservation Measures Partnership (CMP, 2004) and the framework for assessing management effectiveness of protected areas (Hockings *et al.*, 2006). Assessing adherence to such standards enables organizations to identify projects working according to best practice and to work to improve those that are not (O'Neil, 2007; Stolton *et al.*, 2008). However, applying standards and assessing adherence do not in themselves ensure positive conservation outcomes, evaluate their achievement, or provide a basis for systematizing or analyzing conservation experience. Thirdly, 'results-based' approaches have been used to evaluate the success of groups of conservation interventions.

These include: complex analysis of quantitative data (e.g. Andam *et al.*, 2008; Linkie *et al.*, 2008); scorecard approaches that have been used to evaluate the predicted outcomes and impacts of agri-environment scheme agreements in relation to their objectives (Carey *et al.*, 2003, 2005); species conservation projects and programmes (Jepson & Canney, 2003; Gratwicke *et al.*, 2007); and assessments of the impacts of a diverse range of zoo-funded projects (the Zoo Measures Group: Mace *et al.*, 2007). The latter study expressed the conservation impact of projects as a product of the 'importance' of their conservation targets, the 'volume' or scale of the potential effect on the targets, and the actual effect of the project (Mace *et al.*, 2007). While this is a promising approach, it has proved difficult to implement consistently (Walter, 2005), and the 'effect' portion of the assessment was particularly problematic.

A need remains for tools that can help in consistent evaluation of the effectiveness of a wide range of conservation actions and thus support synthesis of conservation experience and evidence-based conservation (Sutherland *et al.*, 2004) as well as analysis of factors affecting conservation outcomes. The Cambridge Conservation Forum[1] (CCF), a network of conservation organizations based in and around Cambridge, UK, has worked collaboratively to develop tools that meet this need for a range of conservation organizations.

A new approach to measuring conservation success

Building on the work of the Zoo Measures Group (Mace *et al.*, 2007), CCF concentrated principally on developing ways to assess 'conservation effect', project-scale impacts on target ecosystems, habitats, species or populations, and the outcomes relating to it. It agreed to define conservation success as '*increasing the likelihood of persistence of native ecosystems, habitats, species and/or populations in the wild (without adverse effects on human well-being)*'. CCF also recognized that conservation projects and programmes characteristically involve several different types of activity, which may each have different outcomes and appropriate measures.

Based on existing categorizations of conservation action (IUCN-CMP, 2006: 12; Salafsky *et al.*, 2002) and as a result of broad consultation within CCF, seven broad categories of conservation activity have been defined (Kapos *et al.*, 2008) that together encompass most of the work that CCF members and other

---

[1] www.cambridgeconservationforum.org.uk.

conservation organizations undertake. These types of action split roughly into two groups: two categories involving direct management of conservation targets such as species or sites, and five that influence conservation status indirectly, through work on policy and legislation, enhanced or alternative livelihoods, capacity building, education, or research (Figure 5.1). Typically, conservation projects include several activity types. An *ad hoc* set of 22 projects offered by CCF member organizations as test subjects ranged in

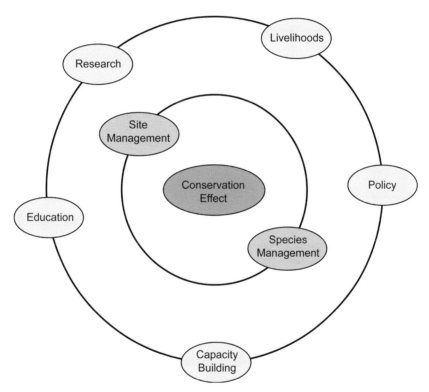

Figure 5.1   **A simple representation of the relationship between different types of conservation activity and conservation impact. Species and site management are more closely linked to conservation impact (see Figure 5.2) than the other, more distant categories of conservation action. For activity types on the outer ring, the links between implementation and conservation impact are more complex (see Figures 5.3–5.7). (From Kapos *et al.* 2008 and reproduced by kind permission of ©2008 Wiley Periodicals Inc.)**

complexity from a single activity type (mostly capacity building or education) to projects involving all seven, and the majority (17) consisted of three to five activity types.

Thematic working groups comprising in total over 50 individuals drawn from more than 20 organizations across CCF focused on each of these activity types to identify common issues in its evaluation and develop a framework and key questions to assess its outcomes. A steering group provided direction and helped to harmonize outputs among the different groups.

Despite the broad diversity of 'communities of practice' represented by the different activity types and working groups, they identified some common obstacles encountered in efforts to evaluate the success of conservation projects (Kapos *et al.*, 2008), including:

- Lack of clear objectives and explicitly articulated assumptions.
- Conservation impacts mostly occur outside project time frames.
- Scarcity of adequate resources and capacity.
- Variation in emphasis among objectives.
- Problems with information management.
- Limited incentives and motivation for evaluation.

The CCF framework and evaluation tool

To help address these issues, CCF working groups developed a framework for evaluation that addresses each type of conservation action in turn (Figures 5.2–5.7). For each action, a conceptual model was developed of the likely relationships between its successful implementation and conservation impact, making explicit the linkages that are often assumed (Box 5.1). These generic 'results chains' (Salafsky *et al.*, 2001) provide a useful framework for assessing the outcomes and effectiveness of conservation projects and for planning the monitoring required to track project outcomes. The models differ strikingly in complexity, between the activity types most directly linked to effects on conservation targets, and those less directly linked to conservation effect. Thus, species and site management tend to act directly on either the threats to a conservation target or the ability of the target to resist or respond to those threats (Figure 5.2). For each of the activity types conceptually more remote from effects on conservation targets, comprising the outer 'orbit' in Figure 5.1, the chain of links between implementation and conservation effect is longer (Figures 5.3–5.7) than for species and site management (Figure 5.2). All of these complex models include 'key outcomes' that occur earlier in the sequence than, and are fundamental to, reducing threats to and/or improving the responses of conservation targets, as follows:

- Livelihoods-related projects (Figure 5.3): abandonment of damaging practices by the target group (or maintenance of sustainable practices).
- Policy work (Figure 5.4): implementation of the policies or legislation promoted.
- Capacity building (Figure 5.5): increases in the quantity and/or quality of conservation action.
- Education and awareness raising (Figure 5.6): change in behaviour by the audience ultimately targeted by the work.
- Research (Figure 5.7): application of research results to conservation practice.

Box 5.1 Conceptual models forming CCF's framework for evaluating success of conservation actions

Conservation actions represent the generic processes by which different types of conservation activity can lead to conservation effects. The components of the flow charts can be divided broadly into implementation, outcomes and effects, separated by horizontal lines (Figures 5.2–5.7). Hexagons and grey arrows denote inputs from other types of conservation activity, but these are indicative rather than exhaustive. While engagement with stakeholders is principally depicted at the early stages of each process, it is fundamental throughout, as are the iterative feedback loops characteristic of adaptive management, which have been omitted for visual clarity. In each of the more complex models (Figures 5.3–5.7) a star indicates the 'key outcome' that is fundamental to reducing threats to, or improving responses of, conservation targets.

Using this framework, CCF has developed a questionnaire-based tool designed to help practitioners assess the outcomes and conservation effect of

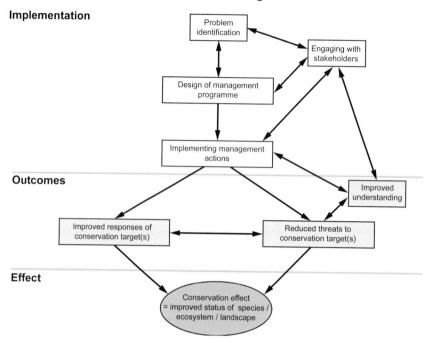

Species or Site Management

Implementation

Problem identification

Engaging with stakeholders

Design of management programme

Implementing management actions

Outcomes

Improved understanding

Improved responses of conservation target(s)

Reduced threats to conservation target(s)

Effect

Conservation effect = improved status of species / ecosystem / landscape

Figure 5.2 **Conceptual model for species or site management projects. A single model is shown for species and site management, which are directly analogous, and have the simplest models. For these activities, implementing management actions leads directly to reduction of threats to the conservation target(s) and/or an improvement in the responses of the conservation target to those threats. (Based on Kapos** *et al.* **2008 and reproduced by kind permission of ©2008 Wiley Periodicals Inc.)**

different sorts of projects in a systematic and consistent manner (Kapos *et al.*, 2008, 2009). For each activity type a single carefully worded questionnaire poses key questions on implementation and outputs, outcomes and conservation effect, using a scorecard approach, and includes a background section on project development and process. The questionnaire also requires that the project-specific meaning of the answer and the evidence on which it is selected are made explicit, and that the evidence is categorized as: (i) opinion, (ii) supported opinion, or (iii) hard evidence. Consistent options address

Livelihoods Enhancement and Alternatives

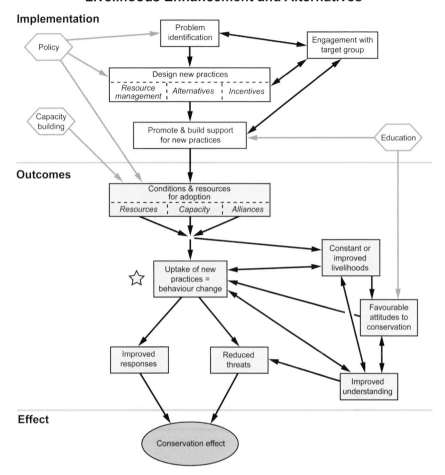

Figure 5.3   Conceptual model for livelihoods-related projects. These projects act by helping to develop sustainable management and use regimes for important natural resources, by encouraging the development of alternative sources of income and improved livelihoods, or by direct incentives. If adequate support for the targeted practices exists or is built, and the necessary conditions and resources for their implementation are in place, changes in practice lead to reduced pressure on, or improved responses of, the focal ecosystems, habitats, species and/or populations. This may or may not be accompanied by real improvements in livelihoods and/or attitudes, which affect conservation status only through their effects on the targeted behaviours or practices. (Based on Kapos *et al.* 2008 and reproduced by kind permission of ©2008 Wiley Periodicals Inc.)

Policy and Legislation

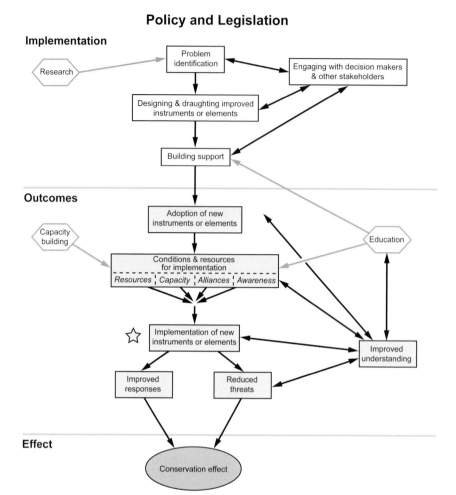

Figure 5.4 **Conceptual model for policy and legislation projects. The basis for efforts to change or enact policy or legislation is that the existing frameworks and/or instruments do not adequately support conservation aims or, indeed, that they positively promote activities that endanger or damage biodiversity. If changes are appropriately designed and drafted, and adequate support is built, improved instruments and frameworks may be adopted. If the necessary conditions are in place, implementation of the new instruments or elements may lead to reduced threat and/or improved responses and thence to conservation effect. (Based on Kapos *et al.* 2008 and reproduced by kind permission of ©2008 Wiley Periodicals Inc.)**

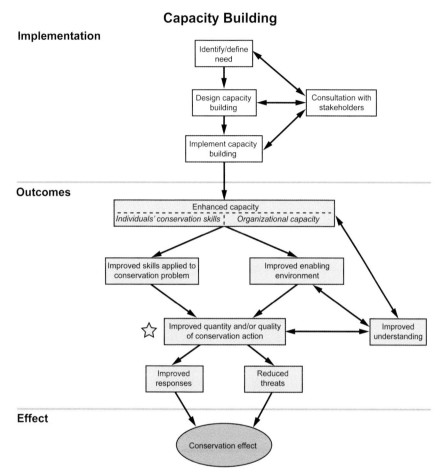

Figure 5.5   Conceptual model for capacity-building projects. Capacity building aims either to improve individuals' skills or to enhance aspects of organizational capacity, or both. If these improvements are applied in conservation, they should lead to more and/or better conservation action to address the problems of interest, leading through reduced threat and/or improved responses to improved status of the conservation targets. (Based on Kapos *et al.* 2008 and reproduced by kind permission of ©2008 Wiley Periodicals Inc.)

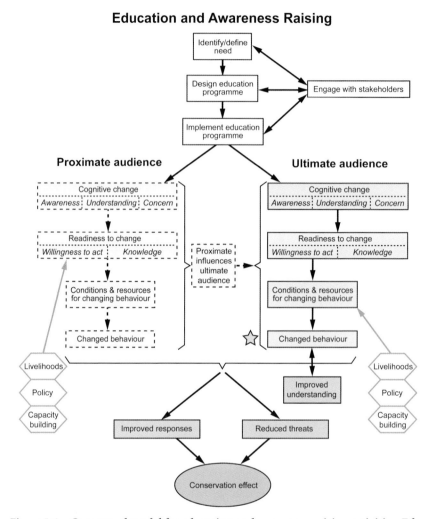

Figure 5.6   Conceptual model for education and awareness-raising activities. Education and awareness-raising activities aim to improve understanding and ultimately to influence behaviour among people not directly involved in conservation action. They do so either by addressing a target audience directly or by using a 'proximate' audience to influence the 'ultimate' audience. In either case the audience advances through increased awareness of an issue, and improved access to information about it, to understanding, concern, knowledge and ultimately willingness to act. If the correct conditions and resources are in place, this leads to behaviour change, which may reduce threats to, or in a few cases increase the responses of, the conservation target, thereby improving its conservation status. (Based on Kapos *et al.* 2008 and reproduced by kind permission of ©2008 Wiley Periodicals Inc.)

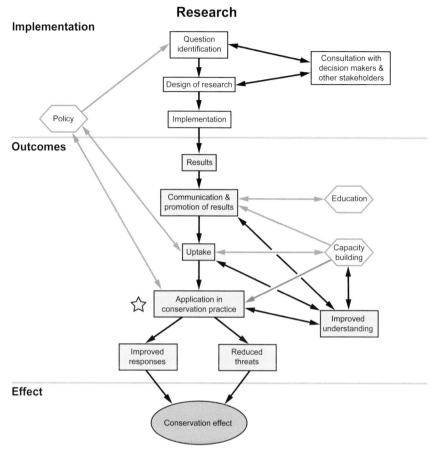

Figure 5.7   **Conceptual model for research projects. Research aims to improve the information base on which conservation action is taken. If the research question is properly identified and the research is well designed and implemented, then it is likely to produce good results. However, these results will only affect conservation if they are appropriately communicated and promoted to the right audiences, if those audiences understand them, and if they apply them in conservation practice. If these things happen then research may play a role in reducing threat and/or improving responses and thus in improving conservation status. (Based on Kapos *et al.* 2008 and reproduced by kind permission of ©2008 Wiley Periodicals Inc.)**

lack of information, distinguishing between the lack of information that will eventually become available and information gaps for which no resolution is foreseen. Consistent rules determine where these kinds of answers are acceptable. For example, it is essential to answer substantively all questions about implementation, while some lack of information about later outcomes and effects is to be expected, especially for recently initiated projects and for activity types less closely linked to conservation effect.

To refine the questionnaires, consistency of interpretation was tested several times by exposing a 'captive audience' to the details of a project and its outcomes through a presentation and some limited questioning. Participants then filled in the relevant questionnaire and we analyzed their answers to identify and improve questions that were not interpreted consistently (Figure 5.8). These trials also confirmed that it is harder to obtain consistent answers to questions that pertain to the parts of the models nearest to conservation effect. Trial use of the questionnaire by individual project leaders further identified where wording and concepts needed to be clarified. Such improvements and adjustments to ensure internal consistency were made iteratively to ensure a practical and high-quality final questionnaire. The finalized evaluation tool is currently in the form of a series of interlinked Excel spreadsheets, with scorecard answers provided in drop-down menus[2]. A more user-friendly version is under development.

When conservation professionals self-evaluated 22 complex conservation projects comprising a total of 56 activities, the results showed that the implementation of a given activity type is a poor predictor of its conservation effect. In contrast, the achievement of 'key outcomes' predicts conservation effect much better (Kapos et al., 2009). This suggests that such outcomes, for which information is often more available than it is for conservation effect (Kapos et al., 2009), can be used to indicate likely conservation effectiveness and so might provide a readily measured metric for cross-project analysis of factors determining conservation success.

Using a framework such as the CCF one to identify such key outcomes can help to identify indicators and monitoring needs that can support adaptive management, facilitate experience sharing and help to improve conservation practice. The need for such tools is highlighted by patterns of experience sharing identified in the CCF sample (Figure 5.9). As noted previously (Redford & Taber, 2000), experiences of successful activities within conservation projects

---

[2] www.cambridgeconservationforum.org.uk/projects/measures/outputs/.

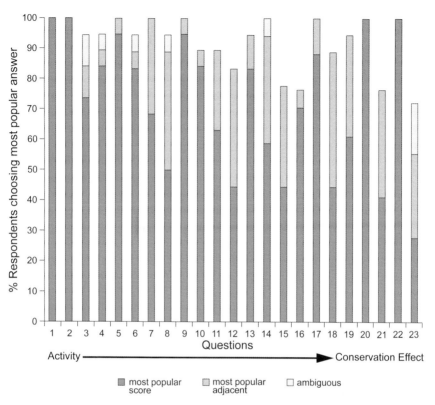

Figure 5.8 Consistency of responses by 19 independent evaluators to questions about conservation of the stone curlew *Burhinus oedicnemus* by the Royal Society for the Protection of Birds. Details of the project were presented in workshop format. Dark bars represent the percentage of respondents choosing the most frequently selected answer. Mid-grey bars show the percentage that chose the next most popular adjacent answer. For example, if the most commonly chosen answer was 'b', then the mid-grey bars show the percentage of respondents choosing whichever was the more popular of 'a' or 'c', and the pale bars show where responses were ambiguous (e.g. 'b/c'). The tallest bars with the greatest proportion of solid area indicate the questions with the greatest degree of agreement among evaluators. The bars that are shortest and those that are multishaded indicate questions that were more difficult to answer and required further revision. As expected, the consistency of answers broadly declined as the questionnaire addressed the more difficult issues of outcomes and conservation effect, rather than activity and output.

Figure 5.9   **Experience-sharing of 22 conservation projects undertaken by 10 organizations. Successful experiences (solid bars) are far more likely to be documented and actively shared both within and outside the implementing organization than less successful ones (pale bars).**

are far more likely to be actively communicated to other practitioners than less successful experiences, which are often poorly disseminated even within organizations. While few organizations are comfortable openly declaring whole projects to be complete failures or sharing their failures widely, we believe there will be greater interest in evaluating projects by their component activity types to identify more and less successful approaches. Analysis of those evaluations may prove a useful mechanism for promoting documentation and sharing of unsuccessful, as well as successful, experiences in conservation.

Many organizations have welcomed the CCF framework and evaluation tool as rigorously developed innovations that may help them to: (i) evaluate and analyze existing conservation experience; (ii) plan future activities and associated monitoring; and (iii) identify steps or actions that could enhance their conservation impact. Users have found them straightforward to apply and have also commented enthusiastically about the usefulness of both the framework and the questionnaires for clarifying project objectives, for planning new work, and for prioritizing new foci and the application of scarce resources for the monitoring.

The tools developed by CCF build on and support much existing thinking about the importance of using conservation experience to inform future conservation actions (Kleiman *et al.*, 2000; Salafsky *et al.*, 2002; Saterson *et al.*, 2004; Ferraro & Pattanayak 2006). Like planning-based approaches such as the logical framework and results chains (Salafsky et al., 2001) and tools for implementing them, like Miradi (Miradi, 2007), the CCF tools help users to make explicit the assumed linkages between their activities and the desired conservation outcomes. By identifying and assessing intermediate or 'key' outcomes that precede, and may be easier to measure than, changes in threat or biological status, the CCF framework and evaluation tool help practitioners to evaluate the likely impacts of their actions. This is so even for projects still in progress and for interventions such as capacity building or policy-related work where biological impacts are not commonly measured. CCF's unified framework for synthesizing existing conservation experience complements the growing drive for evidence-based conservation (Sutherland *et al.*, 2004; Sutherland, 2005) by providing a means to assess the effects of whole projects as well as single management interventions, and helping to ensure that the data needed to build an evidence base are collected in future. The utility of such analytical frameworks for identifying and generating quantitative and qualitative data for analysis has been a key component of efforts to build the evidence base in a wide variety of fields, including library services (Markless & Streatfield, 2006; Streatfield & Markless, 2008) complementary medicine (Verhoef *et al.*, 2006) and humanitarian assistance (Reaves *et al.*, 2008).

Looking ahead, wider application of the CCF framework and especially the evaluation tool could help to provide substantive data on the outcomes and effects of conservation projects, which are generally hard to come by (Kleiman *et al.*, 2000; Saterson *et al.*, 2004; Brooks *et al.*, 2006). Such a dataset would be invaluable for promoting conservation learning and for testing hypotheses about predictors and determinants of conservation success both within, and external to, projects. A prototype scoring system for use with the evaluation tool will be tested and further developed to facilitate statistical analyses. Such a dataset could be used to validate further the degree to which conservation effect can be predicted from earlier stages in the models, and to investigate what project characteristics and context variables help determine success of a given activity type. Questions of interest to be addressed with such analyses might include: Are particular types of intervention more effective in large and/or complex projects or in small and/or simple ones? What components of capacity in partner organizations are most important in determining the success of particular activities?

Finally, by combining the results of such project evaluations with assessments of the importance of the conservation targets and the scale of the interventions, it may also be possible to extend the work of the Zoo Measures Group (Mace *et al.*, 2007), and begin building a picture of overall conservation impacts.

Conclusions

We believe the CCF tools, now made available for wider use on the CCF website[3], will help circumvent the shortage of experimental interventions and the reluctance of many organizations to share failures (Redford & Taber, 2000). The CCF tools can help to overcome many of the problems commonly associated with the evaluation of conservation projects by clarifying conservation objectives and identifying key outcomes that can be assessed in relatively short time frames. These tools can help to make more efficient use of scarce monitoring and evaluation resources, and may also serve as a useful basis for assembling, managing and using information about project outcomes and existing conservation experience. With wide application, the CCF framework and evaluation tool can provide a powerful platform for drawing on the experience of past and ongoing conservation projects to identify quantitatively factors that contribute to conservation success. Conservation practitioners are encouraged to try these tools and thereby help build a systematic catalogue of conservation experience that can be used to establish ways of improving performance. Collaborators interested in using such evaluations to test hypotheses about determinants of conservation effectiveness would be welcomed. We believe that adopting a better measure of success is crucial to deciding not just what to save and where, but what to do in conservation and how best to do it.

Acknowledgments

This work was supported by a grant from the John D. and Catherine T. MacArthur Foundation to the Cambridge Conservation Forum via the

---

[3] www.cambridgeconservationforum.org.uk/projects/measures/outputs.

University of Cambridge. The member organizations of the Cambridge Conservation Forum (see footnote 1) generously supported the participation of many of their staff members in the working groups, workshops and discussion events of this project. Earthwatch and the British American Tobacco Biodiversity Partnership were also instrumental in testing and refining the evaluation tool. Members of the Conservation Measures Partnership and IUCN, the International Union for the Conservation of Nature Programme Evaluation Group also provided helpful input. The University of Cambridge Department of Zoology and United Nations Environment Programme World Conservation Monitoring Centre were especially helpful in accommodating project staff and many project meetings.

We are grateful to the following individuals, who participated in the project's working groups and discussions, and in testing the evaluation tool: Muhtari Aminu-Kano, Malcolm Ausden, Esther Ball, Sue Barnard, Alan Bowley, Evan Bowen-Jones, Rob Brett, Mike Brooke, Pete Brotherton, Paul Buckley, Nadia Bystriakova, David Coomes, Barrie Cooper, Barney Dickson, Julian Doberski, Edwin van Ek, Jon Ekstrom, Lianne Evans, David Gibbons, Michael Green, Rhys Green, Annelisa Grigg, Monica Harris, Dawn Hawkins, Peter Herkenrath, Alex Hipkiss, Graham Hirons, Ruth Hossain, Francine Hughes, Julian Hughes, Jon Hutton, Claudia Ituarte, David Kingma, Paul Laird, Annette Lanjouw, Phyllis Lee, Chris Magin, Tom Milliken, Roger Mitchell, David Noble, Sheila O'Connor, Thomasina Oldfield, Kay O'Regan, Emma Papworth, Ana Rodrigues, Rondang Sinegar, Per Stromberg, Bill Sutherland, David Thomas, Rosie Trevelyan, Graham Tucker, Sue Wells, James Williams and Michael Wright.

References

Andam, K.S., Ferraro, P.J., Pfaff, A., Sanches-Azofeifa, A. & Robalino, J.A. (2008) Measuring the effectiveness of protected area networks in reducing deforestation. *Proceedings of the National Academy of Sciences of the USA*, 105, 16089–16094.

Brooks, J.S., Franzen, M.A., Holmes, C.M., Grote, M.N. & Borgerhoff Mulder, M., (2006) Testing hypotheses for the success of different conservation strategies. *Conservation Biology*, 20, 1528–1538.

Carey, P.D., Manchester, S.J. & Firbank, L.G. (2005) Performance of two agri-environment schemes in England: a comparison of ecological and multi-disciplinary evaluations. *Agriculture Ecosystems and Environment*, 108, 178–188.

Carey, P.D., Short, S., Morris, C. *et al.* (2003) The multi-disciplinary evaluation of a national agri-environment scheme. *Journal of Environmental Management*, 69, 71–91.

Christensen, J, (2003) Auditing conservation in an age of accountability. *Conservation in Practice*, 4, 12–19.

CMP (Conservation Measures Partnership) (2004) *Open Standards for the Practice of Conservation*. Conservation Measures Partnership, Washington, DC.

Ferraro, P.J. & Pattanayak, S.K. (2006) Money for nothing? A call for empirical evaluation of biodiversity conservation investments. *PLoS Biology*, 4, 482–488.

Gratwicke, B., Seidensticker, J., Shrestha, M., Vermilye, K. & Birnbaum, M. (2007) Evaluating the performance of a decade of Save The Tiger Fund's investments to save the world's last wild tigers. *Environmental Conservation*, 34, 255–265.

Gusset, M., Ryan, S.J., Hofmeyr, M. *et al.* (2007) Efforts going to the dogs? Evaluating attempts to reintroduce endangered wild dogs in South Africa. *Journal of Applied Ecology*, 45, 100–108.

Hockings, M., Stolton, S., Leverington, F., Dudley, N. & Courrau, J. (2006) *Evaluating Effectiveness: a framework for assessing management effectiveness of protected areas*, 2nd edn. IUCN World Commission on Protected Areas and James Cook University, Brisbane.

International Union for the Conservation of Nature and Conservation Measures Partnership (IUCN-CMP) (2006) *Unified Classification of Conservation Actions*. Available at: http://conservationmeasures.org/CMP/Site_Docs/IUCN-CMP_Unified_Actions_Classification_2006_06_01.pdf.

Jepson, P. & Canney, S. (2003) *The State of Wild Asian Elephant Conservation in 2003: an independent audit for elephant family*. Elephant Family and Conservation Direct, London and Oxford.

Kapos, V., Balmford, A., Aveling, R. *et al.* (2008) Calibrating conservation: new tools for measuring success. *Conservation Letters*, 1, 155–164.

Kapos, V., Balmford, A., Aveling, R. *et al.* (2009) Outcomes, not implementation predict conservation success. *Oryx*, 43, 336–342.

Kleiman, D.G., Reading, R.P., Miller, B.J. *et al.* (2000) Improving the evaluation of conservation programs. *Conservation Biology*, 14, 356–365.

Linkie, M., Smith, R.J., Zhu, Y. *et al.* (2008) Evaluating biodiversity conservation around a large Sumatran protected area. *Conservation Biology*, 22, 683–690.

Mace, G.M., Balmford, A., Leader-Williams, N. *et al.* (2007) Measuring conservation success: assessing zoos' contribution. In *Zoos in the 21st Century: catalysts for conservation?*, eds A. Zimmermann, M. Hatchwell, L.A. Dickie & C.D. West, pp. 322–342. Cambridge University Press, Cambridge.

Markless, S. & Streatfield, D. (2006) Gathering and applying evidence of the impact of UK university libraries on student learning and research: a facilitated action research approach. *International Journal of Information Management*, 26, 3–15.

Miradi (2007) *Miradi: adaptive management software for conservation projects*. Conservation Measures Partnership and Benetech, Washington, DC.

O'Neil, E. (2007) *Conservation Audits: auditing the conservation process: lessons learned, 2003–2007*, p. 42. Conservation Measures Partnership, Washington, DC.

Reaves, E.J., Schor, K.W. & Burkle, F.M. Jr. (2008) Implementation of evidence-based humanitarian programs in military-led missions: Part II. The impact assessment model. *Disaster Medicine and Public Health Preparedness*, 2, 237–244.

Redford, K.H. & Taber, A. (2000) Writing the wrongs: developing a safe-fail culture in conservation. *Conservation Biology*, 14, 1567–1568.

Salafsky, N. & Margoluis, R. (1999) Threat reduction assessment: a practical and cost–effective approach to evaluating conservation and development projects. *Conservation Biology*, 13, 830–841.

Salafsky, N., Margoluis, R. & Redford, K.H. (2001) *Adaptive Management: a tool for conservation practitioners*. Biodiversity Support Program, Washington, DC.

Salafsky, N., Margoluis, R., Redford, K.H. & Robinson, J.G. (2002) Improving the practice of conservation: a conceptual framework and research agenda for conservation science. *Conservation Biology*, 16, 1469–1479.

Saterson, K.A., Christensen, N.L., Jackson, R.B. *et al.* (2004) Disconnects in evaluating the relative effectiveness of conservation strategies. *Conservation Biology*, 18, 597–599.

Stem, C., Margoluis, R., Salafsky, N. & Brown, M. (2005) Monitoring and evaluation in conservation: a review of trends and approaches. *Conservation Biology*, 19, 295–309.

Stolton, S., Hockings, M., Dudley, N. *et al.* (2008) *Management Effectiveness Tracking Tool: reporting progress at protected area sites*, 2nd edn. WWF, Gland, Switzerland.

Streatfield, D. & Markless, S. (2008) Evaluating the impact of information literacy in higher education: progress and prospects. *Libri*, 58, 102–109.

Sutherland, W.J. (2005) How can we make conservation more effective? *Oryx*, 39, 1–2.

Sutherland, W.J., Pullin, A.S., Dolman, P.M. & Knight, T.M. (2004) The need for evidence-based conservation. *Trends in Ecology and Evolution*, 19, 305–308.

Verhoef, M.J., Vanderheyden, L.C., Dryden, T., Mallory, D. & Ware, M.A. (2006) Evaluating complementary and alternative medicine interventions: in search of appropriate patient-centered outcome measures. *BMC Complementary and Alternative Medicine*, 6, 38. Available at: doi 10.1186/1472-6882-6-38.

Walter, O. (2005) *Are we making a difference? Evaluating the effectiveness of the Mace* et al. *(in press) scoring system to evaluate the impact of conservation effort.* Unpublished MSc dissertation, University of Kent, Canterbury, UK.

Part II
Influence of Value Systems

6

Conserving Invertebrates: How Many can be Saved, and How?

Michael J. Samways

Department of Conservation Ecology and Entomology, University of Stellenbosch, Matieland, South Africa

Introduction

Invertebrates probably make up at least 99% of all species on Earth, making them a major component of the current biodiversity crisis (Lawton & May, 1995). For insects alone, it has been estimated that at least 100 000 species will become extinct over the next 300 years (Mawdsley & Stork, 1995). Some suggest that this figure may be too conservative, and that a quarter of all insect species, which could be ∼3 million species, are under imminent threat of extinction (McKinney, 1999). Furthermore, the extinction crisis is also likely to be affecting marine insects where the number of undescribed species still remains enormous.

With such potential losses to biodiversity, a constructive plan needs to be in place to soften the extinction crisis. Such a plan needs to operate at various spatial scales, from global down to local. Yet time is short. A major trade-off in current invertebrate conservation is, on the one hand, acting quickly, and on the other, knowing how to do this based on very little knowledge of the focal taxa. This raises the critical question: how do we save what we do not know? Consequently, this chapter first seeks some generalizations about invertebrate conservation. Then it moves to consider the value and importance of invertebrates. Third, the great challenges of insect conservation

Trade-offs in Conservation: Deciding What to Save, 1st edition.   Edited by N. Leader-Williams, W.M. Adams and R.J. Smith.   © 2010 Blackwell Publishing Ltd.

are considered, and how approaches to triage may help guide actions to implement species- and landscape-level approaches.

Seeking generalizations for invertebrate conservation

Establishing where the global hotspots are in the world (Mittermeier *et al.*, 2004) is important for invertebrate conservation as these are home to many irreplaceable, narrow-range endemics. A trade-off at this large spatial scale is between whether to focus conservation efforts in these hotspot areas or to spread activities more evenly across Earth (Wilson *et al.*, this volume, Chapter 2; Murdoch *et al.*, this volume, Chapter 3). As it stands, there is a disproportionate focus per species in the economically more powerful northern hemisphere than at lower latitudes.

At the smaller spatial scale, there is merit in establishing where protected areas should be situated within a region, through systematic conservation planning (Margules & Pressey, 2000). However, protected areas lie in a matrix of variably transformed landscapes (Wilson *et al.*, this volume, Chapter 2; Murdoch *et al.*, this volume, Chapter 3). A local protected area, so long as it has not undergone ecological relaxation, then becomes the natural control against which other landscapes can be compared in terms of their invertebrate compositional and functional diversity. These trans-landscape comparisons then form a baseline from which to establish general management principles with applicability across the world. In other words, as the subject of invertebrate conservation is so vast, research undertaken in one geographic area needs to be maximized by translating the findings as guidelines for other parts of the world.

Valuing invertebrates

As with all conservation action, invertebrate conservationists need to be very clear both on the ethical foundation of their action and on the precise conservation goal, so as to move forward with both speed and efficiency. One of the questions that often arises is just how important are invertebrates? The issue of intrinsic value is particularly important for invertebrates, as there are so many species and individuals. In other words, insects have much additive worth, gene upon gene, individual upon individual, and species upon species. According to the principle of intrinsic value, all organisms have an equal right to exist, with one having the same value as the next: a bee equals a bee-eater. This is illustrated by IUCN, the International Union for the Conservation of Nature's Red List (IUCN, 2008) where the entry of each species has equivalence, i.e. each threatened species receives equal coverage without prejudice to its taxonomic status.

A more difficult issue, and one that is perhaps more relevant to most conservation planners, is assessing the utilitarian value of invertebrates. How important are invertebrates to humans? One simple spectrum of activity lies with direct/proximal/single issues at one end, and indirect/distal/multiple issues at the other. This may be translated into particular service provision, such as the production of medicines from a particular species, or the harvesting of honey from another species. Also, the services may be provided by the invertebrate community as a whole, such as components of food webs and recycling of nutrients (Beattie & Ehrlich, 2001). These direct versus indirect approaches, which are not mutually exclusive, provide some major challenges and careful consideration of trade-offs for invertebrate conservationists. Additionally, these challenges require a more open debate between invertebrate conservationists and other taxonomic sectors of the conservation community. The first group must think harder about mainstreaming invertebrates into conservation agendas, while those interested in other taxa need more open-mindedness as to the value of invertebrates for maintaining healthy ecosystems on Earth.

The importance of invertebrates

In terms of ecosystem function, some invertebrates are virtually redundant, while others are major players, and some are even ecosystem engineers. Certain termites can locally influence structural, compositional and functional biodiversity. West African termites, by modifying water balance and organic matter status, increase local tree diversity (Abbadie *et al.*, 1992), while some ants allow certain plants to exist (Folgarait, 1998). Grasshoppers in the arid regions of South Africa produce faeces that are finely divided and rapidly provide important nutrients to plants (Milton & Dean, 1996). On the small Seychelles island of Cousine, giant millipedes *Seychelleptus seychellarum* are so abundant that their biomass is equivalent to one elephant per hectare, and these millipedes play a major role in the formation of soil cover on the sterile granitic bedrock (Lawrence & Samways, 2003). Indeed, Coleman and Hendrix

(2000) emphasize the role of invertebrates in ecosystem processes by calling their book *Invertebrates as Webmasters in Ecosystems*.

While many invertebrates, particularly insects, mites and nematodes, can cause a reduction in agricultural production, others can be used as control agents to reduce the former's impact on cropping systems. This is an enormous service provided by these small animals, which often achieve their changing of ecosystem function by being numerous rather than large. Predatory mites, barely visible to the naked eye, can rapidly increase and very efficiently control harmful spider mites that would otherwise cause a major loss of production (Pringle & Heunis, 2006).

The interrelationships between invertebrates and other components of the ecosystem are often complex (Memmott *et al.*, 2007), and even go beyond the boundaries of particular ecosystems (Knight *et al.*, 2005). These interactions may provide important ecosystem services, such as pollination, and these interactions and services may, in some cases, also be threatened (Kremen & Chaplin-Kramer, 2007). These interactions and contingent services can have great monetary value. Estimates suggest that 'wild' insects which control pests, pollinate flowers, bury dung and provide nutrition for other wildlife are worth $57 billion per year in the USA alone (Losey & Vaughan, 2006).

These sorts of interactions and figures are a direct manifestation of invertebrate diversity and abundance. Yet the functional value of invertebrates is not in isolation. Insect communities are largely associated with plants and their products. Interestingly, and perhaps meaningfully, about 99% of compositional biodiversity comprises animals, mostly invertebrates, while the bulk of structural biodiversity is almost the exact converse ratio, with 99% of the world's biomass in plants (Samways, 1983). Yet through interactions such as pollination and nutrient recycling, much of this biomass is maintained by invertebrates. Thus, the huge part of structural, compositional and even functional biodiversity lies with the plants and associated invertebrates. Yet there are two great challenges in conserving this important group of animals.

The 'great challenges'

The conservation of invertebrates is arguably confronted by two 'great challenges'. First is the 'perception challenge', that bugs are ugly, not nice and unimportant. Second is the 'taxonomic challenge', how can so many species be conserved when most of them do not even have names? Both these challenges have some negative and some positive aspects in common. The challenges of both lie at the species level, while the major opportunities arising from them lie at the landscape level.

One perception is that insects and other invertebrates are there simply to annoy us, with little appreciation that most ecological and evolutionary invertebrate personalities are actually sustaining us. However, that is not an exciting enough notion for most people. Quite simply, the world around us, and its maintenance, are taken for granted, at least as long as people do not appreciate the ecosystem services that biodiversity provides (Costanza *et al.*, 1997; Goldman *et al.*, this volume, Chapter 4). Invertebrates remain largely unseen and unappreciated. Of course, not all invertebrate species are unappreciated. Butterflies, corals and sea shells have great aesthetic value, but regrettably are admired more often dead than alive. Nevertheless, there is a sector of humanity that is very appreciative of the living animal. Most encouragingly, Butterfly Conservation, a UK-based society for all who are interested in the conservation of butterflies, has ~14 000 members.

Like the perception challenge, the taxonomic challenge is also at its greatest at the species level. Yet we do not know how many species there are, and possibly, at best, only 10% have scientific names. Then, in addition, we have only sparse biological information on a tiny minority of those. Yet time is also short, with the anthropogenic extinction crisis already well upon us (Lawton & May, 1995). Without doubt, there are going to be many Centinelan extinctions, those extinctions of species that occur before they are even scientifically described. One species of bush cricket that was endemic to the United States, and was described only after it became extinct, even bears an extinction epithet: *Neduba extincta* (Rentz, 1993).

The perception challenge and the taxonomic challenge can be addressed at the species level, although this does not have the same degree of positive impact as landscape-level approaches. The perception challenge can be tackled by using icons, where single, charismatic species are singled out for special attention (Figure 6.1). This might appear to be a purely invertebrate challenge. However, mammal conservationists also face the same challenge, and charismatic megafauna are inevitably preferred over the cryptic and small species: tiger *Panthera tigra* over tenrec Tenrecidae (Leader-Williams & Dublin, 2000).

Human bias in favour of pretty invertebrates is based on utilitarian ethics, even though non-consumptive. Fluttering butterflies, buzzing bees and chirping crickets make people feel good, and a sense of well-being pervades surrounding landscapes. When taken away, there is a loss: the extinction of experience (Samways, 2007a). At least when ecologically managed, urban parks go some way towards returning an experience of nature, which is particularly important for urban dwellers, especially children (Suh & Samways, 2001). However, overcoming the extinction of experience relative to invertebrate conservation, a parallel argument arises when considering conserving the function of ecosystems. That is, as well as functional species, the integrity of ecological systems must be recognized in terms of the actual species that constitute the ecosystems. Theoretically, there could be a 'functioning system' but composed of alien organisms. Similarly, nature can be experienced in an urban setting, but that nature may well be largely foreign. Equally, when those urban species are indigenous, they are likely to be the common and widespread species that are well able to live in the disturbed urban context.

When assessing a functioning system, species integrity should be considered. Similarly, the experience of nature must recognize all indigenous species, including the rare endemics that may be far away out of sight, out of mind, in a remote reserve, but brimming with intrinsic value.

Sometimes these threatened rarities are also charismatic and then they immediately curry special favours in terms of their conservation: hence the large membership of Butterfly Conservation. Notwithstanding, of course, that in some cases their charisma may contribute to their demise through over-collecting. Rare and threatened fritillaries Nymphalidae, blues Lycaenidae, apollos and swallowtails Papilionidae, indeed attract a relatively disproportionate amount of funding, especially when they have the good fortune to live in the recovering post-glacial, depauperate yet monetarily wealthy north.

Among invertebrates, the taxonomic challenge is also relevant to the large and glamorous, such as butterflies and dragonflies (see Figure 6.1). Both these groups of species can have their own protected areas (Figure 6.2), especially in countries where they have cultural significance, as with dragonflies in Japan (Primack et al., 2000). Therefore, if invertebrate conservation is to be taken seriously, an assessment is required of just how effective these well-known, charismatic and favored icons are as umbrellas or surrogates for a whole host of other small, 'little brown jobs' that have been molded by natural selection to be cryptic and unimportant in the human psyche – such as already undertaken for mammals (e.g. Berger, 1997). This becomes interesting from an ethical perspective because, effectively, the species of utilitarian importance protect others that in turn have only intrinsic value. By conserving the

Figure 6.1 **Dragonflies are large and conspicuous, and so are flagship species for invertebrate conservation. The gilded presba *Syncordulia legator* shown here is a Red List species that occurs in a specific part of the Cape Floristic Region, and is a flagship for many other rare, threatened and endemic species. (Photograph by Michael Samways.)**

charismatic megafauna, even if they are without backbones, the trade-off is that the uncharismatic meso- and microfauna can also be conserved. However, concordance between the large and charismatic, and the cryptic multitude, cannot be assumed and requires focused research (but see Bried *et al.*, 2007).

So, at the species level, or fine-filter level, both the perception challenge and the taxonomic challenge can be addressed, with the hope that other species will be automatically conserved in their wake. However, there is another approach, the coarse-filter or landscape approach, which considers conserving whole landscapes in an informed way so that a vast array of structural, compositional and functional biodiversity, at both ecological and evolutionary time scales, is conserved. This is possibly even more important in the marine realm, where the moving fluid medium inevitably creates much more connectivity than across the land surface. Let us now look at this larger-scale, landscape (or seascape) approach in more detail.

Figure 6.2 **Dragonfly reserves play an important traditional role in Japan, where they raise awareness of other insects as well as dragonflies. One of the foremost is Nakamara Dragonfly Reserve, with a sophisticated information centre on the left. (Photograph by Michael Samways.)**

After systematic conservation planning, then what?

Systematic conservation planning is a process whereby geographic areas are selected for protection. Together, these selected areas aim to conserve as much irreplaceable biodiversity as possible (Margules & Pressey, 2000; Murdoch *et al.*, this volume, Chapter 3). After such planning exercises, which effectively generate a wish list of important areas for conservation, there is the issue of which areas, in reality, can be procured and which species can actually be conserved. In practice, new land and many water bodies are added opportunistically, while there are possibly other important areas that should also be actively sought for acquisition. These new areas are often of lesser conservation value, having been anthropogenically transformed. Choice of which areas are worth acquiring involves triage (Box 6.1), where areas for selection fall into three categories: (i) so badly damaged that successful restoration is unlikely; (ii) in largely good shape so that they can be incorporated into the protected area by simply 'taking down the fence';

or (iii) areas where, with minimal management, one can obtain maximum conservation value (rolled out in Figure 6.3). The most highly desirable areas, arguably, are those of great biodiversity value where little or no conservation management is required to restore them to their original naturalness. However, such areas are often few outside existing protected areas. Often the reality is that the newly acquired area requires some remediation.

Newly acquired areas become effective when they increase the size of an existing protected area, while also adding value as source, rather than sink, habitats. Such an increase in area is generally always beneficial (e.g. in terms of risk aversion to adverse weather, improving a species' genetic base) for biodiversity, including invertebrates. These add-on areas include new tracts of land, corridors (Hilty *et al.* 2006), ecological networks (Samways, 2007b), field sports areas (Oldfield *et al.*, 2003) and set-aside land, as well as the instigation of marine protected areas. The important point for invertebrates is that these add-on areas should be high-quality habitat, with appropriate restoration, so that conditions are then optimal for the rare, specialist and often sensitive species.

As regards invertebrates, maintenance of protected areas, set-aside land and restoration of landscapes instigates the precautionary principle (Cooney & Dickson, 2005) where, to cite Principle 15 of the 1992 Rio Declaration on Environment and Development, '*full scientific certainty shall not be used as a reason for postponing cost-effective measures to prevent environmental degradation*'. In other words, the huge complexities of structural, compositional and functional biodiversity – which largely concerns invertebrates

Box 6.1 **Explaining a conceptual model of ecosystem restoration triage (see Figure 6.3)**

The optimal trade-off requires maximizing biodiversity conservation against the cost of management input required for effective restoration of ecological integrity and ecosystem health (resistance and resilience). There is a gradient from ecosystems with intense-and-frequent, mostly anthropogenic, disturbance and low ecological integrity, to ecosystems with intense, intense-and-very-infrequent or mild-at-any-time natural events and high ecological integrity. Urbanization lies at one end of this spectrum and the 'original' state at the other. 'Original' is in parentheses because (i) it depends how far back we go in time, and (ii) because there is never any certainty over what all the original structural, compositional and functional biodiversity was at any one time in the past. The lost 'original' state is the pristine state, which no longer exists, as anthropogenic impacts reach all parts of the world. Restoration here is a biocentric, deep-ecology view, where there is a genuine aim to bring back all aspects of ecological integrity.

The starting point in the decision process is whether to restore, or not to restore, ecological integrity. It does not invoke decisions on whether to re-green, ecologically landscape or rehabilitate. These three have various cultural, aesthetic and engineering components, and not just a biocentric one. There are two extremes of 'doing nothing': (i) where ecological integrity is irretrievably lost (e.g. a harbour for large ships), and (ii) where ecological integrity is intact. Where ecological integrity is irretrievably lost, only re-greening, rehabilitation or ecological landscaping, but not restoration, is possible. The third prong of triage is the one where ecological restoration is restorable, and is the highest level of biocentricity.

Re-greening is simply putting back a vegetation cover with more consideration to aesthetics and engineering value than to ecological integrity (e.g. the grass cover of road cuttings). The maximal ecological integrity value for re-greening is roughly at the level of recreational areas, with disturbance ranging from intense and frequent (e.g. mowing) to infrequent and mild. Rehabilitation aims to recover some ecological integrity but has major aesthetic and/or human cultural components combined with ecological considerations (e.g. mine dump rehabilitation, removal of pollutants from a stream). Like re-greening, the maximal ecological integrity value achievable through rehabilitation is low. This contrasts with ecological landscaping, which deliberately aims to recreate what we believe to be a 'natural' ecosystem, which may be aesthetic (deliberately or inadvertently anthropocentric) or not (purely biocentric). Carefully planned planting of indigenous trees along roadsides is an example of ecological landscaping. Researched well, ecological landscaping can have great ecological integrity value, at least over time after indigenous biodiversity returns. Ecological landscaping is also of value to greenways, ecological networks and protected areas with management. We are then finally left with restoration, which can normally only be done on minimally degraded ecosystems (hence the dashed line in the upper left of Figure 6.3). Restoration aims for the 'original' state, but this is rarely actually achievable, e.g. because of invasive aliens (hence the dashed line in the lower right of Figure 6.3).

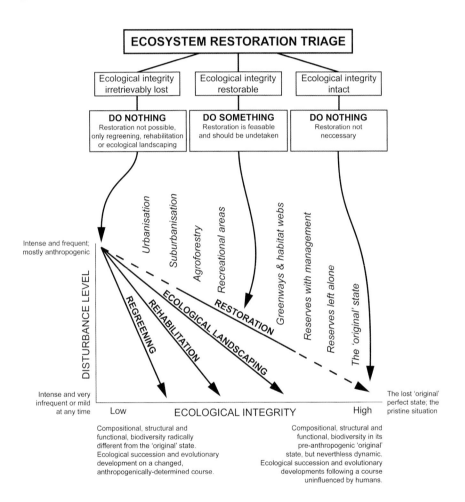

Figure 6.3 **A conceptual model of ecosystem restoration triage. A detailed explanation of this model can be found in Box 6.1. (From Samways, 2000 and reproduced by kind permission of ©2000 Kluwer Academic Publishers.)**

and their interactions – should not be a knowledge impediment to action. Landscape-scale conservation is an umbrella approach to cover intact or restoring ecosystems as a precaution against invertebrate and other biodiversity extinction, local or global. This precautionary approach increases the area of occupancy, and thus reduces the risk of local extinction that might otherwise come about through stochastic environmental effects and genetic erosion in a smaller physical area.

This restoration trajectory of increased conservation area and increased quality aims at some idealized state as perceived by us humans. We idealize it as the best approach for conserving all aspects of ecological integrity, having all natural species present, and ecosystem health, having a fully functional and resilient system. However, there may be components, or mesofilters, that have not been recognized for their value or have been forgotten, which are very important for particular species groups. Saproxylic beetles may, for example, require particular types of dead wood, and potter wasps need muddy pools for making their larva-rearing chambers.

Nevertheless, the most sensible way forward is to use a local, fully natural area as a control or model, and as a goal towards which to aim. Anything short of this, with, for example, an absence of certain key components, such as megaherbivores that modify the landscape and provide suitable conditions for certain invertebrates, arguably is not restoration but a form of rehabilitation or re-greening. The trade-off here is that to identify the idealized state, and to manage towards it, requires financial resources. This, in turn, points to the importance of restoring ecosystems adjacent to intact protected areas, which then act as source habitats from which biotic propagules can spread and establish in the area being restored.

Synthetic management approach for invertebrates and other biodiversity

Now that the field of invertebrate conservation is beginning to mature, certain principles or premises are emerging for conservation at the landscape level (Samways, 2007b). These principles have wider applicability than just to invertebrates, and may apply just as well to many other taxonomic components of biodiversity. The six interrelated landscape-level principles have a golden thread running through them that applies at the population level (Figure 6.4). For ease of reference, this thread may be termed the 'meta-population trio',

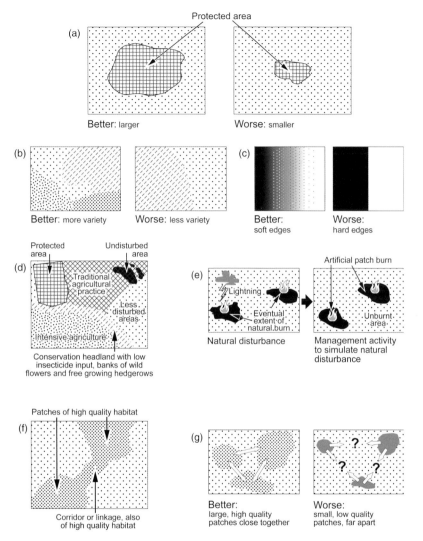

Figure 6.4 **Landscape approaches to insect conservation (and biodiversity conservation more generally), showing the importance of different management approaches: (a) maximize the size of protected areas; (b) maintain habitat heterogeneity; (c) reduce contrast between disturbed areas and adjacent natural areas; (d) maintain as much undisturbed or minimally disturbed land as possible outside protected areas; (e) simulate natural disturbance through management actions; (f) link patches of high-quality habitat; and (g) maintain metapopulations through encouraging large patch sizes, good patch quality and reducing isolation between patches.**

comprising 'large patch size', 'good patch quality' and 'reduced patch isolation'. The tenet here is not to argue over which of these three is the most important, but rather consider all three as inextricably linked as an ideal conservation goal. The six interrelated approaches that are pivotal for conservation success at the landscape, coarse-filter level are as follows.

Maintain natural protected areas

Protected areas should be as large as possible and as many as possible (Figure 6.4a). 'How large' depends on the organisms in question. Protected areas are particularly important for specialist species.

Maintain quality habitat heterogeneity

To maximize biodiversity conservation, the aim is to maintain as much natural variety as possible at various spatial scales (Figure 6.4b). This is important so as to maintain a diversity of opportunities for as many indigenous species as possible. This also involves the removal or suppression of invasive aliens.

Reduce contrast between disturbed and adjacent natural areas

Softening hard, contrasting edges allows for more variety of habitat conditions, especially for small ectothermic organisms like insects. Soft edges, and gentle ecotones, also improve connectivity within the landscape (Figure 6.4c).

Undisturbed or minimally disturbed land outside protected areas

It is important to maintain as much undisturbed or minimally disturbed land as possible outside protected areas. Set-aside effectively comprises instigating conservation headlands, conservancies, agri-environment schemes and other management activities that provide undisturbed, or at least less disturbed, areas to complement natural reserve areas. In this way, the area of occupancy and connectivity are improved, so improving chances of survival (Figure 6.4d).

Simulate natural disturbance

Most landscapes experience some disturbance, whether abiotic, such as fire or flooding, or biotic, such as grazing impact from large herbivores. These disturbances may be patchy and at small spatial scales (e.g. trampling around waterholes) or large scale (e.g. extensive grassland fires). The aim is to simulate these natural events (Figure 6.4e), with particular attention to extent, intensity and timing of the management activity.

Link patches of quality habitat

It is essential to maintain the ecological *status quo* and evolutionary potential of landscapes (Figure 6.4f). This means allowing movement of organisms, maintenance of high population levels and dispersal of propagules to buffer adverse conditions and promote genetic vigour.

The 'golden thread'

The six principles above apply a landscape-level approach to insect (and other biodiversity) conservation. At the population level, they also apply as the principle of maintaining the metapopulation trio of large patch size, good patch quality and reduced patch isolation (Figure 6.4g). To maintain metapopulation viability, it is essential that a large population is maintained through good-quality habitat and sufficient habitat size. These large, quality habitats also need to be close together to ensure movement back and forth between them. This population-level approach is the essence of single-species or fine-filter conservation.

The question that now arises is what are the trade-offs for instigating these principles? The overall trade-off is essentially willingness versus cost. There is no reason why the six principles cannot be put into practice across many landscapes. The inhibition is usually financial. Nevertheless, there are financial incentives for engaging these six principles, moving a trade-off into a symbiosis (Figure 6.5).

Such a six-principled synthetic management approach, besides being a contingency activity that aims to conserve as many invertebrates as possible,

Figure 6.5   **A trade-off between maximizing biodiversity conservation and optimizing agro-forestry production moving into a symbiosis through the introduction of large-scale ecological networks (ENs) of remnant landscape elements between patches of planted timber. This photograph shows one of the large ENs in South Africa, within a pine plantation setting where roughly one-third of the land is set aside to maintain hydrological processes and maintain biodiversity. In addition to ENs of intrinsic value to biodiversity conservation, ENs also confer a financial advantage in that the timber, when produced in an ecologically sensitive manner, can attract Forest Stewardship Council ratification, giving the timber products access to high-value, ecologically structured markets. (Photograph by Michael Samways.)**

is also a means of maintaining ecosystem services. These may be natural services for their own sake (i.e. of intrinsic value), as with the formation of soil by millipedes, or for the benefit of humans (Díaz *et al.*, 2006). The restoration process may not necessarily be taken through to re-establish the full complement of the natural species and their interactions, but modified at some point to deliberately provide a service with minimal impact on surrounding natural processes. This is manifested by techniques such as ecological engineering (Gurr *et al.*, 2004), where the originally impoverished ecosystem is ecologically landscaped to provide a particular service or services, such as improved biological control of pests.

The various trade-offs are very difficult to measure, although direct financial benefits can be approximated as a generalized natural service (Losey & Vaughan, 2006) and as a targeted service, such as biological control (Kremen & Chaplin-Kramer, 2007). However, a huge number of interactions take place in any particular ecosystem at any one time (Memmott *et al.*, 2007), so the concept of trade-offs becomes more and more of a grey area and less black or white. The reason for this, which rests on the precautionary principle, is that the more we delve, the more natural systems reveal themselves to be complex. In effect, we have to conserve what we do not know.

How many invertebrates are going to be saved?

One of the ironies of current biodiversity conservation is that some narrow-range endemics have a good chance of survival in small, remnant patches. The reason is that many were already adapted to living in small areas, long before humans began to impact upon their survival. In other words, some of these narrow-range species are safe, so long as the human footprint, quite literally, does not land exactly on their entire population. Where a severe impact does coincide with the full extent of a population, that species is doomed. In contrast, where the footprint does not coincide, they may well survive in the fragment, on the condition that the quality of their remnant habitat patch stays in good condition and does not suffer from edge impacts and invasion from alien species (Samways, 2006).

At the other extreme are those species that require large geographic areas to survive in the long term. They may be susceptible to widespread land transformation. Each continent has at least one locust species, and once upon time, so did North America. The Rocky Mountain locust *Melanoplus spretus* was once so abundant that when the first railways were laid out across the central plains, they were said to be so numerous that their squashed bodies caused the wheels of the locomotives to slip. Yet by the early 20th century, the species was extinct, having been unable to survive the agricultural development that took place in its egg-laying sites (Lockwood & DeBrey, 1990).

Conclusions

Clearly, with so much habitat loss, particularly in the tropics, many invertebrate species will be lost, mostly habitat specialists. However, such losses can be traded off by implementing land management principles that will be an umbrella for conserving as much invertebrate diversity as possible, and putting the precautionary principle into practice. But in doing so, some realism is required, and triage will have to be implemented. This will involve setting aside land with restoration as a necessity. However, many of the land mosaics outside protected areas are far from devoid of significant biodiversity, and there can be low-impact utilization of wild lands (e.g. Oldfield *et al.*, 2003) through to more intensive, but still valuable in terms of biodiversity, land use such as ecological engineering in transformed landscapes.

International initiatives, such as the Forest Stewardship Council and Marine Stewardship Council certification processes are going a long way towards promoting invertebrate conservation worldwide. In the other direction, any support that invertebrate conservationists can give to these initiatives by way of research, management and monitoring recommendations will inevitably

Figure 6.6   View from the top of Table Mountain, from a protected area surrounded by the city of Cape Town. Many rare and threatened invertebrate and plant species occur in this reserve, and conserving them gives support to the conservation of a whole host of other invertebrate species, many of which are still not scientifically described. (Photograph by Michael Samways.)

lead to saving a considerable amount of biodiversity. In some cases, this will involve conserving small pieces of land or sea, which although perhaps not always suitable for large animals with large home ranges, may nevertheless have considerable value for a wide range of invertebrates. An example is the conservation of Table Mountain, Cape Town (Figure 6.6), where a small protected area in the heart of the city has considerable importance for the conservation of invertebrates, as well as plants (Pryke & Samways, 2008), though it is too small for certain large vertebrates.

The complexities and perceptions surrounding the conservation of invertebrates may at first sight appear to be almost insurmountable. However, conservationists must not lose sight of the great contribution that systematic conservation planning and synthetic management has for conservation of invertebrates. Nevertheless, single-species approaches overlain on the landscape-scale approach should not be ignored. Certain iconic invertebrate species, although receiving conservation in their own right, can also form a conservation umbrella for many other, less charismatic, small species with smaller home ranges.

Acknowledgments

Thanks to Mondi for supporting, in practice, some of the principles outlined here.

References

Abbadie, L., Lepage, M. & Le Roux, X. (1992) Soil fauna at the forest–savanna boundary: role of termite mounds in nutrient cycling. In *Nature and Dynamics of Forest-Savanna Boundaries*, eds P.A. Furley, J. Proctor & J.A. Ratter, pp. 473–484. Chapman & Hall, London.

Beattie, A. & Ehrlich, P.R. (2001) *Wild Solutions*. Yale University Press, New Haven, CT.

Berger, J. (1997) Population constraints associated with the use of black rhinos as an umbrella species for desert herbivores. *Conservation Biology*, 11, 69–78.

Bried, J.T., Herman, B.D. & Ervin, G.N. (2007) Umbrella potential of plants and dragonflies for wetland conservation: a quantitative case study using the umbrella index. *Journal of Applied Ecology*, 44, 833–842.

Coleman, D.C. & Hendrix P.F. (eds) (2000) *Invertebrates as Webmasters in Ecosystems*. CAB International, Wallingford, UK.

Cooney, R. & Dickson, B. (eds) (2005) *Biodiversity and the Precautionary Principle: risk and uncertainty in conservation and sustainable use.* Earthscan, London.

Costanza, R., d'Arge, R., de Groot, R. *et al.* (1997) The value of the world's ecosystem services and natural capital. *Nature*, 387, 253–260.

Díaz, S., Fargione, J., Chapin, F.S. III & Tilman, D. (2006) Biodiversity loss threatens human well-being. *PLoS Biology*, 4, 1300–1305.

Folgarait, P.J. (1998) Ant biodiversity and its relationship to ecosystem functioning: a review. *Biodiversity and Conservation*, 7, 1221–1244.

Gurr, G.M., Wratten, S.D. & Altieri, M.A. (eds) (2004) *Ecological Engineering for Pest Management: advances in habitat manipulation for arthropods.* CSIRO, Collingwood, Australia.

Hilty, J.A., Lidicker, W.Z. Jr. & Merenlender, A.M. (2006) *Corridor Ecology: the science and practice of linking landscapes for biodiversity conservation.* Island Press, Washington, DC.

International Union for the Conservation of Nature (IUCN) (2008) *The IUCN Red List of Threatened Species.* IUCN, Gland, Switzerland and Cambridge, UK.

Knight, T.M., McCoy, M.W., Chase, J.M. *et al.* (2005) Trophic cascades across ecosystems. *Nature*, 437, 880–883.

Kremen, C. & Chaplin-Kramer, R. (2007) Insects as providers of ecosystem services: crop pollination and pest control. In *Insect Conservation Biology*, eds A.J.A. Stewart, T.R. New & O.T. Lewis, pp. 349–382. CAB International, Wallingford, UK.

Lawrence, J.M. & Samways, M.J. (2003) Litter breakdown by the Seychelles giant millipede and the conservation of soil processes on Cousine Island, Seychelles. *Biological Conservation*, 113, 125–132.

Lawton, J.H. & May, R.M. (eds) (1995) *Extinction Rates.* Oxford University Press, Oxford.

Leader-Williams, N. & Dublin, H.T. (2000) Charismatic megafauna as flagship species. In *Priorities for the Conservation of Mammalian Diversity: has the Panda had its day?*, eds A. Entwistle & N. Dunstone, pp. 53–81. Cambridge University Press, Cambridge.

Lockwood, J.A. & DeBrey, L.D. (1990) A solution for the sudden and unexplained extinction of the Rocky Mountain grasshopper (Orthoptera: Acrididae). *Environmental Entomology*, 19, 1194–1205.

Losey, J.E. & Vaughan, M. (2006) The economic value of ecological services provided by insects. *BioScience*, 56, 311–323.

Margules, C.R. & Pressey, R.L. (2000) Systematic conservation planning. *Nature*, 405, 243–253.

Mawdsley, N.A. & Stork, N.E. (1995) Species extinctions in insects: ecological and biogeographical considerations. In *Insects in a Changing Environment*, eds R. Harrington & N.E. Stork, pp. 321–369. Academic Press, London.

McKinney, M.L. (1999) High rates of extinction and threat in poorly studied taxa. *Conservation Biology*, 13, 1273–1281.

Memmott, J., Gibson, R., Carvalheiro, L.G. *et al.* (2007) The conservation of ecological interactions. In *Insect Conservation Biology*, eds A.J.A. Stewart, T.R. New & O.T. Lewis, pp. 226–244. CAB International, Wallingford, UK.

Milton, S.J. & Dean, W.R.J. (1996) Rates of wood and dung disintegration in arid South African rangelands. *African Journal of Range and Forage Science*, 13, 89–93.

Mittermeier, R.A., Robles-Gil, P., Hoffmann, M. *et al.* (2004) *Hotspots Revisited: Earth's biologically richest and most endangered terrestrial ecoregions.* CEMEX, Mexico City.

Oldfield, T.E.E., Smith, R.J., Harrop, S.R. & Leader-Williams, N. (2003) Field sports and conservation in the United Kingdom. *Nature*, 423, 531–533.

Primack, R., Kobri, H. & Mori, S. (2000) Dragonfly pond restoration promotes conservation awareness in Japan. *Conservation Biology*, 14, 1553–1554.

Pringle, K.L. & Heunis, J.M. (2006) Biological control of phytophagous mites in apple orchards in the Elgin area of South Africa using the predatory mite, *Neoseiulus californicus* (McGregor) (Mesostigmata: Phytoseiidae): a cost–benefit analysis. *African Entomology*, 14, 113–121.

Pryke, J.S. & Samways, M.J. (2008) Conservation of invertebrate biodiversity on a mountain in a global biodiversity hotspot, Cape Floral Region. *Biodiversity and Conservation*, 17, 3027–3043.

Rentz, D.C.F. (1993) Orthopteroid insects in threatened habitats in Australia. In *Perspectives on Insect Conservation*, eds K.J. Gaston, T.R. New & M.J. Samways, pp. 125–138. Intercept, Andover, UK.

Samways, M.J. (1983) Insects in biodiversity conservation: some perspectives and directives. *Biodiversity and Conservation*, 2, 258–282.

Samways, M.J. (2000) A conceptual model of ecosystem restoration triage based on experiences from three remote oceanic islands. *Biodiversity and Conservation*, 9, 1073–1083.

Samways, M.J. (2006) Insect extinctions, and insect survival. *Conservation Biology*, 20, 245–246.

Samways, M.J. (2007a) Rescuing the extinction of experience. *Biodiversity and Conservation*, 16, 1995–1997.

Samways, M.J. (2007b) Insect conservation: a synthetic management approach. *Annual Review of Entomology*, 52, 465–487.

Suh, A. & Samways, M.J. (2001) Development of a dragonfly awareness trail in an African botanical garden. *Biological Conservation*, 100, 345–353.

7

Trade-offs between Animal Welfare and Conservation in Law and Policy

Stuart R. Harrop

Durrell Institute of Conservation and Ecology, University of Kent, Canterbury, UK

Introduction

Reconciling the views of animal ethicists and scientific conservationists meets epistemological problems (Perry & Perry, 2008). Indeed, bridging the divide between animal ethics and animal welfare science can be a difficult enough task, even without the added complication of animal conservation (Fraser, 1999). Once animals reach a minimum level of phylo-genetic sophistication that is assumed to give rise to a capacity to suffer, they can be attributed rights. Furthermore, their status can be measured with reference to their sentience, which in turn can require animals to be considered on an individual basis.

By contrast, the conservation scientist often has a very different perspective and only needs to focus attention on individual animals when deploying research methods that require counting or tagging individual animals, or where a population has been reduced to very small numbers. Consequently, when conservation scientists descend below habitats or ecosystem diversity, or rise above genetic diversity, their lowest common denominator is most likely to be at the species, subspecies or population levels. Given these different perspectives, this chapter first analyzes the relationship between animal welfare concerns and conservation strategies and, second, seeks to prescribe scope for trade-offs where the two perspectives are in conflict.

Trade-offs in Conservation: Deciding What to Save, 1st edition.　Edited by N. Leader-Williams, W.M. Adams and R.J. Smith.　© 2010 Blackwell Publishing Ltd.

Animal welfare and conservation

As populations of high-profile species of wild mammals are lost, and as more of their habitats are encroached, conservationists increasingly need to encourage positive interventions to ensure the persistence of biodiversity. Examples of interventions that are likely to raise welfare issues include: captive breeding and reintroduction, with their concomitant husbandry and transport requirements; potentially invasive capture and radio collaring research techniques; ranching; practical incentive-driven human–animal conflict resolution techniques; translocation; culling; dealing with the hybridization of wild species by domesticates; and so on. Moreover, international legislation and policy continues to place emphasis on sustainable *use* of natural resources (Rosser & Leader-Williams, this volume, Chapter 8). Such use can range from trade in living or dead animals, to subsistence hunting and through to recreational hunting and wildlife viewing tourism. These uses represent only a small subset of the many human interventions that involve disturbance of free-living species, and that consequently raise welfare issues.

It is well understood that welfare plays a key role in the husbandry of farm animals. However, it is not always appreciated that welfare has a growing role in animal conservation that is reflected in some aspects of international, regional and national conservation regulation (Harrop, 1997). The number of non-governmental organizations (NGOs) dealing with wild animal welfare issues and advocating change is growing, particularly in the USA and Europe. This probably reflects both an increasing desire to incorporate compassion into the science of conservation and growing media interest in the issue (Bremner & Park, 2007). Whereas interdisciplinarity is now regarded as crucial for effective conservation, it is rare to find those holding strong ethical views working closely with pragmatic conservation scientists. Indeed, there is much room for misunderstanding and conflict.

Clarification of terms

In order to understand the issues, it is important to distinguish between terms used in welfare from those used in conservation. However, it must be borne in mind that classifications often force the creation of stark differences that fail to reflect the fluid nature of human motives. For example, a conservation scientist may also be a closet welfarist, while a welfare proponent may be an archetypal conservationist at heart. Furthermore, there is much scope for grey areas between an act designed to preserve an animal, within a strictly defined conservation regime, and an act designed to preserve its welfare without any thought given to the continued survival of the species. Finally, there are also distinct approaches to making value judgments in conservation and welfare, which themselves generate differences.

Animal welfare, animal liberation and animal rights

The phrases 'animal welfare' and 'animal rights' are regularly used interchangeably. Such use is inaccurate as both terms have distinct meanings. Moreover, 'animal rights' may be broken into two further philosophical subsets of 'rights' and 'liberation'. Welfarists accept animal exploitation but seek to increase welfare standards, whereas those who promote animal liberation or animal rights seek to end exploitation (Francione, 1996). For the conservationist, welfare provides some scope for trade-offs. However, rights and liberation perspectives allow little room to manoeuvre.

Rights and liberation

The ethical basis for the animal liberation and rights approach is diverse. Thus one manifestation of the liberationist derives the term *specieism*, which thereby argues that all animals capable of suffering should be treated equally rather than permitting humans to reside within a unique case (Singer, 1975; Ryder, 2005). Another rights-based approach contends that sentient animals should be granted parallel rights to those enjoyed by humans because those rights are allocated on the basis of cognition (Regan, 2004). The purpose here is not to debate further the animal rights or liberation philosophies. Nor is it to critically analyze the types of *direct action* performed in their name, which have included legal and illegal acts to 'liberate' animals from research establishments and so on, and to persuade persons concerned with such acts to cease their activities[1] Instead, the purpose here is merely to note that these concerns are founded on a philosophical basis. However, it should be borne in mind that a rights or liberation approach generally opposes animal exploitation with very few

---

[1] See, for example, www.animalliberationfront.com.

exceptions. Consequently, from the conservation perspective, unequivocal opposition may arise from this quarter when recreational hunting (Dickson, 2009), culling or intrusive methods of research are aspects of conservation strategy. Animal liberation/rights decisions are often based on an ethical proposition without the requirement of a corroborating scientific proposition deriving from that ethic. Although there are many gradations of perspectives, and welfarists and liberationists can regularly assume the role of the other, this chapter now concentrates on welfare perspectives.

Animal welfare

The ambit of *animal welfare* does not entirely escape contention (Sztybel, 2006). This chapter avoids analyzing the precise breadth of the concept and approaches the subject descriptively. Whereas this description will certainly reflect most of the constituents of animal welfare in any part of the world, it is based on its operation within Europe, and particularly in Britain. Animal welfare accepts the human use of animals in principle, but welfarists continuously review the *necessity* of each practice as scientific knowledge develops and moral (or cultural) opinion changes. Thus, many agricultural practices acceptable in Europe in the 19th century are now banned. For example, the use of the once standard method of catching fur-bearing mammals, the leg-hold trap, is now illegal (Harrop, 2000). Likewise, a number of predominantly recreational practices once thought to be acceptable as sport, are now illegal in Britain. These include cock fighting, badger baiting and a contemporary and controversial addition to the list: hunting with dogs. Welfarists also seek to secure regulation of the exploitation of animals to avoid *unnecessary suffering*. Thus, complex regulations in Europe deal with the manner in which animals of various species are transported and slaughtered, while other regulations deal with animals in agriculture.

Additionally, there are a number of international measures dealing with the transport and care of wild animals, particularly in the Convention on International Trade in Endangered Species of Wild Fauna and Flora (CITES). In addition to the complex array of specific laws in Britain dealing, *inter alia*, with agricultural practices, there are general offences prohibiting acts that cause unnecessary suffering (see the Animal Welfare Act 2006 and the Wild Mammals (Protection) Act 1996). Pursuant to these offences, a court determines the acts that cause unnecessary suffering either by application of precedent or on an *ad hoc* basis, usually in response to evidence of government policy, official veterinary stipulations, current scientific understanding or best practice.

The approach to determining necessity, or the level of suffering that is necessary, is based on the propensity of an animal to suffer irrespective of its conservation status, and the benefits of such actions to humans, or even to relevant or other species. Of course, welfare regulations may also be arbitrary or difficult to assess from a rigorously logical perspective because they also arise from a number of shifting factors such as: the result of public feeling at a particular time; political expediency; aesthetic considerations; or the exercise of power of non-governmental lobbying from time to time coinciding with a particular interested party in government (Harrop, 1997).

Conservation and its ethical basis

The *Concise Oxford Dictionary* defines *conservation* as: *'preservation, esp. of the natural environment'* (Oxford, 1996). An informal search through web-based dictionaries reveals a preponderance of the same approach. Some web definitions refine the approach by providing the foundations of an ethical base such as the anthropocentric justification of retaining natural resources in terms of *inter-* and *intra-*generational equity. This reflects, to an extent, the international consensus evidenced in law and policy, which focuses on variants of the concept of sustainable use. However, one international instrument, the World Charter for Nature[2] appears at first glance to go much further than this, by affirming that *'every form of life is unique, warranting respect regardless of its worth to man'*. This remarkable ecocentric ethical base not only appears to build an instant bridge between the animal welfarist and the conservation scientist, but it also ostensibly suggests that we should embrace the *Anopheles* mosquito and other forms of life that are an anathema to humans. However, the statement is merely a component of a preamble that leads to a more pragmatic set of principles for implementation. Indeed, the body of its substantive text is practically qualified by many propositions that support the mainstream institutional agenda describing, *inter alia*, the need to conserve natural resources for the purposes of maintaining generational equity and general principles of optimal sustainable use.

---

[2] See UN GA RES 37/7 at http://www.un.org/documents/ga/res/37/a37r007.htm.

There is obvious scope, however, for commencing a dialogue to define a unified root for conservation, animal welfare and animal liberation/rights philosophies. All of these perspectives share, to varying degrees, a common desire to facilitate wild species living freely according to their natures in healthy ecosystems. Such a root might be the foundation for a trade-offs formula.

Aspects of animal welfare and conservation regulation

The foci for confrontation and for cooperation between conservation and welfare lobbies include the use of wildlife in trade, for consumption and for recreation, and also aspects of practical conservation practices deployed in the field. Most of these foci give rise to regulations or policy-based decisions at some point and this is the key area to examine.

Wild animal legislation in the United Kingdom

Laws in a democratic country reflect negotiated trade-offs as political expediencies or realities. Sometimes the laws may also reflect stand-offs. Wild animal law in the UK tends to be designed to suit either conservation or welfare objectives, but with little attempt to link the two. The UK's Wildlife and Countryside Act 1981[3] is a conservation instrument driven by that agenda. Despite the many consequences of conservation to animal welfare, the Act includes only one obvious welfare stipulation: a requirement that caged birds should be able to spread their wings. In contrast, the UK's Animal Welfare Act 2006 is directed at the welfare of animals under the control of humans and expressly restricts its ambit to animals not living in the wild state[4] The position does not change significantly when welfare legislation deals with wild animals. The UK's Wild Mammals (Protection) Act 1996 is designed to prevent unnecessary suffering to wild mammals, irrespective of their conservation status. The act only contains an implied reference or attempt to link with conservation strategies where 'pest control' may justify a lower standard of welfare.

Some activities, such as hunting and fishing, may be specifically regulated from the prime perspective of recreational hunting. Instruments in this field

---

[3] All British legislation referred to herein may be found at www.opsi.gov.uk/acts/.
[4] See section 2(c).

tend to deal effectively and consistently with both welfare and conservation issues. The UK deer and fisheries legislation[5] contains conservation provisions that establish closed seasons and prohibit non-selective practices. Simultaneously, these instruments prohibit cruel methods of taking and killing and so deal equally with conservation, hunting and welfare agendas. Swedish hunting legislation goes further than this and also regulates the use of dogs in hunting, specifically in order to protect the dogs from suffering while hunting (Harrop & Harrop, 2001).

These few examples only serve as illustrations of some of the widely different approaches to allocating priorities to welfare and conservation in national law, a full analysis of which would be an unnecessary diversion. However, this UK case study shows a lack of uniformity that demonstrates the polarity of this debate. Thus, welfare and conservation law contain stand-offs rather than trade-offs, although hunting legislation appears to demonstrate a more middle way approach. Nevertheless, even in this field there are distinct exceptions when hunting engenders greater welfare feeling. The recent laws in Britain[6] that ban the controversial practice of hunting with hounds ignore highly relevant conservation issues (Harrop, 2005).

European leghold legislation

Sometimes a law may change emphasis, as if it possesses a life of its own, and so alter its original intention. The European Leghold Trap Regulation[7] sought, *inter alia*, to abolish the use of this trap throughout the European Union. The trap has been widely regarded as cruel, at least since Charles Darwin proclaimed it as such (Harrop, 2000), and this regulation has always been heralded as an example of animal welfare legislation. However, the regulation claimed its legal base from provisions in the European Union's Habitats Directive[8]

---

[5] See, for example, the Deer Act 1991, Deer (Scotland) Act 1996, Salmon and Freshwater Fisheries Act 1975 and Salmon and Freshwater Fisheries (Consolidation) (Scotland) Act 2003.

[6] Hunting Act 2004 and Protection of Wild Mammals (Scotland) Act 2002.

[7] Council Regulation (EEC) No. 3254/91 of 4 November 1991 prohibiting the use of leghold traps in the Community and the introduction into the Community of pelts and manufactured goods of certain wild animal species originating in countries which catch them by means of leghold traps or trapping methods which do not meet international humane trapping standards. *Official Journal* L 308, 09/11/1991, p. 0001–0004 (1991).

[8] Council Directive 92/43/EEC of 21 May 1992 on the conservation of natural habitats and of wild fauna and flora. *Official Journal* L 206, 22/7/1992, p. 7–50 and Council of Europe ETS No. 104: *Convention on the Conservation of European Wildlife and Natural Habitats.*

and in the Berne Convention[9] that were aimed at eradicating non-selective methods of taking animals on the basis of their danger to endangered species in addition to the targeted common species. Consequently, the scientific basis underpinning a conservation measure was claimed to form an ethical base by a completely different lobby.

International whaling law

At the international level there is little welfare legislation but a reasonably sized and focused portfolio of conservation legislation. The original legislative purposes in this body of international law can also shape-shift over time, as may be seen in the context of whaling. The issue of whaling draws a great deal of public interest and the state representatives at meetings of the International Whaling Commission (IWC) are often outnumbered by communities of NGOs. Moreover, whereas the IWC was originally developed to be a vehicle to enable whaling nations to regulate their industry, this regulatory body has been gradually transformed into a conservation organization managing strict delimitations on minimal cetacean harvesting.

The metamorphosis has continued in recent years, as welfare debates have become more frequent, resulting in the IWC at times shape-shifting into other epistemological realms, as it becomes almost exclusively focused on welfare (Harrop, 2003). The IWC is regularly involved in debates concerning the welfare of hunted whales and has prescribed a small number of changes to the schedule of its parent convention[10] to take account of welfare in killing methods. There are, of course, trade-offs that arise from this type of shape-shifting as politics lends a hand. In consequence, welfarists and conservationists can often be seen working together to support each other's arguments.

One catalyst for the increased interest in welfare by conservationists, who might fear that the hunting of whales cannot be executed and controlled as a selective practice, is the possible recovery of some whale stocks to a level that could permit a reasonable off-take. A case in point is the minke whale

---

[9] Council of Europe ETS No. 104: *Convention on the Conservation of European Wildlife and Natural Habitats.*

[10] International Convention for the Regulation of Whaling, 1946 Schedule as amended by the Commission at the 59th Annual Meeting, Anchorage, USA, 28–31 May 2007. The Schedule to the Convention is the operative part of the document and is regularly amended to take account of new regulatory requirements.

Balaenoptera acutorostrata, although controversy surrounds the perceived recovery of the species (Clapham *et al.*, 1999). Setting aside the complexity of the debate concerning subspeciation of the minke whale and so on, if it were conclusively demonstrated that populations could be used sustainably, there are precedents within international policy, and both soft and hard law, to support a case for a reduction in the ambit of the IWC's moratorium (Harrop, 2003). In these circumstances, the welfare arguments concerning the inhumanity of killing whales becomes of paramount importance to those conservationists and others who are unwilling to accept the interpretation of the population figures and thus continue to oppose whaling. This is a useful illustration of welfare and conservation ideals combining forces to operate in confluence rather than conflict.

CITES

The Convention on International Trade in Endangered Species of Fauna and Flora (CITES) remains true to its original volition and its text has always contained very clear trade-offs, particularly where limited trade in threatened species takes place. It expressly prescribes welfare measures where animals in international trade mirror the circumstances in which welfare regulation is applied to domestic and agricultural animals. Thus, facilities for the transport and reception of animals in international trade are required to conform to limited welfare requirements. Although CITES has largely avoided interfering with the pre-trade process, as confirmed when CITES was requested to effect a general ban on trade in animals captured by inhumane means (Harrop, 2003), it has imposed some welfare requirements on 'ranched' wild animals that could potentially enter international trade. Although recently sourced from wild populations, ranched wild animals correspond closely to animals in agriculture and there can be no moral reason to differentiate between the standard of care extended to domestic or wild animals in these circumstances (Harrop, 2003).

In situ *conservation*

Conventions dealing with methods of *in situ* conservation do not prescribe standards of welfare except in very limited circumstances. There are distinct areas that could be relevant, particularly in dealing with human interventions such as hunting, trapping, fishing and taking methods. However, where these are regulated, the relevant legislation tends to apply to endangered and threatened species irrespective of an animal's capacity to suffer. Moreover, even a provision that might be construed as welfare oriented, such as the ban on indiscriminate methods of taking endangered and threatened species in the Berne Convention, was originally included for conservation rather than welfare reasons. Thus, this measure originally had the intent of avoiding non-selective methods of killing that could disproportionately harm the numbers within a population or incidentally catch other endangered species (Harrop, 2003).

Welfare debates at the international level that deal with *in situ* conservation issues can also draw on perspectives beyond conservation. An example is the debate arising from the exemptions for traditional aboriginal subsistence whaling granted by the IWC as exceptions to its moratorium on harvesting large cetaceans. These exemptions permit specific communities to hunt whales in accordance with their traditions. Traditional hunting methods often fall well below the much-criticized standards of the commercial whaling nations such as Japan and Norway. This creates some difficulties for those who take animal and human rights-based moral standpoints, since it is difficult to square the desire to morally support the rights of indigenous peoples and at the same time assert the right of the whale to a quick, clean kill. This dichotomy has been identified by the Japanese delegates to the IWC who have asserted that the trade-off, if there is to be a welfare standard in whaling, should be consistently applied to all whalers (Harrop, 2003).

Confluence and conflict, trade-offs and stand-offs

Animal welfare and conservation appear to have a close working relationship where the treatment of wild animals mirrors closely the husbandry of domesticates. Wild species in transport are captive in a similar manner to domesticates, and so welfare law tends to apply to both. Ranched wild species are curtailed in their behaviour in a similar manner to domestic animals in agriculture. In these circumstances, ranched animals are specifically designated as individuals by their human captors, and these individuals rather than the species receive specific consideration. In consequence, attention has turned to the welfare perspective and our norms regulating behaviour seek to qualify our otherwise unlimited dominion over the husbandry of the animals within our dominion.

However, despite many national laws (Harrop & Harrop, 2001), international regulation largely avoids dealing with the welfare incidences arising from human interaction with species freely living in the wild. The IWC's limited excursions into this area have already been discussed. CITES is reluctant to interfere with the welfare of species prior to the point of shipment. Finally, limited, regional attempts to regulate the welfare incidents of trapping[11] have demonstrated a lowest common denominator approach and have not challenged standards of welfare (Harrop, 1998, 2000). The scenario leading up to that agreement demonstrates the almost insurmountable gulf between welfare and rights arguments, conservation strategies and user interests. Indeed, the agreement was only completed by a deliberate exclusion of the welfare representatives from the negotiations (Harrop, 2000).

Current challenges arise when conservation strategies support hunting, culling or invasive methods of research that disrupt the well-being of wild animals. Culling, when executed solely as a conservation strategy, is not without controversy deriving from welfare concerns, for example the culling of African elephants *Loxodonta africana* in Kruger National Park, South Africa (van Aarde *et al.*, 1999). Culling may be part of a strategy to defeat disease transmission, to prevent hybridization or to control burgeoning populations of a species. The examples are numerous and include: controlling the populations of African elephants within confined protected areas (Morris, 2007); eradicating introduced hedgehogs *Erinaceus europaeus* (Figure 7.1) that threaten nesting sea-birds on Scottish islands (Jackson, 2001); preventing the spread of disease between the Ethiopian wolf *Canis simensis* and domestic dogs (Laurenson *et al.*, 1998); eliminating the hybridization of animals with domesticates, such as the Ethiopian wolf with the domestic dog, or the Scottish wild cat *Felis sylvestris grampia* (Figure 7.2) with feral domestic cats (Beaumont *et al.*, 2001); and hybridization with escapee exotics, such as the threat to the European white-headed duck *Oxyura leucocephala* posed by escapees from collections of the North America ruddy duck *Oxyura jamaicensis rubida* (Rhymer & Simberloff, 1996).

In these cases, culling may be the only effective conservation strategy to preserve either genetic integrity in wild species or ecological balance. However, the welfarist considers the goal of maintaining genetic diversity to be subordinate to securing freedom from suffering. Culling, even when

---

[11] Agreement on international humane trapping standards between the European Community, Canada and the Russian Federation. *Official Journal* L 042, 14/02/1998, p. 0043–0057 (1998).

Figure 7.1 **Culling conducted as part of a conservation strategy can prove controversial when it intersects with welfare concerns, for example the eradication of hedgehogs *Erinaceus europaeus* that threaten nesting sea-birds on Scottish islands. (Photograph by Stuart Harrop.)**

expertly carried out, is likely to cause some suffering. Wild animals cannot be simply pre-stunned and then cleanly and quickly killed like domestic animals in a slaughterhouse. Indeed, from the welfare perspective the destruction of alien species, as a component of conservation strategy, has been described as analogous to ethnic cleansing (Smout, 2003).

Other areas that could generate welfare concerns include intrusive aspects of conservation research that share some characteristics with hunting and capture methods. In the UK such methods of research involving wild animals require licenses and obtaining them can be a difficult business and they may not all be granted, *inter alia*, where the welfare cost outweighs the value of the research.[12]

---

[12] See, for example, the UK's Animals (Scientific Procedures) Act 1986. See also the UK Government's 'Guidance on the Operation of the Animals (Scientific Procedures) Act 1986' where it is stated in paragraph 2.45: '*In deciding whether and on what terms to grant a project licence, the Secretary of State must weigh the likely adverse effects on the animals involved against the benefit (to humans, other animals or the environment) likely to accrue from the programme of work*'.

Figure 7.2 **Welfare concerns can also arise when attempts are made to eliminate the hybridization of wild animals with domesticates, such as the Scottish wild cat** *Felis sylvestris grampia* **with feral domestic cats. (Photograph by Stuart Harrop.)**

One relevant example of a conservation method that to an extent mimics hunting is the increasingly used method of radio/satellite tracking. This can involve significant interference with an animal, first through capture and insertion or attachment of a device and, thereafter, where normal behaviour may be altered or inhibited as a result either of the shock of the capture experience or through some characteristic of the attached device. Disruption of behaviour may be particularly apparent where a large device is required, perhaps to accommodate significant battery power to enable satellite downloading of data where the subject animal's habitat is an inaccessible location. In one case, a leatherback turtle *Dermochelys coriacea*, captured in the Irish Sea was noted, following the attachment of a satellite uplink device to it, to swim without stopping until it was in a subtropical zone before it recommenced its normal feeding routine. This tracked behaviour may have been an immediate response to the experience of capture[13].

---

[13] See Irish Sea Leatherback Turtle Project Populations, Origins and Behaviour INTERREG IIIA Initiative 2003–2006: Final Project Report at www.turtle.ie/publications/6.pdf.

Reconciling animal welfare and conservation in international law

The design of an international legal strategy always requires a trade-off to be made between *laissez-faire* and intervention. Furthermore, establishing an international regime takes considerable time, since altering the *status quo* erodes jealously guarded national sovereignty or threatens existing commercial or social practices. Animal welfare norms can give rise to an even greater list of sensitivities than proposed international conservation provisions, because they derive from widely differing ethical perspectives that are not shared uniformly within social groups, cultures or countries. Indeed, some counties may be forgiven for regarding animal welfare requirements as mere heads on the hydra of western protectionist policies.

Moreover, if welfare requirements are to be become mandatory for less economically prosperous countries, other issues should probably be tackled first. Animal welfare measures cost money, particularly where applied on a commercial scale. This is well understood in the context of farm animal welfare, which adds significant costs to animal production (McGlone, 2001). Nevertheless, there is a limit to how much of these costs the producers and consumers are willing to sustain, even in wealthier countries (Bennett, 1997). In less developed countries, those at the frontline of production have little hope of affording western welfare standards.

That said, three potential approaches are now set out that could have the effect of further embedding welfare interests in international conservation norms. First, there could be general regulation of non-selective, indiscriminate and inhumane hunting/capture methods within an international instrument. Similar measures are already contained in the Berne Convention in relation to the taking of endangered or threatened species. However, in order to satisfy animal welfare philosophy, an international instrument would need to go further than these existing measures and apply them to all species irrespective of their conservation status.

Such a measure could also require minimum training and weapons require-ments for all hunting, whether subsistence, commercial, recreational or scientific. Such measures are common in national conservation instruments, and are usually seamlessly enmeshed with a volition that emanates from a welfare lobby (Harrop & Harrop, 2001). This instrument could also detail regulations for allowed methods of capture to be used in radio tracking and other conservation strategies. It could contain minimum standards for devices used in such strategies or create approval procedures. Nevertheless, the detail required in this sort of legislation might be better suited to a standardization approach, which I discuss next, rather than the slow and less flexible mechanism of international conservation law.

The second approach would involve the creation of international standards designed to deal in detail with hunting and conservation strategies. This mechanism would be specially suited to deal with the many technical aspects of the subject. These would be implemented either through a parent international convention or through their development under the auspices of the International Standards Organization, and implementation through national regulatory or trade measures. If the latter approach is followed, the lessons of previous attempts to create a standard for the humane trapping of animals (at the time known as ISO TC191) should be carefully studied. Despite 10 years of extensive negotiations, that standard failed to come into existence because of a lack of common ground between the welfarists and the trappers (Harrop, 2000). Other debates on the subject should also be carefully scrutinized. For example, CITES also examined the issue as to whether it could ban trade in inhumanely captured animals and chose to reject the proposal (Harrop, 2003). Moreover, if the technical standards route is to be followed, the requirements of the World Trade Organization's Technical Barriers to Trade Agreement may also need to be considered. Whereas that agreement may be a useful mechanism to support otherwise illegal state-supported eco-labels, there is some doubt as to whether a technical standard in place to support animal welfare is permitted by that agreement (Harrop, 2003).

The third proposed approach would be to legislate, either through a convention or standards-based instrument, to take all conflicting priorities in conservation into account through an impact assessment. Such an instrument could require that all conservation projects execute, *inter alia*, a welfare impact assessment or audit. In this way, all conservation interventions such as trapping, radio collaring, culling and so on would become transparent. Such an assessment could also require conservation scientists to specifically address welfare issues in their planning by examining alternate arrangements and measuring them against the central proposal, such as the feasibility of translocation against culling, or the feasibility of physical observation against intrusive radio tracking.

References

Beaumont, M., Barratt, E.M., Gottelli, D. *et al.* (2001) Genetic diversity and introgression in the Scottish wildcat. *Molecular Ecology*, 10, 319–336.

Bennett, R.M. (1997) Farm animal welfare and food policy. *Food Policy*, 22, 281–288.

Bremner, A. & Park, K. (2007) Public attitudes to the management of invasive non-native species in Scotland. *Biological Conservation*, 139, 306–314.

Clapham, P.J., Young, S.B. & Brownell, R.L. (1999) Baleen whales: conservation issues and the status of the most endangered populations. *Mammal Review*, 29, 35–60.

Dickson, B. (2009) The ethics of recreational hunting. In *Recreational Hunting, Conservation and Rural Livelihoods*, eds B. Dickson, J. Hutton & W.M. Adams, pp. 59–72. Blackwell, Oxford.

Francione, G.L. (1996) *Rain Without Thunder: the ideology of the Animal Rights Movement*. Temple University Press, Philadelphia.

Fraser, D. (1999) Animal ethics and animal welfare science: bridging the two cultures. *Applied Animal Behaviour Science*, 65, 171–189.

Harrop, S.R. (1997) The dynamics of wild animal welfare law. *Journal of Environmental Law*, 9, 287–302.

Harrop, S.R. (1998) The Agreements on International Humane Trapping Standards: background, critique and the texts. *Journal of International Wildlife Law and Policy*, 6, 79–104.

Harrop, S.R. (2000) The international regulation of animal welfare and conservation issues through standards dealing with the trapping of wild mammals. *Journal of Environmental Law*, 13, 387–394.

Harrop, S.R. (2003) From cartel to conservation and on to compassion: animal welfare and the International Whaling Commission. *Journal of International Wildlife Law and Policy*, 6, 79–104.

Harrop, S.R. (2005) The role and protection of traditional practices in conservation: an enquiry into the UK's implementation of Article 8 (j) Convention on Biological Diversity. *Environmental Law and Management*, 16, 244–251.

Harrop, S.R. & Harrop, D.F. (2001) Comparing different national regulatory approaches to the practice of hunting wild animals with dogs. *Journal of Wildlife Law and Policy*, 4, 469–481.

Jackson, D. (2001) Experimental removal of introduced hedgehogs improves wader nest success in the Western Isles, Scotland. *Journal of Applied Ecology*, 38, 802–812.

Laurenson, K., Sillero-Zubiri, C., Thompson H., Shiferaw, F., Thirgood, S. & Malcolm, J. (1998) Disease as a threat to endangered species: Ethiopian wolves, domestic dogs and canine pathogens. *Animal Conservation*, 1, 273–280.

McGlone, J.J. (2001) Farm animal welfare in the context of other society issues: toward sustainable systems. *Livestock Production Science*, 72, 75–91.

Morris, E. (2007) Africa conservation: making room. *Nature*, 448, 860–863.

Oxford (1996) *The Concise Oxford Dictionary*, 9th edn. Oxford University Press, Oxford.

Perry, D. & Perry, G. (2008) Improving interactions among animal rights groups and conservation biologists. *Conservation Biology*, 22, 27–35.

Regan, T. (2004) *The Case for Animal Rights*. University of California Press, Berkeley and Los Angeles.

Rhymer, J.M. & Simberloff, D. (1996) Extinction by hybridization and introgression. *Annual Review of Ecology and Systematics*, 27, 83–109.

Ryder, R. (2005) All beings that feel pain deserve human rights: equality of the species is the logical conclusion of post-Darwin morality. *The Guardian*, 6 August.

Singer, P. (1975) *Animal Liberation*. Avon Books, New York.

Smout, T.C. (2003) The alien species in 20th century Britain: constructing a new vermin. *Landscape Research*, 28, 11–20.

Sztybel, D. (2006) The rights of animal persons. *Animal Liberation Philosophy and Policy Journal*, 4, 1–37.

van Aarde, R., Whyte, I. & Pimm, S. (1999) Culling and the dynamics of the Kruger National Park African elephant population. *Animal Conservation*, 2, 287–294.

8

Protection or Use: a Case of Nuanced Trade-offs?

Alison M. Rosser and Nigel Leader-Williams

Durrell Institute of Conservation and Ecology, University of Kent, Canterbury, UK

Introduction

Over 190 national governments around the world have acceded to the Convention on Biological Diversity (CBD), making it the largest United Nations convention in terms of state parties. Through accession, signatory governments recognize that conservation requires a dual strategy, outlined in Article 1 of CBD, that encompasses protection, sustainable use and equitable sharing of the benefits of biodiversity (Leader-Williams *et al.*, this volume, Chapter 1). Nevertheless, the conservation movement continues polarized debates over whether to preserve biodiversity through protected areas and legislation that ban all or most extractive use, so-called 'protection', or through incentives that encourage managed extractive use of wild species that is sustainable. The latter has been characterized as the 'use it or lose it' debate (Freese, 1997). Equally, there is justified concern that efforts to conserve biodiversity, whether through strict protection or through extractive or non-extractive use, are not sustainable in the face of growing pressures on land, on national exchequers and because of the fickle nature of market-dependent approaches, to which the worsening status of biodiversity (Lawton & May, 1995; Butchart *et al.*, 2004) more than bears testament. Even community-based conservation approaches incorporating benefit sharing from resource use appear increasingly inequitable, and do

Trade-offs in Conservation: Deciding What to Save, 1st edition.   Edited by N. Leader-Williams, W.M. Adams and R.J. Smith.   © 2010 Blackwell Publishing Ltd.

not provide individuals and households with control over resources they use (Norton-Griffiths, 2007). In many biodiversity-rich areas, patterns of human population growth are changing (Sachs *et al.*, 2009), livelihood options are lacking, numbers of landless poor are growing (Brockington & Igoe, 2006) and governance structures are ineffective and corrupt (Smith *et al.*, 2003). As local conservation systems become increasingly over-run, and as win–win solutions of local involvement promoted in the 1980s are found wanting, the protectionist lobby has shown a resurgence that advocates strict enforcement of non-extractive use in approaches that return 'back to the barriers' (Hutton *et al.*, 2005).

Consequently, positions and debates over the two paradigms of conservation through 'protection' or through 'use' have become even more polarized, raising questions about trade-offs between different approaches to conservation for the 21st century. Given the apparent international consensus of the CBD, this chapter aims to re-articulate the trade-offs that need to be made between these polarized positions, by examining factors that contribute to the long-term conservation success of both approaches. First, we examine definitions and underlying philosophical questions in debates over protection and use. Second, we outline some of the difficulties that conservationists face in practice when conserving biodiversity through protection or use. Third, we examine the effectiveness of following a single strategy approach that promotes either protection or use, including some case studies. Fourth, we suggest the need for a dual approach in which already nuanced trade-offs between strict protection and unfettered extractive use are made explicit. Finally, we call for an end to sterile arguments over the relative effectiveness of each approach.

Definitions and philosophy

What do conservationists mean by the two terms 'protection' and 'use', and are there clear-cut differences or overlaps between the ways in which these terms are used? As Box 8.1 shows, some conservationists seek to afford protection, either to geographic areas, known generically as 'protected areas', or to particular species. Such protection can be afforded through legislative frameworks that require strict enforcement to ensure their effectiveness. However, law enforcement is costly in terms of manpower and financial resources (Leader-Williams & Albon, 1988), so conservationists often seek to trade-off some protection against non-extractive uses such as tourism (Honey, 1999), or against cultural and aesthetic uses (Homewood, this volume, Chapter 10) and, increasingly, against provision of ecosystem services (Goldman *et al.*, this volume, Chapter 4): this we call 'strict protection'. Equally, some protected areas have been established through legislation that sanctions some forms of extractive use, such as recreational hunting (Dickson *et al.*, 2009). Furthermore, areas that are not formally protected can support private or communally sanctioned removal of resources: this we call 'extractive use' (see Hutton & Leader-Williams, 2003). However, all forms of resource use should be sustainable, following the clear definition for sustainable use encompassed in Article 2 of the CBD (Box 8.1).

Box 8.1 **Philosophy and definitions of protection and use**

Area protection

A 'protected area' comprises a clearly defined geographic space, recognized, dedicated and managed, through legal or other effective means, to achieve the long-term conservation of nature, with associated ecosystem services and cultural values (Leader-Williams *et al.*, 1990a; IUCN, 2008). Protected areas can be designated at different political levels, from local to international. Furthermore, protected areas can include different objectives that may be reflected in names given by different jurisdictions to different categories of protected area. Most protected areas have been designated under systems that assume some form of central control, appropriate to the level at which the protected area is designated. However, there is now increasing recognition of the coverage and role that privately owned and communally controlled areas can play in providing area-based, and indeed species, protection.

Species protection

Species can also be protected directly through species-protection measures enshrined in taboos or local custom (Homewood, this volume, Chapter 10), and in local, national, regional and international legislation. Species protection measures can vary, from international conventions to local by-laws, and may list protected species on different appendices or annexes. In turn, listings can afford different levels of protection, depending on perceived levels of threat to those species. Such species-specific legislation generally seeks to protect species against overuse, usually from extractive use. However, some species-specific legislation may also seek to afford protection against habitat destruction through development.

Sustainable use

According to Article 2 of the CBD, sustainable use comprises '*the use of components of biological diversity in a way and at a rate that does not lead to the long-term decline of biological diversity, thereby maintaining its potential to meet the needs and aspirations of present and future generations*'. The operational part of this definition has a distinctly biological slant, and requires incentives to be in place for resource users to practice sustainable, as opposed to unsustainable, use (Hutton & Leader-Williams, 2003). However, Article 11 of the CBD notes that conservation should: '. . . *as far as possible and appropriate adopt economically and socially sound measures that act as incentives for conservation and sustainable use . . .*'. Incentive-based approaches can include: conservation on private and communal land outside protected area networks; benefits derived from non-extractive and extractive uses of biodiversity, in areas such as private tourist reserves and hunting estates; agricultural set aside; biodiversity-friendly cultivation and farming practices; payments for ecosystem services; and cultural or aesthetic 'uses' of biodiversity.

Protection and use in practice

The previous section showed that protected areas comprise a core strategy for many conservationists. The definition of protected areas suggests they offer a somewhat inflexible approach to conservation. In practice, the means to achieve area or species protection can be very flexible and allow conservationists, stakeholders and relevant authorities to select the best forms of designation or use to deal with the local ecological and socioeconomic needs. IUCN, the International Union for the Conservation of Nature (IUCN, 2003)

Table 8.1   **The series of six management categories defined by the IUCN (2008) for protected areas based on primary management objectives.**

Category	Name	Primary management objective
Ia	Strict Nature Reserve	Science
Ib	Wilderness Area	Wilderness protection
II	National Park	Ecosystem protection and recreation
III	National Monument	Conservation of specific natural features
IV	Habitat/Species Management Area	Conservation through managed intervention
V	Protected Landscape/Seascape	Landscape/seascape conservation and recreation
VI	Managed Resource Protected Area	Sustainable use of natural ecosystems

has developed a global system of management categories (Table 8.1). While not explicitly stated, the three categories of Ia, Ib and II are generally uninhabited by people, implying that these areas are 'strictly protected'. Category Ia and Ib areas do not generally allow any form of use, while category II areas generally allow non-extractive, recreational visitor use. By contrast, protected areas in categories V and VI generally allow human habitation and extractive use that is intended to be sustainable.

Protected areas in categories Ia, Ib and II are usually established on state land managed by state agencies. More recently, various conservation non-governmental organizations (NGOs) and foundations have been buying up private land (Bruner *et al.*, this volume, Chapter 11) to establish strictly protected conservation areas. Thus, organizations from developed countries are working with partners in biodiversity-rich countries to establish private reserves and take over extractive concessions for non-extractive purposes (Carter *et al.*, 2008). Meanwhile, various charitable organizations and businesses have entered private partnerships with national governments to run protected areas that the state can no longer afford. In some cases, they may require the removal of illegal settlers, as with the case of African Parks in Ethiopia. More recently, the role of indigenous people in managing natural resources has been recognized through the category of Indigenous Community Conserved Areas (ICCAs). This increasing variety of governance structures is captured in the IUCN matrix for describing protected areas on the basis of management and governance (IUCN, 2008).

Sustainable use approaches rest on the premise that sustainable and well-managed use of resources can provide incentives that encourage local stakeholders to conserve resources that generate ongoing social and economic benefits (Freese, 1997; IUCN, 2000). Most terrestrial biodiversity remains in the tropics, also inhabited by many of the world's poor who depend on the extractive use of biodiversity for their livelihoods (Roe & Elliott, 2006). From the short-term economic perspective, the greatest monetary benefits generally derive from mining those natural resources (Clark, 1973; Lande et al., 1994). Indeed, many species have been driven to extinction or near extinction by overuse, due to issues such as lack of tenure over the resource, poor scientific understanding of the productivity of resources, high discount rates, lack of appropriate monitoring of the resource and feedback to amend the harvest management, as well as issues of human population growth and poverty (Swanson, 1994; Milner-Gulland & Rowcliffe, 2007). Thus, so-called 'sustainable use' approaches have often been advocated in situations where safeguards are poorly implemented, and the resulting use is usually nowhere near sustainable, yet labelled a failure of 'sustainable use' approaches. Unsurprisingly, many conservationists remain to be convinced of the effectiveness of extractive use as a conservation tool. Meanwhile, the social and ecological impacts of non-extractive uses are attracting increasing attention from both conservationists and those concerned with social development (Hutton & Leader-Williams, 2003).

Despite their apparently separate philosophy and definitions (see Box 8.1), trade-offs have regularly been made between 'protection' and 'use' or vice versa. In many areas, biodiversity conservation is zoned, and strictly protected core areas are surrounded by areas of managed extractive use, as in the design of Biosphere Reserves (Batisse, 1986). However, the rhetoric of many conservationists sees the 'best' or 'strictest' protected areas as those in 'higher' management categories (see Table 8.1) where people are excluded and extractive use is prohibited (Terborgh, 2004). Similarly, in relation to 'use', conservationists tend to show less concern over non-extractive uses like tourism, and payments for ecosystem services, than about extractive uses where individual specimens are purposefully removed from the population. Consequently, the way the different positions blend into each other is inconsistent, even if protagonists pretend that 'protection' and 'use' are separate paradigms. We now review how supposedly different systems of 'protection' and 'use' perform.

Performance of protection and use approaches

The case for adopting strict area-based protection, where human settlement is not allowed, or high levels of species protection, which precludes commercial extractive use, is often based on the realistic concern that much extractive use is not actually sustainable. However, just as use may not be properly regulated, strict protection may also not be fully enforced, often due to funding shortages and lack of political will. We illustrate some extremes in the spectrum of polarized views, using three case studies of our own and colleagues' work (Box 8.2), followed by a review of some general issues.

Box 8.2 **Case studies of protection and use**

These case studies indicate that strict protection, although much more effective than not affording any protective status at all, is not foolproof in and of itself. Equally, sustainable use can provide incentives for careful management under appropriate tenurial arrangements.

Kerinci Seblat National Park

Kerinci Seblat National Park (KSNP) in Sumatra covers ~13 300 km^2 of critically important tropical forest, contains probably the largest remaining population of Sumatran tigers (Linkie *et al.*, 2006) and enjoys high, IUCN category II protection status. However, KSNP lacks direct revenue-generating opportunities through tourism that, coupled with its sheer size and inaccessibility, has resulted in little development compared with surrounding, non-gazetted areas lying within similar forested ecosystems. On this basis, KSNP could be judged something of a success in terms of conserving forest resources. However, thanks to minimal enforcement and despite community involvement through a World Bank-funded Integrated Conservation and Development Project

---

(ICDP), edge encroachment of critically important lowland forest remains a problem (Linkie *et al.*, 2008) (Figure 8.1). Meanwhile, tigers remain threatened by poaching, as limited law enforcement patrols have to cover large areas (Linkie *et al.*, 2003). In this case, neither strict protection nor an ICDP that aimed to generate incentives have been fully successful.

Figure 8.1 **Farmers clearing land in the lowland margins of Kerinci Seblat National Park in Sumatra, Indonesia. Lowland rainforest is the most biodiversity-rich rainforest and its encroachment continues despite the presence of a national park boundary. (Photograph by Nigel Leader-Williams.)**

Masai Mara National Reserve

Masai Mara National Reserve (MMNR) in Kenya is the northernmost extension of the world famous Serengeti–Mara ecosystem. It covers $\sim$1500 km^2 and attracts high levels of tourism, the revenue from which could both result in good management inside MMNR, and provide benefits to share with local communities outside MMNR. Therefore, the protective status of MMNR might be expected to fare better than KSNP, thanks to the high tourist revenues per unit area. However, even in this case, there are issues over encroachment, disturbance to wildlife (Figure 8.2), lack of enforcement of MMNR regulations, issues with disbursing benefits to local communities (Walpole *et al.*, 2003) and embezzlement of tourism revenue by local district councils (Thompson & Homewood, 2002).

Figure 8.2 **Nature-based tourism has been traded off against protection in many high category protected areas, such as in Amboseli National Park, Kenya. However, the impact of tourist vehicles crowding round large carnivores is little known. (Photograph by Nigel Leader-Williams.)**

Guassa Commons

The Guassa Commons in Ethiopia covers an area of ~50 km^2 within the biodiversity-rich highlands of Ethiopia. The commons have been managed for extractive use by local community structures defined by descent groups since the 1700s (Ashenafi & Leader-Williams, 2005). Under this system, the so-called 'guassa grass' *Festuca* spp. provides thatching straw for local houses and supports a rich natural biodiversity that includes the world's second largest population of the critically endangered Ethiopian wolf *Canis simiensis*, which remains far outside any formal state-run protected area (Ashenafi *et al.*, 2005). In this case the area is protected by community norms that allow a managed harvest, and people have traded off year-round unlimited access for a predictable, long-term but restricted harvest. The local community management institution has remained resilient to the political changes that have gone on around it, and have prevented the area becoming an externally imposed protected area.

Is strict protection effective?

Conservationists have persuaded governments worldwide to set aside large areas of their terrestrial land surface as protected areas. Indeed, the initial global target of 10% coverage of the world's terrestrial land area has been exceeded, and coverage is currently $\sim$12% (IUCN, 2003). Nevertheless, the status of biodiversity continues to worsen and many conservationists suggest further protection (Rodrigues *et al.*, 2004). More justifiably, calls have been made to rapidly increase the areas of freshwater and marine habitats included within protected area networks (Roberts *et al.*, 2002), which lag far behind the coverage of terrestrial protected areas (Adams, this volume, Chapter 16).

Establishing a protected area network does not necessarily result in successful protection of biodiversity. For example, biologists working at the site level have raised concerns that many protected areas are not effectively conserving their constituent biodiversity (Curran *et al.*, 2004; DeFries *et al.*, 2005; Southworth *et al.*, 2006). Nevertheless, studies of their effectiveness suggest that protected areas can reduce tropical deforestation (Bruner *et al.*, 2001; Naughton-Treves *et al.*, 2005; Nepstad *et al.*, 2006), as confirmed by recent studies of the counterfactual of the outcome had there been no protected area (Ferraro & Pattanayak, 2006; Gaveau *et al.*, 2009a). Furthermore, marine protected areas have been shown to increase fish yields in areas surrounding no-take zones at the core of protected areas (Roberts *et al.*, 2005).

Despite some encouraging results, studies of protected area effectiveness have demonstrated problems with edge incursions and erosion due to a lack of enforcement amongst other things, raising more questions about the long-term viability of the protected area approach (Gaveau *et al.*, 2009b). There are also questions about the overall success of protected areas in conserving a variety of targets such as large animals, less charismatic species and ecosystems (Gardner *et al.*, 2007; Stoner *et al.*, 2007). Similarly, research on carnivores has shown that protected areas are not sufficiently large to accommodate wide-ranging species like wild dogs *Lyacon pictus* (Woodroffe & Ginsberg, 1998). Thus, discrete protected areas raise issues of connectivity and adaptability in the face of environmental change, which could be better addressed by a landscape approach that includes conservancies (Lindsey *et al.*, 2009).

Strict protection, coupled with non-extractive uses such as photographic tourism, has received widespread support from conservationists as win–win solutions, where revenue generation can help incentivize conservation and enforce regulations (Honey, 1999). However, more recent research has questioned the apparently benign nature of non-extractive uses, particularly of mass tourism and of various wildlife watching ventures, perhaps as some types of nature-based tourism become more successful than originally envisaged (Roe *et al.*, 1997). So, strict protection may not be fully effective either, particularly where no monitoring is undertaken and unregulated uses go undetected during funding shortfalls.

Is sustainable use an effective conservation tool?

The extractive use of species has also produced very mixed results as a conservation tool. The United Nations Food and Agriculture Organization (FAO) now recognizes that over 50% of fish stocks are either overused or fully depleted (FAO, 2006), and ~90% of large predatory fish populations have now been lost (Meyers & Worm, 2003). A very high profile example of overfishing was closing the Atlantic cod *Gadus morhua* fishery on the Grand Banks of Canada, an example of management failure in the waters of a well-resourced, developed country that resulted in much economic hardship amongst local fishing communities (Kurlansky, 1997). Commercial trade has driven overuse in other species including saiga antelope *Saiga tatarica* and beluga sturgeon *Huso huso*, once numbering in the millions and now on the brink of extinction, largely since the demise of the Soviet Union (Milner-Gulland *et al.*, 2003; Pikitch *et al.*, 2005). Meanwhile, many tropical forest species have been locally reduced by bushmeat hunting (Bennett *et al.*, 2007). Thus many conservationists correctly feel nervous about use, as many harvested species have declined, despite being afforded some nominal protection.

Managed use has, however, been traded off against some effective protection in some situations, and has provided the incentives necessary to justify investment in appropriate management. Examples include the ADMADE and CAMPFIRE programmes in Zambia and Zimbabwe, respectively, the case of markhor *Capra falconeri* trophy hunting in Pakistan, the hunting of white rhino *Ceratotherium simum simum* on private lands in South Africa, the sale of seahorses from the Philippines, and a bushmeat management programme in the Peruvian Amazon (Lewis & Alpert, 1997; Leader-Williams, 2002; Martin-Smith *et al.*, 2004; Rosser *et al.*, 2005; Fang *et al.*, 2006; Frost & Bond, 2008).

What are the challenges to effectiveness of protection and use?

The way forward is not to adopt polarized positions about the effectiveness of single approaches, but instead focus on how to improve the effectiveness of both approaches. In the final analysis, cases of effectively managed sustainable use and effectively protected areas depend on effectively functioning socio-political systems. Thus, the once lauded CAMPFIRE programme (Adams, this volume, Chapter 16), established in the newly independent Zimbabwe, now has to operate in an increasingly lawless country. As neither approach of 'protection' or 'use' is working on its own, what are the common challenges?

Importance of enforcement

Enforcement is vital both to effective area-based protection and to managed use of wildlife. Enforcement can be put in place either directly at the national level, through the employment of civil servants whose job is to guard biodiversity owned and/or managed by the state, or indirectly through community conservation projects that empower local people to act as guardians of locally owned or managed resources. In many biodiversity-rich areas, national governments face huge drains on their resources to fund human development, reduce poverty and increase human well-being. In such cases, biodiversity conservation is a low priority, and the ensuing lack of resources and political will means that weak enforcement of protected area boundaries and integrity results in so-called 'paper parks' (Brandon *et al.*, 1998). Consequently, the 7th World Parks Congress focused on 'Benefits Beyond Boundaries', to encourage greater recognition of the benefits of protected areas and to ensure their successful implementation. Equally, lack of enforcement of both protected areas and of wildlife trade regulations has been associated with waves of poaching, which reduced elephants and rhino populations in Africa in the 1980s (Figure 8.3), and which have seen the more recent decimation of Indian, Indonesian and Russian tiger populations (Leader-Williams *et al.*, 1990b; Dinerstein *et al.*, 2007; Jachmann, 2008). To address challenges posed by lack of enforcement, conservationists should consider whether further trade-offs between protection and extractive use could provide incentives to deliver conservation of species and areas, as in the Guassa Commons of Ethiopia (see

Figure 8.3 **Black rhinos killed illegally for their horn in Mkaya, Swaziland. The establishment of extensive networks of protected areas in sub-Saharan Africa has failed to stem the loss of flagship species such as rhinos and elephants. (Photograph by Nigel Leader-Williams.)**

Box 8.2). Because funds for conservation are short, it may become increasingly important for others to undertake conservation action voluntarily and thereby trade-off state for local control.

Increasing local support

If governments do not implement adequate enforcement measures, might those who traditionally live in wildlife areas have more reason to do so? Unfortunately, there is sometimes a lack of support for conservation measures because protected areas have been imposed on the local population, and villagers have been forced to re-settle, for example in Luangwa Valley in Zambia, and even now in Gabon and Ethiopia. As a result, resources are taken out of local control. In other areas, peoples' traditional right to collect forest materials have been curtailed and this too can result in both hardship and resentment, as in the case of Royal Chitwan National Park in Nepal (Straede & Treue, 2006). Therefore, a more effective trade-off might be to work with local people to mitigate the worst overuse (Xu & Melick, 2007).

Reducing poverty around protected areas

Conservationists have often been accused of imposing negative impacts on rural people whose access to natural resources has been restricted through establishing protected areas (Dowie, 2005; Kaimowitz & Sheil, 2007). Further-more, ICDPs and other community-based projects have often not delivered expected benefits at the household and individual levels (Barrett & Arcese, 1995; Gubbi *et al.*, 2008). Recently, restricted access associated with stricter management categories (see Table 8.1) has been linked with possibly increased levels of poverty in the surrounding villages and suggestions of greater child mortality. However, these relationships have not yet been shown to hold on a global scale (de Sherbinin, 2008; Upton *et al.*, 2008).

In contrast, the edges of protected areas have recently been shown to experience higher population growth rates than in surrounding rural areas, suggesting that protected areas attract rather than repel human settlement (Wittemyer *et al.*, 2008). For example, the Wolong Biosphere Reserve, home of the giant panda *Ailuropoda melanoleuca* in China, has attracted several industries including hotels, hydropower and tourism that, together with the jobs in the conservation sector, have directly raised the employment level of local communities (Lu *et al.*, 2006). As these contrasting cases show, addressing poverty in biodiversity-rich areas will require clear long-term management plans developed with local stakeholders that recognize the long-term benefits of biodiversity to stakeholders near and far. Trade-offs between protection and use will be required to equitably distribute costs and benefits between local and distant communities.

Mitigating human–wildlife conflict and land hunger

Human–wildlife conflicts often occur in or around protected areas (Woodroffe *et al.*, 2005). Such conflicts can arise where protection is successful and animals overspill outside protected area boundaries to cause conflict with local people. Human–wildlife conflicts can also arise where human density is growing in once sparsely populated areas, and people now live in a hard edge against national park boundaries, for example in Kenya, Uganda and India. In many countries, protected areas are increasingly encircled by a sea of humanity and land use change that threatens natural habitats as land is converted to agriculture. Indeed, levels of encroachment into national parks may also track world commodity prices (Gaveau *et al.*, 2009b). As human populations continue to increase, conflict seems set to increase, where people become more land hungry in poor areas. In contrast, among increasingly urbanized populations in the developed world, rural areas are becoming depopulated, and previously developed or farmed land is reverting to semi-natural or natural habitats.

Meeting development expectations

Concerns have also been raised that promoting conservation through sustainable use will condemn people to a poverty trap. Thus natural populations and habitats have finite productivity and so do not offer development opportunities as human populations grow. The argument runs that the productivity of wildlife populations within a given area can only support a finite level of harvest, but that humans by nature generally strive to increase their standard of living, and so consumption is likely to increase beyond sustainable levels (Rao & McGowan, 2002). Consequently, the case for promoting use may have less value in future, unless account can be taken of the real value of renewable resources, in order to derive increased income from the same levels of harvest.

Future trade-offs and more nuanced approaches to protection and use?

The previous discussion on challenges to the effectiveness of 'protection' or 'use' approaches, and the case studies, have shown that polarized strategies to conservation through strict protection or sustainable use may not be very helpful. Hence, this section argues that context-specific trade-offs are often needed to deliver a middle and more nuanced approach to conservation. For example, a comparison of the performance of closed marine parks in Kenya with a collaboratively managed multiple-use area in Tanzania concluded that a combined approach to protection and use would provide the most effective outcome (McClanahan *et al.*, 2006). The importance of managing trade-offs between protectionist strategies and extractive uses to support local livelihoods is also highlighted in Nepal, where conservation outcomes of particular national parks are undermined by poor relations with local people. Some villages have harvested thatch and fuel wood illegally within Royal Chitwan National Park for over 30 years. To reduce local resentment towards enforcement of national park regulations, some extractive use of thatch and fuel wood should be legalized and managed in a trade-off with strict protection (Straede & Treube, 2005). In such cases, combining protection and use may provide a better outcome than sticking rigidly to path-dependant policies (Adams, this volume, Chapter 16).

Conclusions

Polarized arguments over whether statutory protected areas or sustainable use remains the most effective approach to conservation appear increasingly futile. Indeed, both approaches operate under suboptimal conditions, as evidenced by the worsening biodiversity crisis (Butchart *et al.*, 2004). Instead, a more nuanced and integrated approach is increasingly required, where protection and use are not seen as mutually exclusive, but where trade-offs between the two approaches can deliver mutual benefits (Redford *et al.*, 2006). Therefore, a key need to underpin future conservation strategies is for pragmatic, site-based research (Smith *et al.*, 2009; Knight & Cowling, this volume, Chapter 15) to determine the factors that predispose towards successful conservation across wider landscapes and with areas under different governance structures (Robinson, 2006). Many states already have far much more conservation management responsibility than they can carry out effectively, with large areas of their land surfaces under some form of, often notional, protection. Therefore, many so-called 'paper parks' are just that and exist only on paper, while many protected areas and protected species are subject to illegal harvest and trade, and remain threatened through lack of funding and political will to enforce protection. Furthermore, the spectre of climate change worsens and, whilst detailed outcomes still remain uncertain (Willis *et al.*, this volume, Chapter 18), it is clear that resilience will require greater landscape connectivity and mainstreaming of biodiversity (Mace, this volume, Chapter 19).

Conservationists and national governments need to work more closely with civil society at large to develop more creative solutions, to empower communities to value and manage their resources and to ensure that people living in developed countries meet the real cost of their consumption patterns (Adams & Jeanrenaud, 2008). The bureaucratic tendency to centralize power and control needs to loosen into an integrative framework that allows local management of local resources to provide resilience in the face of change. Thus, a dual approach to protection and use is required and trade-offs need to be carefully articulated and expected outcomes assessed. Additional gains should be sought and incentives created to ensure optimal conditions for conservation on state, private and communal land so as to create mosaics in the landscape. Optimal conditions are required to support all approaches, rather than arguing about the effectiveness of single approaches under conditions that are not conducive to allowing either approach to function effectively. The level of international buy-in to the CBD, and the importance to mankind of conserving biodiversity, deserves no less.

References

Adams, W.M. & Jeanrenaud, S. (2008) *Transition to Sustainability: towards a humane and diverse world*. Island Press, Washington, DC.
Ashenafi, Z.T. & Leader-Williams, N. (2005) An indigenous common property resource system in the central highlands of Ethiopia. *Human Ecology*, 33, 539–563.

Ashenafi, Z.T., Coulson, T.N., Sillero-Zubiri, C. & Leader-Williams, N. (2005) The behaviour and ecology of the Ethiopian wolf in a human-dominated landscape outside protected areas. *Animal Conservation*, 8, 113–121.

Barrett, C.B. & Arcese, P. (1995) Are ICDPs sustainable? On the conservation of large mammals in sub-Saharan Africa. *World Development*, 23, 1073–1085.

Batisse, M. (1986) Developing and focusing the biosphere reserve concept. *Nature and Resources*, 12, 2–11.

Bennett, E.L., Blencowe, E., Brandon, K. *et al.* (2007) Hunting for consensus: reconciling bushmeat harvest, conservation, and development policy in west and central Africa. *Conservation Biology*, 21, 884–887.

Brandon, K., Sanderson S. & Redford K.H. (1998) *Parks in Peril: people, politics, and protected areas.* Island Press, Washington, DC.

Brockington, D. & Igoe, J. (2006) Eviction for conservation: a global overview. *Conservation and Society*, 4, 424–470.

Bruner, A.G., Gullison, R.E., Rice, R.E. & da Fonseca, G.A.B. (2001) Effectiveness of parks in protecting tropical biodiversity. *Science*, 291, 125–128.

Butchart, S.H.M., Stattersfield, A.J., Bennun L.A. *et al.* (2004) Measuring global trends in the status of biodiversity: Red List indices for birds. *PLoS Biology*, 2, 2294–2304.

Carter, E., Adams, W.M. & Hutton, J. (2008) Private protected areas: management regimes, tenure arrangements and protected area categorization in East Africa. *Oryx*, 42, 177–186.

Clark, C.W. (1973) Profit maximization and the extinction of animal species. *Journal of Political Economy*, 81, 950–961.

Curran L.M., Trigg, S.N., McDonald, A.K. *et al.* (2004) Lowland forest loss in protected areas of Indonesian Borneo. *Science*, 303, 1000–1003.

de Sherbinin, A. (2008) Is poverty highest near parks? An assessment of infant mortality rates around protected areas in developing countries. *Oryx*, 42, 26–35.

DeFries, R., Hansen, A., Newton, A.C. & Hansen, M.C. (2005) Increasing isolation of protected areas in tropical forests over the past twenty years. *Ecological Applications*, 15, 19–26.

Dickson, B.W., Hutton, J. & Adams, W.M. (2009) *Recreational Hunting, Conservation and Rural Livelihoods: science and practice.* Wiley-Blackwell, Oxford.

Dinerstein, E., Loucks, C., Heydlauff, A. *et al.* (2007) *Setting Priorities for the Conservation and Recovery of Wild Tigers: 2005–2015: a user's guide.* WWF-US, WCS, Smithsonian, and NFWF-STF, Washington, DC and New York.

Dowie, M. (2005) Conservation refugees: when protecting nature means kicking people out. *Orion*, 16–27.

Fang, T., Rios, C. & Bodmer, R. (2006) Implementación de un programa piloto de certificación de pieles de pecaríes (*Tayassu tajacu* y *T. pecari*) en la comunidad de Nueva Esperanza, río Yavarí Mirí. Revista electronica. *Manejo de Fauna Silvestre en Latinoamérica*, 1(8), 15 pp.

FAO (Food and Agriculture Organization) (2006) *The State of World Fisheries and Aquaculture 2006*. FAO Fisheries and Aquaculture Department, FAO, Rome.

Ferraro P.J. & Pattanayak S.K. (2006) Money for nothing? A call for empirical evaluation of biodiversity conservation investments. *PLoS Biology*, 4, 482–488.

Freese, C.H. (1997) *Harvesting Wild Species*. John Hopkins University Press, Baltimore.

Frost, P.G.H. & Bond, I. (2008) The CAMPFIRE programme in Zimbabwe: payments for wildlife services. *Ecological Economics*, 65, 776–787.

Gardner, T.A. Caro, T., Fitzherbert, E.B., Banda, T. & Lalbhai, P. (2007) Conservation value of multiple-use areas in East Africa. *Conservation Biology*, 21, 1516–1525.

Gaveau, D.L.A., Epting, J., Lyne, O., Linkie, M., Kanninen, M. & Leader-Williams, N. (2009a) Evaluating whether protected areas reduce tropical deforestation in Sumatra. *Journal of Biogeography*, 36, 2165–2175.

Gaveau, D.L.A., Linkie, M., Suyadi, Levang, P. & Leader-Williams, N. (2009b) Three decades of deforestation in southwest Sumatra: effects of coffee prices, law enforcement, and rural poverty. *Biological Conservation*, 142, 597–605.

Gubbi, S., Linkie, M. & Leader-Williams, N. (2008) Evaluating the legacy of an integrated conservation and development project around a tiger reserve in India. *Environmental Conservation*, 35, 331–339.

Honey, M. (1999) *Ecotourism and Sustainable Development: who owns paradise?* Island Press, Washington, DC.

Hutton, J. & Leader-Williams, N. (2003) Sustainable use and incentive-driven conservation: realigning human and conservation interests. *Oryx*, 37, 215–226.

Hutton, J.M., Adams, W.M. & Murombedzi, J. (2005) Back to the barriers? Changing narratives in biodiversity conservation. *Forum for Development Studies*, 2, 341–370.

IUCN (International Union for the Conservation of Nature) (2000) *The IUCN Policy Statement on Sustainable Use of Wild Living Resources*. IUCN World Conservation Congress, Amman, October 2000.

IUCN (International Union for the Conservation of Nature) (2003) *2003 United Nations List of Protected Areas*. IUCN, Gland, Switzerland.

IUCN (International Union for the Conservation of Nature) (2008) *Guidelines for Applying Protected Area Management Categories*. IUCN, Gland, Switzerland.

Jachmann, H. (2008) Illegal wildlife use and protected area management in Ghana. *Biological Conservation*, 141, 1906–1918.

Kaimowitz, D. & Sheil, D. (2007) Conserving what and for whom? Why conservation should help meet basic human needs in the tropics. *Biotropica*, 39, 567–574.

Kurlansky, M. (1997) *Cod: a biography of the fish that change the world*. Walker Books, London.

Lande, R., Engen, S. & Saether, B-E. (1994) Optimal harvesting, economic discounting and extinction risk in fluctuating populations. *Nature*, 372, 88–90.

Lawton, J.H. & May, R.M. (eds) (1995) *Extinction Rates*. Oxford University Press, Oxford.

Leader-Williams, N. (2002) Regulation and protection: successes and failures in rhinoceros conservation. In *The Trade in Wildlife: regulation for conservation*, ed. S. Oldfield, pp. 89–99. Earthscan, London.

Leader-Williams, N. & Albon, S.D. (1988) Allocation of resources for conservation. *Nature*, 336, 533–535.

Leader-Williams, N., Albon, S.D. & Berry, P.S.M. (1990b) Illegal exploitation of black rhinoceros and elephant populations: patterns of decline, law enforcement and patrol effort in Luangwa Valley, Zambia. *Journal of Applied Ecology*, 27, 1055–1087.

Leader-Williams, N, Harrison, J & Green, M.J.B. (1990a) Designing protected areas to conserve natural resources. *Science Progress*, 74, 189–204.

Lewis, D.M. & Alpert, P. (1997) Trophy hunting and wildlife conservation in Zambia. *Conservation Biology*, 11, 59–68.

Lindsey, P.A., Romanach, S.S. & Davies-Mostert, H.T. (2009) The importance of conservancies for enhancing the value of game ranch land for large mammal conservation in southern Africa. *Journal of Zoology*, 277, 99–105.

Linkie, M., Chapron, G., Martyr, D.J., Holden, J. & Leader-Williams, N. (2006) Assessing the viability of tiger subpopulations in a fragmented landscape. *Journal of Applied Ecology*, 43, 576–586.

Linkie, M., Martyr, D.J., Holden, J. *et al.* (2003) Habitat destruction and poaching threaten the Sumatran tiger in Kerinci Seblat National Park, Sumatra. *Oryx*, 37, 41–48.

Linkie, M., Smith, R.J., Zhu, Y. *et al.* (2008) Evaluating biodiversity conservation around a large Sumatran protected area. *Conservation Biology*, 22, 683–690.

Lu, Y.H., Fu, B.J., Chen, L.D., Xu, F. & Qi, X. (2006) The effectiveness of incentives in protected area management: an empirical analysis. *International Journal of Sustainable Development and World Ecology*, 13, 409–417.

Martin-Smith, K.M., Samoilys, M.A., Meeuwig, J.J. & Vincent, A.C.J. (2004) Collaborative development of management options for an artisanal fishery for seahorses in the central Philippines. *Ocean and Coastal Management*, 47, 165–193.

McClanahan, T.R., Verheij, E. & Maina, J. (2006) Comparing management effectiveness of a marine park and a multiple-use collaborative fisheries management area in East Africa. *Aquatic Conservation: Marine and Freshwater Ecosystems*, 16, 147–165.

Meyers, R.M. & Worm, B. (2003) Rapid worldwide depletion of predatory fish communities. *Nature*, 423, 280–283.

Milner-Gulland, E.J. & Rowcliffe, M. (2007) *Conservation and Sustainable Use*. Oxford University Press, Oxford.

Milner-Gulland, E.J., Bukreeva, O.M., Coulson, T.M. *et al.* (2003) Reproductive collapse in antelope harems. *Nature*, 422, 135–135.

Naughton-Treves, L., Buck, M. & Brandon. K. (2005) The role of protected areas in conserving biodiversity and sustaining local livelihoods. *Annual Review of Environment and Resources*, 30, 219–252.

Nepstad, D.C., Schwartzman, S., Bamberger, B. *et al.* (2006) Inhibition of Amazon deforestation and fire by parks and indigenous reserves. *Conservation Biology*, 20, 65–73.

Norton-Griffiths, M. (2007) How many wildebeest do you need? *World Economics*, 8, 41–64.

Pikitch, E.K., Doukakis, P., Lauck, L., Chakrabarty, P. & Erickson, D.L. (2005) Status, trends and management of sturgeon and paddlefish fisheries. *Fish and Fisheries*, 6, 233–265.

Rao, M. & McGowan, P. (2002) Wild-use, food security, livelihoods and conservation. *Conservation Biology*, 16, 580–583.

Redford, K.H., Robinson, J.G. & Adams, W.M. (2006) Parks as shibboleths. *Conservation Biology*, 20, 1–2.

Roberts, C.M., Bohnsack, J.A., Gell, F.R., Hawkins, J.P. & Goodridge, R. (2002) Marine reserves and fisheries management. *Science*, 295, 1233–1235.

Roberts, C.M., Hawkins, J.P. & Gell, F.R. (2005) The role of marine reserves in achieving sustainable fisheries. *Philosophical Transactions of the Royal Society of London, Series B*, 360, 123–132.

Robinson, J.G. (2006) Conservation biology and real-world conservation. *Conservation Biology*, 20, 658–669.

Rodrigues, A.L., Andelman, S.J., Bakarr, M.I. *et al.* (2004) Effectiveness of the global protected area network in representing species diversity. *Nature*, 428, 640–643.

Roe, D. & Elliott, J. (2006) Pro-poor conservation: the elusive win–win for conservation and poverty reduction? *Policy Matters*, 14, 53–63.

Roe, D., Leader-Williams, N. & Dalal-Clayton, D.B. (1997) *Take Only Photographs, Leave Only Footprints: the environmental impacts of wildlife tourism*. Wildlife and Development Series No. 10. International Institute for Environment and Development, London.

Rosser, A.M., Tareen, N. & Leader-Williams, N. (2005) Trophy hunting and the precautionary principle: a case study of the Torghar Hills population of straight-horned markhor. In *Biodiversity and the Precautionary Principle: risk and uncertainty in conservation and sustainable use*, eds R Cooney & B. Dickson, pp. 55–72. Earthscan, London.

Sachs, J.D., Baillie, J.E.M., Sutherland, W.J. *et al.* (2009) Policy forum: biodiversity conservation and the Millennium Development Goals. *Science*, 325, 1502–1503.

Smith, R.J., Muir, R.J.D., Walpole, M.J., Balmford, A.P. & Leader-Williams, N. (2003) Governance and the loss of biodiversity. *Nature*, 426, 67–70.

Smith, R.J., Verissimo, D., Leader-Williams, N., Cowling, R.M. & Knight, A.T. (2009) Let the locals lead. *Nature*, 462, 280–281.

Southworth, J., Nagendra, H. & Munroe, D.K. (2006) Exploring human–environment tradeoffs in protected area conservation. *Applied Geography*, 26, 87–95.

Stoner, C., Caro, T., Mduma, S. *et al.* (2007) Assessment of effectiveness of protection strategies in Tanzania based on a decade of survey data for large herbivores. *Conservation Biology*, 21, 635–646.

Streade, S. & Treue, T. (2006) Beyond bufferzone protection: a comparative study of park and buffer zone products' importance to villagers living inside Royal Chitwan National Park and to villagers living in its buffer zone. *Journal of Environmental Management*, 78, 251–267.

Swanson, T.M. (1994) *International Regulation of Extinction*. Macmillan, London.

Terborgh, J. (2004) Reflections of a scientist on the World Parks Congress. *Conservation Biology*, 18, 619–620.

Thompson, M. & Homewood, K.M. (2002) Entrepreneurs, elites and exclusion in Maasailand: trends in wildlife conservation and pastoralist development. *Human Ecology*, 30, 107–138.

Upton, C., Ladle, R., Hulme, D., Jiang, T., Brockington, D. & Adams, W.M. (2008) Are poverty and protected area establishment linked at a national scale? *Oryx*, 42, 19–25.

Walpole, M.J., Karanja, G.G., Sitati, N.W. & Leader-Williams, N. (2003) *Wildlife and People: conflict and conservation in Masai Mara, Kenya*. International Institute for Environment and Development, London.

Wittemyer, G., Elsen, P., Bean, W.T. *et al.* (2008) Accelerated human population growth at protected area edges. *Science*, 321, 123–126.

Woodroffe, R. & Ginsberg, J.R. (1998) Edge effects and the extinction of populations inside protected areas. *Science*, 280, 2126–2128.

Woodroffe, R., Thirgood, S.J. & Rabinowitz, A. (eds) (2005) *People and Wildlife: conflict or co-existence?* Cambridge University Press, Cambridge.

Xu, J. & Melick, D.R. (2007) Rethinking the effectiveness of public protected areas in southwestern China. *Conservation Biology*, 21, 318–328.

9

Whose Value Counts? Trade-offs between Biodiversity Conservation and Poverty Reduction

Dilys Roe[1] and Matthew J. Walpole[2]

[1]International Institute for Environment and Development,
London, UK
[2]Fauna & Flora International, Cambridge, UK

Introduction

In recent years there has been much debate about the role of conservation agencies in poverty reduction (Roe, 2008). Critical attention paid to the activities and accountability of big international conservation non-governmental organizations (NGOs), and their impacts on local, and particularly indigenous, communities (e.g. Chapin, 2004) has forced this up the policy agenda. At the same time the often disappointing experience of Integrated Conservation and Development Projects (ICDPs) (Wells & McShane, 2004) and a shift in development assistance policy towards poverty reduction as a priority, emphasized by commitment to the Millennium Development Goals (MDGs) in 2000, and towards direct budget support rather than project funding as an aid delivery mechanism (Roe, 2008), has fuelled divergent voices within the conservation community. On the one hand some question why conservation should shoulder the additional burden of tackling poverty reduction (Robinson & Bennett, 2002; Terborgh, 2004; Oates, 2006), and suggest that those concerned with conservation and those concerned with poverty reduction

Trade-offs in Conservation: Deciding What to Save, 1st edition.   Edited by N. Leader-Williams, W.M. Adams and R.J. Smith.   © 2010 Blackwell Publishing Ltd.

simply carry on their work in two separate policy realms without reference to each other (described in Adams *et al.*, 2004). On the other hand are those who argue for greater alignment of biodiversity conservation and poverty reduction agendas, both through putting conservation at the heart of development assistance (Sanderson & Redford, 2003; Sanderson, 2005; Development and Environment Group, 2006), and by reorienting conservation to embrace a more pro-poor mission (e.g. Roe *et al.*, 2003; Kaimowitz & Sheil, 2007).

Regardless of whether or not development policy towards biodiversity has changed, and whether or not it is within the mandate of conservation organizations to concern themselves with poverty reduction, it is clear that biodiversity conservation and poverty reduction are not mutually exclusive (Goldman *et al.*, this volume, Chapter 4). However, what is less clear is the degree to which current conservation interventions already support poverty reduction and how they might evolve if conservation policy was indeed to truly embrace this additional goal. This chapter seeks to answer this question. More specifically, the chapter first seeks to explore whether the components of biodiversity – and the ecosystem services they provide – that are valued by poor people are the same as those targeted by international conservation action. Second, the chapter considers how trade-offs might be resolved if and when these two sets of priorities are different.

Complex relationships between biodiversity conservation and poverty reduction

That poverty reduction and biodiversity conservation are linked is not a new hypothesis. As early as the 1940s there was increasing recognition that conservation could provide revenue-generating opportunities though trade, hunting and tourism, that could contribute to local economic development in poor countries (Adams, 2004). International policy processes throughout the 1970s, 1980s and into the 1990s sought to emphasize the links but often by more broadly focusing on *environment* and *development* rather than biodiversity conservation and poverty reduction. The 1980 World Conservation Strategy (WCS) recognized potential trade-offs between conservation and development, for example where conservation strategies such as protected areas reduce local access to resources or where development activities result in a loss of biodiversity (IUCN *et al.*, 1980). However, it also emphasized the potential synergies, a new concept at the time (Talbot, 1980).

The 1992 Convention on Biological Diversity (CBD) in its preamble recognizes '*that economic and social development and poverty eradication are the first and overriding priorities of developing countries*'. In 2002, the Parties to the Convention committed themselves '*to achieve by 2010 a significant reduction of the current rate of biodiversity loss at the global, regional and national level* as a contribution to poverty alleviation [with emphasis added] *and to the benefit of all life on Earth*'. This target was subsequently endorsed at the 2002 World Summit on Sustainable Development and incorporated into the MDGs in 2006.

The role of biodiversity in society is articulated in the conceptual framework for the Millenium Ecosystem Assessment (MA), in which biodiversity underpins the provision of ecosystem services that contribute to human well-being and, by implication, to poverty reduction (MA, 2005). At the practical level, however, the nature and extent of the links, as well as the roles and responsibilities of different interest groups in addressing them, remain disputed (Sanderson & Redford, 2003). Does poverty fuel biodiversity decline, or does growth and development do so? Or both? Do the poor rely on biodiversity and, if so, can conservation reduce (or at least help prevent) poverty or does it cause or enhance poverty by denying the poor access to resources? Are conservation and poverty reduction mutually supportive, mutually antagonistic, or neither or both? Is biodiversity in fact irrelevant in the fight to alleviate poverty?

One of the problems fuelling this dispute is the tendency of protagonists on both sides to talk in generalities or at cross-purposes, and the lack of sound evidence on which clear judgments can be made. A case in point is the debate around the impact of conservation activities on poor people. Some commentators have categorically stated that conservation, particularly in the form of protected areas, is bad, for example: '*Conservation policies, in general, and protected areas, in particular, have increased poverty by denying or reducing community access to resources traditionally used for survival and livelihoods*' (Lockwood *et al.*, 2006: 56). Others highlight the benefits of protected areas in sustaining flows of ecosystem services (e.g. Borrini-Feyerabend *et al.*, 2004; BirdLife International, 2007). Overall, protected areas are unlikely to be as damaging, or as beneficial, as these different commentators suggest (Upton *et al.*, 2008). Much depends on how 'protected areas' are defined, the contexts under which they are established and managed (Rosser & Leader-Williams, this volume, Chapter 8) and the ways in which benefits and costs are distributed (Wells, 1992).

It is clear that there is no single, linear relationship between biodiversity and poverty on which generalities can be based. Despite the link posited in the MA, it is not the case that human well-being declines uniformly with biodiversity loss. For example, millions and millions of people have benefited from the expansion of the agricultural frontier and the subsequent increase in food security (MA, 2005). Nevertheless, the MA has found that the benefits have not been evenly or equitably distributed, with the poor as the biggest losers. Political, economic and social institutions govern the ability of different interest groups to access and benefit from ecosystem services. Furthermore, the link between levels and types of biodiversity and ecosystem service provision is not clear (e.g. Ash & Jenkins, 2007) and there is disagreement about 'how much biodiversity is enough?' (e.g. Norton-Griffiths, 1996) and 'which bits are the most important?' (e.g. Vermeulen & Koziell, 2002).

At the local level it is clear that the relationship between biodiversity conservation and poverty reduction is played out differently in different contexts and different localities (Box 9.1). In every context and each location there are likely to be trade-offs, both between conservation and poverty reduction goals and within each of those goals. These may be temporal (e.g. benefits now, costs later), spatial (e.g. benefits here, costs there) or beneficiary-specific (some people/species/ecosystem services win, others lose) (Schei, 2007). How significant these trade-offs are depends largely on how aligned or misaligned the biodiversity interests of conservation and of poor people really are.

Box 9.1 **Five reasons for focusing on human needs in conservation (FFI, 2007)**

1. Biodiversity underpins local livelihoods – so poverty reduction is a rationale for conservation.
2. Poverty damages biodiversity – so poverty reduction is a tool for conservation.

3. Conservation inflicts disproportionate costs on poor people – so conservation hinders poverty reduction.
4. Development fuels biodiversity loss – so conservation is hindered by poverty reduction.
5. Local support for conservation is necessary – reducing poverty helps build goodwill and trust.

An analysis of the project portfolio of Fauna & Flora International (FFI) revealed that 85% of projects were engaging with local livelihoods in their conservation work. Across the portfolio, all five rationales above were cited as motivations for focusing on livelihoods and poverty, with individual projects often citing multiple rationales (Walpole & Wilder, 2008).

Biodiversity priorities of poor people and international conservation organizations

Although international conservation organizations are, without exception, supportive of the CBD's 2010 target, the MA notes that '*for a reduction in the rate of biodiversity loss to contribute to poverty alleviation, priority would need to be given to protecting the biodiversity of particular importance to the well-being of poor and vulnerable people*' (MA, 2005: 15). Which bits of biodiversity have particular importance to the poor, and how does this compare with the priorities of conservation organizations?

Earlier chapters have described the processes that conservation organizations themselves go through in order to decide where, and on what, to focus their interventions (Wilson *et al.*, this volume, Chapter 2), whether through priority areas, priority species or taxa or priority ecosystems (Murdoch *et al.*, this volume, Chapter 3). In many cases these strategies are focused on conserving the rare or threatened species (Brooks *et al.*, 2006) rather than on conserving maximum biodiversity (Samways, this volume, Chapter 6).

Species-focused conservation does have the potential to generate benefits for local people through sustainable use, whether consumptive through hunting, or non-consumptive through photographic tourism (Rosser &

Leader-Williams, this volume, Chapter 8), and conservation organizations are going to considerable effort to demonstrate such values from their species-focused work (WWF, 2006). Yet such benefits are often small scale, vulnerable to external forces and captured by elites rather than genuinely targeting the poor (Walpole & Goodwin, 2000; Walpole & Thouless, 2005). Moreover such species often take up space and resources that could be of value to the poor, and sometimes come into direct conflict with people locally (Woodroffe *et al.*, 2005).

Landscape-scale conservation approaches can also benefit poor people through delivery of local ecosystem service benefits (e.g. Brooks *et al.*, 2006; Goldman *et al.*, this volume, Chapter 4). In other cases, however, poor people lose out, particularly where conservation means exclusion from resources on which they are reliant for their day-to-day livelihoods (MA, 2005; Cernea & Schmidt-Soltau, 2006). Even where they do not, biodiversity often simply provides a safety net for the rural poor rather than a genuine route out of poverty (Wunder, 2001; Shackleton *et al.*, 2007).

Overall, however, international conservation promotes primarily a global, rather than local, agenda. The Global Environment Facility (GEF) recognizes this fundamental difference, having been established to cover the incremental costs that are incurred by developing countries conserving biodiversity assets that have global rather than local value.

Local perspectives of poor people are likely to be very different to those of international conservation organizations and their supporters in wealthy countries. Poor people focus on the direct use values of biodiversity and its cultural associations rather than the continued existence of threatened species or habitats (Box 9.2). An example of the clear differences in the value systems of different stakeholder groups is evidenced by the debate in the literature on conservation and indigenous rights (see, for example, Redford, 1990; Alcorn, 1993; Redford & Stearman, 1993). Distinctions between domesticated or cultivated species, or agro-biodiversity, and the wild species that are the focus of international conservation, are also less meaningful to many rural communities, who mix farming with hunting and harvesting of wild resources as part of their livelihood strategies. Indeed domesticated biodiversity is generally more important to rural livelihoods than wild nature. Although both livestock and crop genetic diversity are included as indicators within the CBD framework for tracking progress towards the 2010 target, they are rarely if ever the focus of international conservation efforts (FAO, 2007).

Box 9.2 **Different views on biodiversity values in Tanzania**

In Tanzania's East Usambara Mountains at least four value systems occur between different stakeholders in relation to wildlife and forest resources: local communities, government, the private sector and the international conservation community. For effective management, these differing perceptions need to be reconciled, and global and local interests should be balanced.

Stakeholder group	Values of wildlife resources
Local communities	Water, timber, fuel, food, medicines, non-timber forest products for subsistence use; land, local climate, small-scale commerce, culture base, tourism
Private sector	Local climate, water catchment, soil conservation, industrial raw materials, timber
National government	Water catchment, regional climatic patterns, power, urban market demands, tourism, export revenues
International community	Rare and unique biodiversity, research and scientific potential, genetic resource, international market demands

Kaimowitz and Sheil (2007) contend that, compared to the focus on saving charismatic species from extinction, relatively little effort is being made to protect the species and habitats that directly meet human needs – for food, medicines, pollinators, nutrient recyclers and so on – and that even less effort has been made to ensure that disadvantaged people retain access to species on which they have traditionally relied. In particular, thousands of medicinal plants are now endangered, supplies of wild meat and fish have been depleted, fuel wood has become scarcer and species of cultural or spiritual significance have been lost.

Trade-offs for 'pro-poor' conservation

Financial resources for conservation are limited (Balmford & Whitten 2003; Bruner *et al.*, 2004). Consequently, conservation organizations have to make choices and allocate priorities (Bruner *et al.*, this volume, Chapter 11) when deciding how and where to intervene financially. Adopting an approach to conservation that emphasizes poverty reduction *as well as* biodiversity would mean shifting allocations of resources to different priorities and different localities, altering the current portfolios of the major international conservation agencies.

Without a detailed analysis of current project portfolios and financial allocations, it is not possible to predict how significant this change would be. Nevertheless, the manner in which current conservation planning and prioritization occurs – which does not take into account the degree of reliance of local people on wild resources or the level of threat to particular species or taxa that are important to poor people – implies there would be a substantial difference. In particular it might imply a decreasing emphasis on charismatic species and an increasing emphasis on medicinal plants, coastal fisheries, bushmeat species (many conservation interventions target threatened primates – particularly apes – but rodents, small antelope and invertebrates can be much more significant (e.g. Barnett, 2000)) and agro-biodiversity including crops and pollinators (FAO, 2007).

It would also mean a change of focus to locations where the majority of poor people live, a list always dominated by countries in sub-Saharan Africa, however poverty is defined. Despite increasing levels of urbanization, the majority (75%) of poor people live in rural areas (WRI *et al.*, 2005). However, Redford *et al.* (2008) note that relatively few poor people live in the undisturbed areas that are of interest to conservation organizations. Currently there is a convergence of interests around forest areas as both a target for conservation interventions and as home – or a source of livelihood – to an estimated 90% of the 1.1 billion people living on less than US$1 per day (World Bank, 2002). But other priority conservation areas for poor people may often be drier, more disturbed and more densely populated than the sites currently targeted by conservation organizations (Kaimowitz & Sheil, 2007).

At this point it must be recognized that addressing poverty reduction by reallocating scarce funds to different priority species and locations is not part of the current conservation agenda and seems unlikely to become so, in part because it subordinates the global values at the heart of international conservation. A more pragmatic question is not whether conservation should be geared more directly towards poverty reduction but rather how conservation and poverty reduction interests can be better aligned (Robinson, 2006).

Conservation organizations clearly do not have wholesale disregard for the needs and values of poor people. Conservation organizations have recently responded to the call to engage in poverty reduction, and a review of the websites of the major international conservation organizations reveals a plethora of policy documents and public statements about the importance of working for and with poor people (Box 9.3). The challenge is translating the rhetoric of policy into meaningful impacts in practice, by adopting different approaches to doing conservation.

Box 9.3 **Public policy statements of some major international conservation organizations**

BirdLife International

BirdLife works together for birds and people: We recognize that it is vital to integrate environmental conservation with social development and livelihood security. Projects in some of the world's most important areas for biological diversity help communities to achieve sustainable livelihoods through the managed use of natural resources, so increasing economic growth and reducing social inequalities.

Fauna & Flora International

Fauna & Flora International (FFI) works to ensure that conservation initiatives will generate tangible community benefits as well as conserving biodiversity.

International Union for the Conservation of Nature

IUCN, the International Union for the Conservation of Nature wants to make sure that all its programmes are responsive to the need for addressing poverty issues; hence poverty and livelihood security concern the whole organization.

The way forward: managing trade-offs, maximizing synergies

Conservation and poverty reduction are both legitimate societal goals, but where they overlap they are not always perfectly aligned. The immediate needs for food, fuel and shelter may clash with the imperative to protect the forest habitat of a critically endangered species. The arguments for protecting large-scale ecosystems may be at odds with those for roads and infrastructure development that are part of national poverty reduction strategies. Even where conservation can provide tangible benefits it is unlikely to have large-scale, immediate impacts on poverty for the reasons discussed above. However, this does not mean that it is irrelevant: local benefits (largely in terms of provisioning and cultural services) can be significant to certain stakeholder groups, and the broader societal benefits of conservation (largely in terms of supporting and regulating services) will affect the poor as well as others (e.g. Wilder, 2008).

The trade-offs between the benefits afforded by conservation and those arising from mainstream approaches to poverty reduction are inevitable and will be played out differently in different situations. It is naïve to expect win–win solutions everywhere (Goldman *et al.*, this volume, Chapter 4). Yet ensuring a less antagonistic and more supportive relationship between the two is possible. It is often the way in which conservation is carried out, as much as the species and habitats that are targeted, that can determine the effect on poverty. For example, protected areas are highlighted time and again as a potential threat to poor people, but there are many different (but often under-recognized) types of protected areas that do not exclude people and provide benefits for poor communities (Pathak *et al.*, 2005; Redford *et al.*, 2006).

Table 9.1 provides examples of different approaches to conservation that have greater or lesser poverty impacts. These are not necessarily mutually exclusive, and all conservation organizations are being encouraged to adopt at the very least a 'do no harm' approach, which requires engaging with poor, local stakeholders and understanding local livelihoods, their interaction with biodiversity and the social impacts of proposed conservation interventions (Mapendembe *et al.*, 2008). This in turn may yield solutions that are beneficial rather than poverty neutral (Box 9.4).

Table 9.1 A typology of pro-poor conservation (updated from IIED, 2003; Roe & Elliott, 2006).

Approach	Description	Examples
Poverty reduction as a tool for conservation	Recognition that poverty is a constraint to conservation and needs to be addressed in order to deliver on conservation objectives	Alternative income-generating projects; many integrated conservation and development projects; many community-based conservation approaches
Conservation that 'does no harm' to poor people	Recognition that conservation can have negative impacts on the poor and that compensation is required where these occur and/or to mitigate their effects	Social impact assessments prior to protected area designations; compensation for wildlife damage; provision of *locally acceptable* alternatives or compensation when access to resources lost or reduced
Conservation that generates benefits for poor people	Conservation still seen as the overall objective but designed so that benefits for poor people are generated	Revenue-sharing schemes around protected areas; employment of local people in conservation jobs; community conserved areas
Conservation as a tool for poverty reduction	Poverty reduction and social justice issues are the overall objectives. Conservation is seen as a tool to deliver on these objectives	Conservation of medicinal plants for healthcare, wild species as food supplies, sacred groves; pro-poor wildlife tourism

Overarching objective moves from conservation to poverty reduction

The nature and scale of trade-offs will be affected by the scale at which conservation interventions occur. Adopting a landscape-level approach to conservation enables trade-offs to be more effectively managed, balancing strictly protected areas at one end of a continuum, in which poor people are compensated for any loss of access to resources, with production areas at the other, recognizing that much of the world's biodiversity occurs outside protected areas (e.g. McNeely & Scherr, 2003). It also allows for better identification of potential synergies: mapping local peoples' priorities as an addition to the current set of global prioritization strategies and seeing where the overlaps are would be a great starting point to identifying where mutual interests lie, where they can be created (e.g. through payments for conservation services provided) and where negotiated solutions – potentially involving compensation for strict protection – need to be developed.

Box 9.4 Pro-poor conservation in practice

Fauna & Flora International (FFI, 2006) has adopted a policy that commits it to ensuring that '*its conservation activities do not disadvantage or undermine poor, vulnerable or marginalised people that are dependent upon or live adjacent to natural resources, and wherever possible will seek to conserve biodiversity in ways that embrace local wellbeing and social equity*'. Underpinning this is a set of principles and procedures to guide project staff and partners. These include a commitment to local participation, to understanding and acting on social impacts, to cross-sectoral partnerships that bring livelihoods expertise to bear, and to lesson learning through multidisciplinary monitoring and evaluation. With support from the Dutch government's international development agency Directoraat Generaal Internationale Samenwerking (DGIS), a range of projects exemplify this approach.

Conserving crocodiles and improving food security in Cambodia

The last stronghold of the critically endangered Siamese crocodile *Crocodylus siamensis* (Figure 9.1) is in the Cardamom Mountains of Cambodia, an area of outstanding biodiversity conservation value, but with scattered villages blighted by extreme poverty and hunger. Aid

Figure 9.1 **Siamese crocodiles are critically endangered, but FFI has established a project that links food security and conservation in the Cardomom Mountains in Cambodia. (Photograph by kind permission of Tom Dacey.)**

agencies were proposing to turn the crocodiles' marshland habitat into rice fields when FFI entered into partnership with the local agricultural development NGO CEDAC (Cambodian Center for Study and Development in Agriculture) to address people's needs as part of a strategy to enhance crocodile and wetland protection. The establishment of the O'Som Natural Resources Management Committee has empowered the local population and increased social cohesion. The introduction of organic intensification methods has improved food security, tripling rice yields on existing farmlands and so decreasing pressure on crocodile habitat. Combining conservation and livelihoods agendas demands greater effort in time, effort and resources than conventional approaches to either, but it enables synergies, and a more sustainable solution, to be found.

Given the extent to which many poor, rural communities depend directly on biodiversity for their day-to-day livelihoods, albeit not necessarily the same bits of biodiversity as those valued by the conservation community, potential synergies are also to be found in adopting a more proactive approach that addresses the root causes and drivers of biodiversity loss rather than the rather more reactive approach of trying to protect what is left. This means focusing on climate change, unsustainable consumption and production, macroeconomic policies and even population growth, which conservation organizations are increasingly doing[1].

Synergies need to be explored, however, not just between different value sets, but also between different stakeholders. Conservation organizations – and NGOs in particular – seem to either have been shouldered with, or have taken on the burden of, trying to link biodiversity conservation with poverty reduction. But other agencies also have a stake. Yet there are few calls for development NGOs to address biodiversity loss or for scientific research or training establishments to play their part. Botanic gardens, for example, have a huge potential to address the conservation of medicinal plants, wild crops and other species of importance to poor people (Waylen, 2006) but are rarely included in calls on conservation organizations to do more.

Redford *et al.* (2003) highlight the need for collaboration amongst conservation groups if the 'race to conserve nature' is not to be lost. Equally as important is collaboration amongst groups that have either conservation or poverty reduction at the core of their mission (Walpole, 2006). Given the scarce resources for both biodiversity conservation *and* for poverty reduction (the United Nations target of 0.7% GDP for development assistance is still to be met by the majority of countries), working together and sharing responsibility for both agendas is likely to be the most effective way to balance trade-offs and ensure different voices are heard and different priorities met.

References

Adams, W.M. (2004) *Against Extinction: the story of conservation*. Earthscan, London.
Adams, W.M., Aveling, R., Brockington, D. *et al.* (2004) Biodiversity conservation and the eradication of poverty. *Science*, 306, 1146–1149.

---

[1] For example see Conservation International's programme on population at http://www.conservation.org/discover/wellbeing/Pages/population.aspx; WWF's macroeconomics programme at http://www.panda.org/about_wwf/what_we_do/policy/macro_economics/index.cfm, and The Nature Conservancy's work on climate change at http://www.nature.org/initiatives/climatechange/.

Alcorn, J.B. (1993) Indigenous peoples and conservation. *Conservation Biology*, 7, 424–427.

Ash, N. & Jenkins, M. (2007) *Biodiversity and Poverty Reduction: the importance of ecosystem services*. UNEP/World Conservation Monitoring Centre, Cambridge.

Balmford, A. & Whitten, T. (2003) Who should pay for tropical conservation, and how could these costs be met? *Oryx*, 37, 238–250.

Barnett, R. (2000) *Food for Thought: the utilization of wild meat in eastern and southern Africa*. TRAFFIC East/Southern Africa, Nairobi.

BirdLife International (2007) *Livelihoods and the Environment at Important Bird Areas: listening to local voices*. BirdLife International, Cambridge.

Borrini-Feyerabend, G., Pimbert, M., Farvar, M.T., Kothari, A. & Renard, Y. (2004) *Sharing Power: learning by doing in co-management of natural resources throughout the world*. International Institute of Environment and Development, London and International Union for the Conservation of Nature, Gland, Switzerland.

Brooks, T.M., Mittermeier, R.A., da Fonseca, G.A.B. *et al.* (2006) Global biodiversity conservation priorities. *Science*, 313, 58–61.

Bruner, A.G., Gullison, R.E. & Balmford, A. (2004) Financial costs and shortfalls of managing and expanding protected area systems in developing countries. *BioScience*, 54, 1119–1126.

Cernea, M.M. & Schmidt-Soltau, K. (2006) Poverty risks and national parks: policy issues in conservation and resettlement. *World Development*, 34, 1808–1830.

Chapin, M. (2004) A challenge to conservationists. *World Watch*, November/December, 17–31.

Development and Environment Group (2006) *A BOND Development and Environment Group response to the UK Department for International Development White Paper 2006: 'Eliminating World Poverty: making governance work for the poor'*. British Overseas NGOs for Development, London.

FAO (Food and Agriculture Organization) (2007) *The State of Food and Agriculture: paying farmers for environmental services*. FAO, Rome.

FFI (Fauna & Flora International) (2006) *The Case for Integrating Conservation and Human Needs*. Livelihoods and Conservation in Partnership Series No. 1. Fauna & Flora International, Cambridge.

FFI (Fauna & Flora International) (2007) *Addressing Human Needs in Conservation*. Livelihoods and Conservation in Partnership Series No. 2. Fauna & Flora International, Cambridge.

IIED (International Institute for Environment and Development) (2003) *A Typology of Pro-poor Conservation*. Mimeo prepared for the World Parks Congress, Durban, September 2003.

IUCN (International Union for the Conservation of Nature), UNEP (United Nations Environment Programme) & WWF (1980) *World Conservation Strategy: living resource conservation for sustainable development*. IUCN, Gland, Switzerland.

Kaimowitz, D. & Sheil, D. (2007) Conserving what and for whom? Why conservation should help meet basic needs in the tropics. *Biotropica*, 39, 567–574.

Lockwood, M., Worboys, G. & Kothari, A. (2006) *Managing Protected Areas: a global guide*. Earthscan, London.

Mapendembe, A., Thomas, D. & Dickson, B. (2008) *Conservation and Poverty: a review of existing commitments*. Briefing note, Fauna & Flora International and BirdLife International, Cambridge.

McNeely, J. & Scherr, S. (2003) *Ecoagriculture*. Island Press, Washington, DC.

MA (Millennium Ecosystem Assessment) (2005) *Ecosystems and Human Well-being: synthesis*. Island Press, Washington, DC.

Norton-Griffiths, M. (1996) Property rights and the marginal wildebeest: a cost benefit analysis of wildlife conservation options in Kenya. *Biodiversity and Conservation*, 5, 1557–1577.

Oates, J.F. (2006) Conservation, development and poverty alleviation: time for a change in attitudes. In *Gaining Ground: in pursuit of ecological sustainability*, ed. D. Lavigne, pp. 277–284. International Fund for Animal Welfare, Guelph, Canada.

Pathak, N., Kothari, A. & Roe, D. (2005) Conservation with social justice? The role of community conserved areas in achieving the Millennium Development Goals. In *How to Make Poverty History: the central role of local organisations in meeting the MDGs*, eds T. Bigg & D. Sattherthwaite, pp. 55–78. International Institute for Environment and Development, London.

Redford, K.H. (1990) The ecologically noble savage. *Orion Nature Quarterly*, 9, 24–29.

Redford, K.H. & Stearman, A.M. (1993) On common ground? Response to Alcorn. *Conservation Biology*, 7, 427–428.

Redford, K.H., Coppolillo, P., Sanderson, E.W. *et al.* (2003) Mapping the conservation landscape. *Conservation Biology*, 17, 116–131.

Redford, K.H., Levy, M.A. Sanderson, E.W. & de Sherbinin, A. (2008) What is the role for conservation organisations in poverty alleviation in the world's wild places? *Oryx*, 42, 516–528.

Redford, K.H., Robinson, J.G. & Adams, W.M. (2006) Parks as Shibboleths. *Conservation Biology*, 20, 1–2.

Robinson, J.G. (2006) Conservation biology and real-world conservation. *Conservation Biology*, 20, 658–669.

Robinson, J.G. & Bennett, E.L. (2002) Will alleviating poverty solve the bushmeat crisis? *Oryx*, 36, 332.

Roe, D. (2008) Documenting the origins and evolution of the conservation–poverty debate: a review of key literature, events and policy processes. *Oryx*, 42, 491–503.

Roe, D. & Elliott, J. (2006) Pro-poor conservation: the elusive win–win for conservation and poverty reduction? *Policy Matters*, 14, 53–63.

Roe, D., Hutton, J.M., Elliott, J., Saruchera, M. & Chitepo, K. (2003) In pursuit of pro-poor conservation: changing narratives . . . or more? *Policy Matters*, 12, 87–91.

Sanderson, S. (2005) Poverty and conservation: the new century's 'peasant question'. *World Development*, 33, 323–332.

Sanderson, S. & Redford, K.H. (2003) Contested relationships between biodiversity conservation and poverty alleviation. *Oryx*, 37, 1–2.

Schei, P (2007) Chairman's report. *The Trondheim/UN Conference on Ecosystems and People – biodiversity for development – the road to 2010 and beyond*. 29 October to 2 November, 2007. Norwegian Directorate for Nature Management, Trondheim, Norway.

Shackleton, C.M., Schackleton, S.E., Buiten, E. & Bird, N. (2007) The importance of dry woodlands and forests in rural livelihoods and poverty alleviation in South Africa. *Forest Policy and Economics*, 9, 558–577.

Talbot, L. (1980) The world's conservation strategy. *Environmental Conservation*, 7, 259–268.

Terborgh, J. (2004) Reflections of a scientist on the World Parks Congress. *Conservation Biology*, 18, 619–620.

Upton, C., Ladle, R., Hulme, D. *et al.* (2008) Are poverty and protected area establishment linked at a national scale? *Oryx*, 42, 19–25.

Vermeulen, S. & Koziell, I. (2002) *Integrating Global and Local Values: a review of biodiversity assessment*. International Institute of Environment and Development, London.

Walpole, M.J. (2006) Partnerships for conservation and poverty reduction. *Oryx*, 40, 245–246.

Walpole, M.J. & Goodwin, H.J. (2000) Local economic impacts of dragon tourism in Indonesia. *Annals of Tourism Research*, 27, 559–576.

Walpole, M.J. & Thouless, C.R. (2005) Increasing the value of wildlife through non-consumptive use. In *People and Wildlife: conflict or coexistence?*, eds R. Woodroffe, S. Thirgood & A. Rabinowitz, pp. 122–139. Cambridge University Press, Cambridge.

Walpole, M.J. & Wilder, E. (2008) Disentangling the links between conservation and poverty reduction in practice. *Oryx*, 42, 539–547.

Waylen, K. (2006) *Botanic Gardens: using biodiversity to improve human well-being*. Botanic Gardens Conservation International, Richmond, UK.

Wells, M.P. (1992) Biodiversity conservation, affluence and poverty: mismatched costs and benefits and efforts to remedy them. *AMBIO*, 21, 237–243.

Wells, M.P. & McShane, T.O. (2004) Integrating protected area management with local needs and aspirations. *AMBIO*, 33, 513–519.

Wilder, E. (2008) *A Compendium of Case Studies, Lessons and Recommendations: sharing FFI's experience of linking biodiversity conservation and human needs*. Fauna & Flora International, Cambridge.

Woodroffe, R., Thirgood, S. & Rabinowitz, A. (2005) *People and Wildlife: conflict or coexistence?* Cambridge University Press, Cambridge.

World Bank (2002) *The Environment and the Millennium Development Goals.* World Bank, Washington, DC.

WRI (World Resources Institute), UNDP (United Nations Development Programme), UNEP (United Nations Environment Programme) & World Bank (2005) *World Resources 2005: the wealth of the poor – managing ecosystems to fight poverty.* WRI, Washington, DC.

Wunder, S. (2001) Poverty alleviation and tropical forests: what scope for synergies? *World Development*, 29, 1817–1833.

WWF (2006) *Species and People: linked futures.* WWF International, Gland, Switzerland.

(10)

The Power of Traditions in Conservation

Katherine M. Homewood

University College London, Department of Anthropology,
London, UK

Introduction

Many non-western traditions and practices have relevance to conservation, from sacred sites through to traditional ecological knowledge and practice. Conservationists seek to harness such traditions for their enterprises. However, institutions that have produced protected sites or sustainable use in small-scale societies may not be a robust basis for western-style conservation. Divergent underlying aims, changing contexts, and the diverse and continually evolving nature of traditions make this a dubious strategy that may undermine rather than reinforce outcomes for environments and communities.

This chapter looks briefly at some non-western conservation traditions, and at attempts to harness potential synergies for western conservation. It proceeds to develop a specific case, summarizing the findings of a multi-site study on reserve-adjacent resource use by Kenya and Tanzania Maasai, and examining the role of wildlife and tourism in land use and livelihoods. The study raises questions about the interplay of conservation with traditions of resource use and sustainable management in Maasailand, and the role of tradition more generally in conservation.

Trade-offs in Conservation: Deciding What to Save, 1st edition. Edited by N. Leader-Williams, W.M. Adams and R.J. Smith. © 2010 Blackwell Publishing Ltd.

Co-option of tradition in conservation

Conservation-relevant traditions and practices of non-western societies range from secret knowledge around, for example, sacred groves, which may form the nuclei of protected forest sites (e.g. Ntiamoa-Baidu, 1993; Dossou-Glehouenou, 1999; Falconer, 1999); through reverence for particular species seen as symbolizing particular lineages (Ntiamoa-Baidu, 1993) or embodying particular deities including monkeys as Lord Hanuman (Box 10.1) and blackbuck as Lord Shiva (Kothari & Das, 1999); and down to the shared knowledge around day-to-day practices and institutions that govern natural resource use in groups relying on forest products, grazing, fishing and other natural resource-related livelihoods (Posey, 1999). Such local ecological knowledge and land use practices have produced valued heritage landscapes (Posey, 1999). Examples include: fire ecology in Australia (Yibarbuk *et al.*, 2001) and West Africa (Laris, 2003); pastoralist grazing and burning in Tanzania (Homewood & Rodgers, 1991); North American Indian 'parklands' (Neumann, 2004); and the traditional land use practices shaping landscapes such as France's Cévennes, UK's Lake District and the English garden (Tilley, 2009). There is a perennial interest in harnessing such conservation-compatible systems of knowledge and belief for the conservation enterprise. Conversely, some conservation traditions and beliefs paradoxically operate to hinder species survival. Campaigns by animal welfare organizations to ban hunting (Harrop, this volume, Chapter 7) may drive biodiversity loss in UK farm landscapes (Leader-Williams *et al.*, 2002) and wildlife decline in Kenya through their removal of incentives for sustainable use (Parker, 2006; Norton-Griffiths, 2007).

Box 10.1 **India's sacred monkeys**

Hanuman is a prominent Hindu deity, and his manifestation as the monkey Hanuman is the key actor in India's national epic, the Ramayana, popular throughout South Asia. The Ramayana praises God Rama and his beautiful spouse Sita. When the demon king abducts Sita to the island of Lanka, Hanuman masterminds her rescue by commanding a victorious army of monkeys. Hanuman is captured and set on fire

---

before escaping and burning Lanka. For orthodox Hindu believers, langurs are the 'true' holy monkeys, as they have charcoal black faces, hands and feet symbolizing the burns Hanuman suffered, and they are called *kala bandhar* or black monkeys. The Ramayana refers repeatedly to Hanuman's long tail, and the Sanskrit word *langulin* from which the English name of 'langur' derives means 'having a long tail'. However, many statues depict Hanuman with a reddish face typical of the (short-tailed) rhesus macaque *Macaca mulatta* called in Hindi *lal bandhar* or red monkey. Grey langurs (also called Hanuman langurs or Indian langurs, *Presbytis (Semnopithecus) entellus*) and rhesus macaques are the two most common forms, but all kinds of monkey species are revered as holy, depending on which occur in any given locality, from Nilgiri langurs *Trachypithecus johnii* to Javanese macaques *Macaca fascicularis*, stumptail macaques *Macaca arctoides*, and various others. Indeed, any monkey that roams in temple compounds (Figure 10.1) is held sacred on the Indian subcontinent and fed by local people (Sommer, 1996, 2002).

Figure 10.1 **Sacred languar monkeys are fed and revered around many temples in India (see Box 10.1). (Photograph by kind permission of Volker Sommer.)**

It is tempting to romanticize the aims, practices and aspirations of societies remote from our own. However, their underlying aims and interests may differ from those of western conservation. Theoretical analyses and empirical studies warn that institutions that have produced protected sites or sustainable use in small-scale societies may not be a robust basis for conservation (Alvard, 1993; Smith & Wishnie, 2000). Formal definitions of conservation practice not only require a conservation effect to be observed, but also require evidence of intent to conserve through giving up a short-term profit in favour of a longer term, delayed pay-off. The resource management practices of a subsistence society may result in 'epiphenomenal conservation' (Alvard, 1993) through underlying decision rules that diverge from conservation criteria. For example, optimal foraging theory predicts the way foragers either continue searching for food in a given patch or move to search a fresh patch. Patch switching maximizes returns and may also lead to quasi-conservation outcomes, whereby prey recover from intensive harvesting in a temporarily abandoned patch, as part of a sustainable cycle of abandonment and re-use (Thomas, 2007). However, under conditions of overall resource scarcity, the underlying decision rules driving such behaviour may lead to over-exploitation rather than conservation (Smith & Wishnie, 2000), as has been argued for African pastoralist grazing systems (Ruttan & Borgerhoff-Mulder, 1999, but see Homewood, 1999).

Local communities may maintain outcomes consistent with conservation through the operation of small groups with long-term reciprocal interactions, with well-defined user groups, through common property resource management of common pool resources, and established institutions for monitoring and enforcement. However, such systems are easily overwhelmed by: external markets, often facilitated by the state and/or outside investors; the difficulty of monitoring and enforcement across increasing scales; and the proliferation of alternative, potentially higher yielding investment opportunities (Ostrom et al., 1999; Smith & Wishnie, 2000), particularly where slow-recovering species are concerned (Caughley, 1993; May, 1994).

Traditions do not represent lasting sets of rules, held in common and universally respected by the many members of a particular group, providing a robust basis on which to build synergies with western conservation enterprises. Traditions in practice are continually evolving, and are continually drawn upon and added to in different ways by different people and interest groups. They are less an expression of unity or homogeneity within a community, or of unchanging, prescriptive behaviour; instead, they more represent a diverse, shifting bundle of internally contradictory postulates that evolve by accreting new layers, again in often contradictory ways (see Caughley *et al.*, 1987, for an analysis of just such contradictory and contested strands within the western conservation tradition). Religious beliefs may simultaneously advocate stewardship while justifying over-exploitation. For example, Islam is a cornerstone of Turkish regard for trees (Tont, 1999: 392), while in Senegal, the Islamic Mourides sect's destruction of forests to expand commercial cultivation is publicly proclaimed as an act of religious virtue (Schoonmaker-Freudenberger, 1991).

Brockington (2005) analyzed multilayered environmentalism in Tanzania. Village people and district and central government make remarkably consistent associations between trees and rainfall despite strong tensions and diverse roots. Alongside strands derived from western scientific discourse, and the state's strategic use of environmental narratives to deflect criticism from its own actions onto rural resource users it accuses of damaging practices, local people adopt conservation and government rhetoric to further their own political ends. Conservation funds and idiom are co-opted as instruments in local power struggles. Lack of contradictions in local and state discourses may not indicate a shared tradition, but rather, in the context of unequal power relations, discourse '. . . *works not through contradiction and opposition but through co-option, manipulation and subversion*' (Brockington, 2005: 116).

Declarations about traditions thus commonly represent politically loaded statements about identities and aspirations: less a credo or framework guiding action, than an arena within which people manoeuvre for political advantage. Focusing on one element within that notional space to build channels for conservation inevitably puts new pressures on everyday processes of negotiation, whereby local power plays over access to and use of resources are continually contested (Walpole & Leader-Williams, 2001). It provides new openings for some well-placed folk at the expense of others, and accordingly fosters vested interests, allegiances, coalitions and resistance. Leaders seen as exploiting such opportunities for their own ends may lose respect and authority (cf. Conklin & Graham, 1995; Igoe, 2003; but see Brockington, 2005: 105–6).

Power-plays thus impact on the evolution of grassroots environmental entitlements and also on political structures and processes at higher levels, creating alternative channels of power and decision making, distorting the operation of systems of representation and government already straining to accommodate conflicting customary and national frameworks of power and authority (Ribot, 2006). Apparently simple choices of local partners representing some synergy with outside conservation goals can lead to major political consequences, either fostering more democratic representation or alternatively deepening exclusion, inequities and conflict (Ribot, 2006).

Historically, there are sinister precedents of powerful outsiders co-opting environment-related traditions for political purposes. In colonial Rhodesia, the government sought to co-opt sacred sites and prophetic cults to pre-empt their forming nuclei for guerrilla resistance movements (Ranger, 2003). In Zambia, the loss of sacred sites to the flooding of Lake Kariba, and to associated tourist developments, meant collapse and loss of respect for, and the power of, traditional leaders and ancestral cults (MacGregor, 2003). In Mozambique, in an attempt to destroy the black African Mozambican democracy emerging on their borders, RENAMO forces supported by apartheid South Africa targeted clinics and hospitals for destruction, while encouraging traditional healers and practitioners of witchcraft and sorcery, causing damage to the point where FRELIMO initially banned all such traditional practices (West, 2005).

The attempt to co-opt traditions for conservation thus invokes a potentially cynical and damaging dynamic with a long and dubious history. Amazonian Indian groups offer a prominent example (Conklin & Graham, 1995). High-profile publicity campaigns have represented Amazonian Indians as 'ecologically noble savages', indigenous first peoples who are natural guardians and stewards of their environment. Western environmentalists created an imagined ideal of people living in harmony with nature and identified these remote groups as living exemplars of their ideal. The alliance worked well for a time, both for human rights and for environmental ends. However, it eventually fractured, as global pressures corrupted local spokespersons and undermined their legitimacy (Conklin & Graham, 1995; cf. Igoe, 2003). Local development initiatives led to Amazonian Indian groups granting logging concessions and other extractive enterprises, and, according to some, transnational enterprises used their financial muscle to pressure conservation agencies to act against the interests of both environment and local people (Chapin, 2004).

This chapter now moves to develop a case study around conservation traditions among the Maasai.

Tradition and conservation in Maasailand

Conventional wisdom has often queried the conservation benefits of pastoralist land use practices (Lamprey, 1983; Behnke *et al.*, 1993; Illius & O'Connor,

1999; Ruttan & Borgerhoff-Mulder, 1999; Vetter, 2005; Gillson & Hoffmann, 2007). It is now widely accepted among conservation and development agencies that pastoralist land use in East Africa is environmentally sound and economically rational (Homewood *et al.*, 2001; WISP, 2008), and one that will assume increasing importance with climate change (Hesse & MacGregor, 2006). At intermediate levels of density, pastoralism enhances biodiversity, although at higher densities this gives way to competition between livestock and wildlife (Western & Gichohi, 1993; Lamprey & Reid, 2004).

Maasai have been widely hailed as natural conservationists (Parkipuny, 1982) (Box 10.2). Conservation initiatives build on this, both by siting protected areas in the landscapes created by millennia of pastoralist use, and by invoking Maasai practices as the basis for community conservation. Maasai are portrayed as quasi-pure pastoralists, for whom livestock are central to their culture and livelihoods (Figure 10.2), as living in harmony with savanna ecosystems and wildlife, and as perennially returning to pastoralist lifestyles. The corollary is held to be that Maasai dislike farming, although poverty makes farming an increasingly common necessity.

Box 10.2 **Maasai pastoralism as conservative and conservation compatible**

'*The Maasai have traditionally lived in harmony with their land, which for centuries was defended from outsiders by majestic and fierce Maasai warriors. Thus, Maasailand has remained one of the few unspoiled African ecosystems still in existence.*' http://www.maasaitrust.org/wildlife_conservation/index.html

'*. . .the Maasai are recognized the world over for their lifestyle centred around their cattle, the tenacity to their culture and their traditional way of dressing.*' Maundu *et al.* (2001: 6)

Maasai contrast their conservation ethic with official conservation agencies

'*. . . the world should know that we are not people who eat the soil until it is finished. We manage the land so that we make sure our small cultivation disturbs neither domestic nor wild animals. The world should learn from* us how we Maasai manage our lands. They shouldn't see us as destroyers of the land.'

'The only time we might kill an animal is when one attacks us.'

'Just look around. The parts of the world left with wildlife are peopled by pastoralists. Why is it so? How is it that supposed 'experts' and 'guardians of nature' come here after having failed to conserve trees and wildlife in their places of origin?'

'In one of the previous NCAA meetings I told them about the name of this place – Ramat – and Ramat means healthy habitat for people and all animals. They said it should be called 'Crater'. There were 125 rhinos when you arrived here and now they're all gone, and it is you who are consuming them.'

Spokespersons for Ngorongoro Maasai, reported in Lane (1996: 15–16)

Figure 10.2   Maasai pastoralists and their cattle live in southern Kenya and northern Tanzania, and were the subject of a multi-site study of their livelihoods and patterns of land use (Homewood *et al.*, 2009). (Photograph by Nigel Leader-Williams.)

Elements of Maasai conservation management are cited by various observers as depending on:

- communal land ownership, with mobile, transhumant herding on open rangelands (Potkanski, 1994);
- grazing and forest reserves delineated and maintained by elders (Stephenson, 1999; Maundu *et al.*, 2001);
- policing and enforcement of exclusion by warrior age sets (Box 10.2);
- traditional leaders' support for conservation as compatible with pastoralism and pastoralist ideals (Box 10.2); and
- tolerance of and compatibility with wildlife (Kipury, 1983).

Conservation workers seeking to build on local traditions nevertheless face a complex history of interactions between Maasai and conservation. Much of the land that, towards the end of the 19th century, was primarily under pastoral Maasai control now constitutes formal conservation estate (Homewood & Rodgers, 1991; Berger, 1993; Igoe & Brockington, 1999; Hughes, 2006). Maasai communities have been displaced by conservation and by settler, state and outside investor enterprises (Igoe & Brockington, 1999; Anderson, 2002). The Maasai Moves of the early 20th century cajoled and coerced Kenya's northern Maasai sections into moving to the Southern Reserve, now Kajiado and Narok Districts (Waller, 1990; Spear & Waller, 1993; Anderson, 2002; Hughes, 2006). Neumann (2004) sees adjudication and later subdivision of group ranches in Kenya Maasailand as a state project to contain the Maasai, separating them from an artificially created wilderness requiring subsequent preservation. Similar progressive dispossession in Tanzania is rendered more powerful by state denial of ethnicity (Shivji & Kapinga, 1998; Igoe & Brockington, 1999; Homewood *et al.*, 2004). Dispossession has fostered hostility towards conservation among many Maasai, with protest killings of wildlife in both Amboseli during the1980s (Lindsay, 1987) and in Tarangire–Simanjiro during the early 2000s (Sachedina, 2008). At meetings of the conservation lease-back Kitengela Landowners Association in 2005, prominent local Maasai said forcefully that they would be better off without wild animals on their land. The impact of conservation on land conflicts forms part of Maasai daily political discourses, for example in Amboseli in Kenya and in Ngorongoro Conservation Area in Tanzania.

Conservation makes appeal to Maasai traditions of stewardship, but may only be heard by many in the context of a history of dispossession.

Communities are diverse assemblages of people of different ages, sexes, occupations, assets, incomes and opportunities. The political positions of different interest groups diverge and mesh with traditions in different ways. For some, conservation holds real appeal and the narrative of stewardship speaks more strongly. For some, it may involve re-working conservation messages as 'local tradition'. For others, dispossession, economic grievance and development needs outweigh possible synergies with conservation.

Actions speak louder than words: Maasai decisions over land use and livelihoods are the clearest expression of the extent to which conservation traditions hold power. The following section outlines the background and findings of a multi-site study of livelihoods and land use in Maasailand.

Multi-site study of Maasai livelihoods

Maasailand comprises some $150\,000\,\text{km}^2$ of arid and semi-arid rangelands straddling the Kenya–Tanzania border (Figure 10.3). There are strong ecological, ethnic and microeconomic continuities throughout Maasailand, including a strong shared identity. These continuities are cross-cut by major contrasts between the two countries. Kenya experienced greater settler impact, more investment and a more highly developed infrastructure at Independence. Post-Independence, Kenya followed a clear-cut capitalist route, moving rapidly to privatize land and other resources. By contrast, Tanzania became a socialist state under Nyerere, imposing collectivization on dispersed pastoralist populations. Despite economic liberalization from the mid-1980s, the state retains strong central control.

Land tenure and hunting policies constitute two of the main contrasts relevant to conservation. Kenya Maasailand has mostly been subdivided into group ranches and more recently into private land holdings. In Tanzania, villages have only relatively recently become able to acquire somewhat tenuous title to their land. The state can reallocate land away from villages for agribusiness or conservation enterprises. Grazing land is classed as 'empty' and is particularly susceptible to alienation (Nelson et al., 2009). In Kenya, hunting is banned. In Tanzania, hunting operators lease blocks of land from the state and purchase tourist licenses to hunt specified wildlife quotas.[1] In

---

[1] There are rare experimental resident hunting schemes where licensing has been controlled by individual villages, such as the Department for Internation Development-supported MBOMIPA

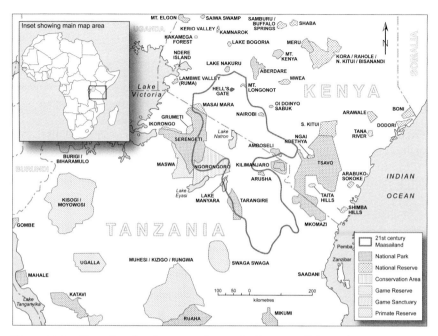

Figure 10.3   **Maasailand and protected areas in Kenya and Tanzania.**

both countries, the economic potential of wildlife, the prospect of building sustainable livelihoods and bringing about poverty reduction through sustainable use of wildlife resources inform policy, including poverty reduction strategy papers (e.g. URT, 2005; Roe & Walpole, this volume, Chapter 9). However, none of this has translated into positive support for pastoralism. No matter how conservation compatible, pastoralism is seen to be economically unproductive and environmentally damaging by both governments, which aim to transform livestock production into sedentary, intensive, 'modern' enterprises (URT, 1997; MoLF, 2006; Homewood *et al.*, 2009).

Our collaborative study pulled together in-depth fieldwork carried out in five main sites (Figure 10.4), each involving protected area-adjacent populations (Homewood *et al.*, 2009) (Table 10.1). In each site, data were collected using standardized variables and a common approach allowing comparative

scheme, now superseded by the Pawaga-Idodi Wildlife Management Area (Walsh, 2003; Coppolillo & Dickman, 2007). These are not at issue here.

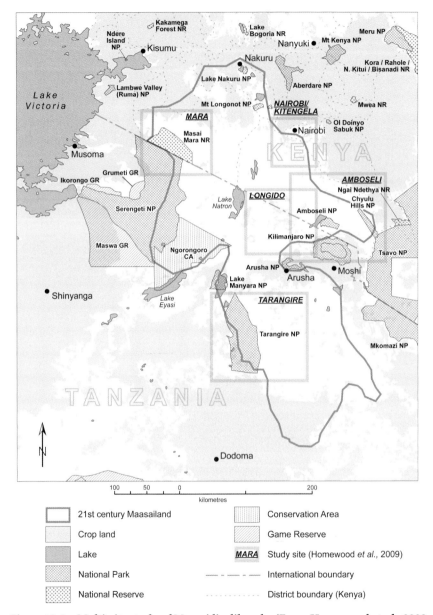

Figure 10.4   **Multi-site study of Maasai livelihoods. (From Homewood *et al.*, 2009 and reproduced by kind permission of Springer Science and Business Media.)**

Table 10.1 **Study sites (see also Figure 10.4) and key issues.**

Country	Protected area	Reserve-adjacent population	Tenure and issues	Principal researcher
Kenya	Masai Mara National Reserve	Mara group ranches	Wildlife viewing income: wildlife associations, private campsites/lodges; conservancies	M. Thompson
	Nairobi National Park	Kitengela landowners	Private landowners + conservation leaseback, urban property values, off-farm work	D. Nkedianye
	Amboseli National Park/Reserve	Kajiado group ranches/landowners	Key resources excised for cultivation or conservation. Amboseli National Park downgraded to (contested) national reserve status by ministerial decree in 2005	S. BurnSilver
Tanzania	Enduimet Wildlife Management Area	Longido villages	Poor agropastoralists + communal village lands; grazing lands open to alienation. Hunting/game viewing revenues centrally controlled despite Wildlife Management Area status	K. Homewood
	Tarangire National Park	Simanjiro villages	Gemstone mining wealth secured land titles. Large-scale cultivation as tenure strategy forestalling extension of conservation area	H. Sachedina

analysis. We focused on what people do, and with what success, exploring the factors which influence what economic activities people pursue and which determine how well they do from those activities (Homewood *et al.*, 2009). This chapter only presents a brief summary of findings across all sites to make a central point about continuity, change and conservation potential in Maasai natural resource use.

Maasai remain pastoralists centred on livestock. Some 95–100% households own livestock in all sites, and livestock contribute >50% of the income for all but the very poorest households, with livestock holdings being the best predictor of total annual household income. That central activity is combined with others in many ways, and with variable success. Around half of all households cultivate, but in most cases this contributes little to incomes. Instead, it may be as or more important as tenure strategy than as food security. Off-farm work is extremely important: across the board, Maasai have diversified away from natural resource-based livelihoods.

Conservation earnings are trivial at the household level in most sites. Community-level earnings may be less trivial, but are very prone to elite capture. There is little evidence of conservation partnering pastoralism. On the contrary, wildlife represent a fast changing and hard-to-capture income stream in most sites, accessed by only a small proportion of households (3–14%) and contributing <5% of mean annual income. Most people make little or nothing from wildlife, and those who do have in the past invested the returns in conservation-incompatible enterprises such as large-scale cultivation, further driving wildlife decline (Homewood *et al.*, 2001; Thompson *et al.*, 2002). Masai Mara in Kenya is the exception. Around 60% of Mara households earn wildlife-related revenue, contributing on average around 20% of household income across the wealth spectrum, although the well-off inevitably capture the lion's share (60–70% to the top quartile: Homewood *et al.*, 2009).

Maasai traditions, livelihoods and conservation

What are the implications of these results for organizations seeking to build on Maasai traditions of conservation-compatible land uses on open rangelands? First, livestock are certainly central to Maasai culture and livelihoods. However, our data show that many Maasai now farm. Together with increasing reliance on off-farm work, many do not return to pastoralist lifestyles. Second, communal land ownership has dwindled. Most of Kenya Maasailand is now subdivided as private holdings, and the remaining group ranch and trust lands are designated for subdivision. In Tanzania, communal land tenure is weak, particularly for grazing land, and pastoralist communities are losing land rapidly to outside investors. Where possible, Maasai continue to practice mobile, transhumant herding on open rangelands, but this is getting harder to maintain. Where communal ownership remains a possibility, Maasai elders continue to set aside grazing reserves, and warriors continue to police and enforce them. However, traditional leaders are by no means unanimously supportive of conservation as compatible with pastoralism. Some are linked into conservation enterprises, and their personal interests align with conservation (Thompson & Homewood, 2002; Thompson *et al.*, 2009), but they invest their few wildlife earnings in conservation-incompatible activities (Thompson *et al.*, 2009). Others see conservation as rampant dispossession to be resisted at all costs (Nelson, 2004, 2007; Sachedina, 2008; Nelson *et al.*, 2009). Tolerance of wildlife is increasingly stretched as poverty or profit drives people to cultivate, and wildlife damage becomes an increasing issue. Extensively reared livestock are broadly compatible with plains wildlife, and few Maasai hunt for food. However, continuing wildlife declines across Maasai rangelands suggest that broad tolerance and compatibility are not enough to outweigh competition for water and forage, alongside disease interactions and wildlife damage.

Conservation resource flows buy compliance from gatekeepers, as the well-known case of Ololosokwan shows (DeLuca, 2004; Nelson, 2004, 2007). However, these flows rarely benefit local communities. Instead, they favour susceptible individuals to whom personal benefits outweigh others' losses, and they distort the grassroots democratic process (Ribot, 2006). Local traditions of ecological knowledge are part of a knowledge system that also encompasses histories of dispossession, alongside current examples of displacement in the name of conservation. Like western traditions of conservation, Maasai perceptions of, and attitudes towards, conservation/pastoralism interactions are diverse, heterogeneous, multilayered and contradictory. It is naïve to try to cherry-pick individual 'congenial' elements as a basis for conservation initiatives. Working with the wider group, consulting and respecting local priorities despite their divergence from conservation goals, and building local democracy that will resist the tendency of susceptible individual gatekeepers to be picked off for short-term personal gain, may build better synergies in the long term.

References

Alvard, M. (1993) Testing the 'ecologically noble savage' hypothesis. *Human Ecology*, 21, 355–387.

Anderson, D. (2002) *Eroding the Commons: the politics of ecology in Baringo, Kenya 1890–1968*. James Currey, Oxford.

Behnke, R., Scoones, I. & Kerven, C. (eds) (1993) *Range Ecology at Disequilibrium: new models of natural variability and pastoral adaptation in African savannas*. Overseas Development Institute, London.

Berger, D. (1993) *Wildlife Extension: participant conservation by Maasai of Kenya*. African Centre for Technical Studies, Nairobi.

Brockington, D. (2005) The politics and ethnography of environmentalisms in Tanzania. *African Affairs*, 105, 97–116

Caughley, G. (1993) Elephants and economics. *Conservation Biology*, 7, 943–945.

Caughley, G., Shepherd, N. & Short, J. (1987) *Kangaroos: their ecology and management on the sheep rangelands of Australia*. Cambridge University Press, Cambridge.

Chapin, M. (2004) A challenge to conservationists. *World Watch Magazine*, November/December, 17–31.

Conklin, B. & Graham, L. (1995) The shifting middle ground: Amazonian Indians and eco-politics. *American Anthropologist*, 97, 695–710.

Coppolillo, P. & Dickman, A. (2007) Livelihoods and protected areas in the Ruaha landscape: a preliminary review. In *Protected Areas and Human Livelihoods*, eds K.H. Redford & E. Fearn, pp. 17–26. Working Paper No. 32. Wildlife Conservation Society, New York.

DeLuca, L. (2004) *Tourism, conservation, and development among the Maasai of Ngorongoro District, Tanzania: implications for political ecology and sustainable livelihoods*. Unpublished PhD thesis, University of Colorado, Boulder, CO.

Dossou-Glehouenou, B. (1999) Cultural and spiritual values of biodiversity in West Africa. In *Cultural and Spiritual Values of Biodiversity: a complementary contribution to the Global Biodiversity Assessment*, ed. D. Posey, pp. 370–371. UNEP and ITDG Publishing, London.

Falconer, J. (1999) Non-timber forest products in southern Ghana: traditional and cultural forest values. In *Cultural and Spiritual Values of Biodiversity: a complementary contribution to the Global Biodiversity Assessment*, ed. D. Posey, pp. 366–370. UNEP and ITDG Publishing, London.

Gillson, L. & Hoffman, M.T. (2007) Rangeland ecology in a changing world. *Science*, 315, 53–54.

Hesse, C. & MacGregor, J. (2006) *Pastoralism: drylands' invisible asset? Developing a framework for assessing the value of pastoralism in East Africa*. Drylands

Issue Paper No. 142. International Institute for Environment and Development, London.

Homewood, K.M. (1999) Pastoralists and payoffs. Comment on L. Ruttan & M. Borgerhoff-Mulder, are East African pastoralists truly conservationists? *Current Anthropology*, 40, 641–642.

Homewood, K.M. & Rodgers, W.A. (1991) *Maasailand Ecology*. Cambridge University Press, Cambridge.

Homewood, K., Kristjanson, P. & Chenevix Trench, P. (eds) (2009) *Staying Maasai: livelihoods, conservation and development in East African rangelands*. Springer, New York.

Homewood, K., Lambin, E.F., Coast, E. *et al.* (2001) Long-term changes in Serengeti-Mara wildebeest and land cover: pastoralism, population or policies? *Proceedings of the National Academy of Sciences of the USA*, 98, 12544–12549.

Homewood, K., Thompson, M. & Coast, E. (2004) In-migrants and exclusion: tenure, access and conflict in east African rangelands. *Africa*, 74, 567–610.

Hughes, L. (2006) *Moving the Maasai: a colonial misadventure*. Palgrave Macmillan, Basingstoke and New York.

Igoe, J. (2003) Scaling up civil society: donor money, NGOs and the pastoralist land rights movement in Tanzania. *Development and Change*, 34, 863–885.

Igoe, J. & Brockington, D. (1999) *Pastoral Land Tenure and Community Conservation: a case study from North-east Tanzania*. Pastoral Land Tenure Series No. 11. International Institute for Environment and Development, London.

Illius, A. & O'Connor, T. (1999) On the relevance of non-equilibrial concepts to arid and semi-arid grazing systems. *Ecological Applications*, 9, 798–813.

Kipury, N. (1983) *Oral Literature of the Maasai*. Heinemann, Nairobi.

Kothari, A. & Das, P. (1999) Local community knowledge and practices in India. In *Cultural and Spiritual Values of Biodiversity: a complementary contribution to the Global Biodiversity Assessment*, ed. D. Posey, pp. 185–192. UNEP and ITDG Publishing, London.

Lamprey, H. (1983) Pastoralism yesterday and today: the overgrazing problem. In *Tropical Savanna*, ed. F. Bourlière, pp. 643–666. Elsevier, Amsterdam.

Lamprey, R.H. & Reid, R.S. (2004) Expansion of human settlement in Kenya's Maasai Mara: what future for pastoralism and wildlife? *Journal of Biogeography*, 31, 997–1032.

Lane, C. (1996) *Ngorongoro Voices*. Forests, Trees and People Programme, Food and Agriculture Organization, Rome.

Laris, P. (2003) Grounding environmental narratives. In *African Environment and Development: rhetoric, programs, realities*, eds W. Moseley & B. Logan, pp. 63–85. Ashgate Publishing, Aldershot, UK.

Leader-Williams, N., Oldfield, T., Smith, R. & Walpole, M. (2002) Science, conservation and fox-hunting. *Nature*, 419, 878.

Lindsay, K. (1987) Integrating parks and pastoralists: some lessons from Amboseli. In *Conservation in Africa*, eds D. Anderson & R. Grove, pp. 149–167. Cambridge University Press, Cambridge.

MacGregor. J. (2003) Living with the river: landscape and memory in the Zambezi Valley, north west Zimbabwe. In *Social History and African Environments*, eds W. Beinart & J. MacGregor, pp. 87–106. James Currey, Oxford.

Maundu, P., Berger, D., Ole Saitibau, C. *et al.* (2001) *Ethnobotany of the Loita Maasai: towards community management of the Forest of the Lost Child*. People and Plants Working Paper No. 8. UNESCO, Paris.

May, R.M. (1994) Resource exploitation and the economics of extinction. *Nature*, 372, 42–43.

MoLF (Ministry of Livestock and Fisheries) (2006) *Draft National Livestock Policy 2006*. Ministry of Livestock and Fisheries Development, Nairobi.

Nelson, F. (2004) *The Evolution and Impacts of Community-based Ecotourism in Northern Tanzania*. Drylands Issue Paper No. 131. International Institute for Environment and Development, London.

Nelson, F. (2007) *Emergent or Illusory? Community management in Tanzania*. Drylands Issues Paper No. 146. International Institute for Environment and Development, London.

Nelson, F. Gardner, B., Igoe, J. & Williams, A. (2009) Community-based conservation and Maasai livelihoods in Tanzania. In *Staying Maasai? Livelihoods, conservation and development in East African rangelands*, eds K.M. Homewood, P. Kristjanson & P. Chenevix Trench, pp. 299–333. Springer, New York.

Neumann, R. (2004) Nature–state–territory: toward a critical theorization of conservation enclosures. In *Liberation Ecologies*, 2nd edn, eds R. Peet & M. Watts, pp. 195–217. Routledge, London and New York.

Norton-Griffiths, M. (2007) How many wildebeest do you need? *World Economics*, 8, 41–64.

Ntiamoa-Baidu, Y. (1993) Indigenous protected areas in Ghana. In *African Biodiversity: foundation for the future*, eds USAID/WWF/TNC/WRI, pp. 66–67. Biodiversity Support Program, Professional Printing Inc., Beltsville, MD.

Ostrom, E., Burger, J., Field, C., Norgaard, R. & Policansky, D. (1999) Revisiting the commons: local lessons, global challenges. *Science*, 284, 278–282.

Parker, I., (2006) *Kenya: the example not to follow*. Available at: www. Africanconservation.org/dcforum/DCForumID21/203.html (also published in *Sustainable*, Newsletter of IUCN Sustainable Use Specialist Group, July 2005).

Parkipuny, L.M.S. (1982) *On behalf of the people of Ngorongoro*. Background paper for the 1981 Management Plan of the Ngorongoro Conservation Area. Unpublished manuscript.

Posey, D. (ed.) (1999) *Cultural and Spiritual Values of Biodiversity: a complementary contribution to the Global Biodiversity Assessment*. UNEP and ITDG Publishing, London.

Potkanski, T. (1994) *Property Concepts, Herding Patterns and Management of Natural Resources among the Ngorongoro and Salei Maasai of Tanzania*. International Institute for Environment and Development, London.

Ranger, T. (2003) Women and environment in African religion. In *Social History and African Environments*, eds W. Beinart & J. MacGregor, pp. 72–86. James Currey, Oxford.

Ribot, J. (2006) Choose democracy: environmentalists' socio-political response. *Global Environmental Policy*, 16, 115–119.

Ruttan, L. & Borgerhoff-Mulder, M. (1999) Are East African pastoralists truly conservationists? *Current Anthropology*, 40, 621–652.

Sachedina, H. (2008) *Wildlife is our oil: conservation, livelihoods and NGOs in the Tarangire ecosystem*. Unpublished DPhil thesis, University of Oxford, Oxford.

Schoonmaker-Freudenberger, K. (1991) *Mbégué: the disingenuous destruction of a Sahelian forest*. IIED Drylands Network Programme Issues Paper No. 29. International Institute for Environment and Development, London.

Shivji, I. & Kapinga, W. (1998) *Maasai Rights in Ngorongoro, Tanzania*. International Institute for Environment and Development, London, and Hakiardhi, Dar es Salaam.

Smith, E.A. & Wishnie M. (2000) Conservation and subsistence in small-scale societies. *Annual Review of Anthropology*, 29, 493–524.

Sommer, V. (1996) *Heilige Egoisten. Die Soziobiologie indischer Tempelaffen*. C.H. Beck, Munich.

Sommer, V. (2002) In divine company: India's temple monkeys. *BBC Wildlife* January, 50–56.

Spear, T. & Waller, R. (eds) (1993) *Being Maasai*. James Currey, London,Mkuki na Nyota, Dar es Salaam, and East African Educational Publishers, Nairobi.

Stephenson, D. (1999) The importance of the Convention on Biological Diversity to the Loita Maasai of Kenya. In *Cultural and Spiritual Values of Biodiversity: a complementary contribution to the Global Biodiversity Assessment*, ed. D. Posey, pp. 531–533. UNEP and ITDG Publishing, London.

Thomas, F. (2007) The behavioural ecology of shellfish gathering in western Kiribati, Micronesia. 2: Patch choice, patch sampling, and risk. *Human Ecology*, 35, 515–526.

Thompson, D.M. & Homewood, K.M. (2002) Elites, entrepreneurs and exclusion in Maasailand. *Human Ecology*, 30, 107–138.

Thompson, M., Serneels, S. & Lambin, E.F. (2002) Land-use strategies in the Mara ecosystem (Kenya): a spatial analysis linking socio-economic data with landscape variables. In *Remote Sensing and GIS Applications for Linking People, Place and Policy*, eds S.J. Walsh & K.A. Crews-Meyer, pp. 39–68. Kluwer Academic Publishers, Boston.

Thompson, M., Serneels, S., Ole Kaelo, D. & Chenevix Trench, P. (2009) Maasai Mara land privatization and wildlife decline: can conservation pay its way? In *Staying Maasai? Livelihoods, conservation and development in East African rangelands*, eds K.M. Homewood, P. Kristjanson & P. Chenevix Trench, pp. 77–114. Springer, New York.

Tilley, C. (2009) What gardens mean. In *Ethnographic Approaches to Material Culture: ethnographic approaches*, ed. P. Vannini, pp. 171–192. Peter Lang Publishing, New York.

Tont S.A. (1999) Of dancing bears and sacred trees: some aspects of Turkish attitudes toward nature, and their possible consequences for biological diversity. In *Cultural and Sprdtual Values of Biodiversity. A complementary contribution to the Global Biodiversity Assessment*, ed. D.A. Posey, pp. 392–393. UNEP, Kenya and ITDG Publishing, London.

URT (United Republic of Tanzania) (1997) *Livestock and Agriculture Policy*. Government of Tanzania Publications, Dar es Salaam.

URT (United Republic of Tanzania) (2005) *Mkukuta: Tanzania's national strategy for growth and reduction of poverty*. Vice-President's Office, United Republic of Tanzania, Dar es Salaam.

Vetter, S. (2005) Rangelands at equilibrium and non-equilibrium: recent developments in the debate. *Journal of Arid Environments*, 62, 321–341.

Waller, R. (1990) Tsetse fly in western Narok, Kenya. *Journal of African History*, 31, 81–101.

Walpole, M.J. & Leader-Williams, N. (2001) Masai Mara tourism reveals partnership benefits. *Nature*, 413, 771.

Walsh, M. (2003) *MBOMIPA. From Project to Association and from Conservation to Poverty Reduction: final project report*. Ministry of Natural Resources and Tourism, Wildlife Division and Tanzania National Parks, in collaboration with Iringa District Council, Iringa, Tanzania.

West, H. (2005) *Kupilikula: governance and the invisible realm in Mozambique*. University of Chicago Press, Chicago.

Western, D. & Gichohi, H. (1993) Segregation effects and impoverishment of savanna parks: the case for ecosystem viability analysis. *African Journal of Ecology*, 31, 269–281.

WISP (World Initiative for Sustainable Pastoralism) (2008) *A Global Perspective on the Total Economic Value of Pastoralism*. International Livestock Research Institute for GEF/UNDP/IUCN, Nairobi.

Yibarbuk, D., Whitehead, P.J., Russell-Smith, J. *et al.* (2001) Fire ecology and Aboriginal land management in central Arnhem Land, northern Australia: a tradition of ecosystem management. *Journal of Biogeography*, 28, 325–343.

Part III
Economics and Governance

11

Misaligned Incentives and Trade-offs in Allocating Conservation Funding

Aaron Bruner, Eduard T. Niesten and Richard E. Rice

Center for Applied Biodiversity Science, Conservation International, Arlington, VA, USA

Introduction

Current funding for environmental conservation is not sufficient to protect Earth's biological diversity and the benefits it provides to society (James *et al.*, 2001). At the same time, a set of perverse incentives suggests that the conservation community may also be achieving far less than it could be with available funds. This chapter offers reflections on these incentives and the trade-offs they create for decisions about how to allocate scarce funds for conservation in developing countries.

The chapter begins by describing the importance of assessing the effectiveness of conservation investments. We then offer some economic concepts to help clarify both the challenges facing efforts to achieve greater effectiveness in conservation spending, and the means to overcome them. We return to these concepts to help assess efficiency throughout the chapter. Third, we discuss the context for conservation in developing countries and the perverse incentives for funding allocations that this context appears to have created. We describe in some detail a set of entirely logical reactions to these incentives that do, indeed, suggest serious inefficiencies in funding allocation. Finally, we discuss several important shifts in conservation practice that we believe are now helping to address inefficiencies. We conclude that, although a significant

Trade-offs in Conservation: Deciding What to Save, 1st edition.   Edited by N. Leader-Williams, W.M. Adams and R.J. Smith.   © 2010 Blackwell Publishing Ltd.

shift in funding allocation may be necessary, the conservation community is increasingly equipped to change, and that doing so would result in both more effective conservation and a greater contribution to human well-being.

Evaluating trade-offs in allocating scarce conservation funds

While we prepared this chapter, millions of conservation dollars will have been spent on workshops, consultancies and plans to address a host of pressing issues, ranging from climate change to water scarcity. At the same time, the management of protected area systems in developing countries is so poorly funded that many lack fuel for vehicles or even boots for staff (van Schaik *et al.*, 1997) (Figure 11.1). The wide array of threats to Earth's biodiversity and ecosystems justifies many different types of conservation investments. However, when what is arguably the bedrock of the world's conservation efforts is systematically underfunded in favour of planning for new challenges, an evaluation of efficiency is clearly warranted.

Figure 11.1   **A guard post at the entrance to Pacaya Samiria National Reserve in the Peruvian Amazon. While donors may provide infrastructure for guard and ranger posts, all over the world guards and rangers often lack for uniforms, transport, salaries and logistical support to undertake their routine duties at the front line of conservation. (Photograph by Nigel Leader-Williams.)**

Such an evaluation is not straightforward. Due to the nature of financial as well as impact reporting on most conservation initiatives, existing information is inadequate even to permit a useable breakdown of how the majority of international conservation funds are spent, much less to facilitate a comprehensive analysis of the cost-effectiveness of each type of spending in contributing to overall conservation impact. Therefore, improving the flow of relevant information is an obvious priority (Kapos *et al.*, this volume, Chapter 5).

Nevertheless, we believe that even in the in the absence of such information, it is possible to substantiate the hypothesis that funds are allocated inefficiently. Here, we seek to demonstrate the existence of inefficiencies by combining two perspectives on incentives for conservation action. First, we characterize the basic dynamics that drive decisions by resource owners/resource users about whether or not to choose conservation. We propose that evaluating the degree to which conservation actions affect resource use choices provides a simple yet concrete way to identify whether those conservation actions, in turn, are likely to be effective. Second, to overcome the lack of information about how funding is actually used, we examine the incentives that drive funding allocation decisions within the conservation community. The following sections compare the conservation actions that are the logical reaction to this incentive structure to those that effectively influence resource use decisions.

Effective conservation in theory and practice

A fundamental economic challenge must be overcome to deliver effective conservation outcomes at a significant scale. Specifically, like national defense or clean air, biodiversity is in many ways a 'public good' from which people benefit independent of their contribution to its provision. Providing public goods typically involves significant individual or local costs, for example in the form of risking life or limb, or from not clearing forested land to grow crops, while the benefits derive to society more broadly. As a result, individuals typically do not choose to provide public goods in the absence of outside intervention.

Indeed, in the case of conservation, few resource owners in any country choose to conserve their lands on an entirely voluntary basis because they are not willing or able to forego the potential earnings from other income-generating options. Even where particular individuals or groups favour conservation, they often require outside financial support to protect their natural resources from significant external pressures (Zimmerman *et al.*, 2001). Ultimately, as with other public goods, if the people who benefit from conservation do not share the costs of providing it with the smaller set of people who bear those costs, conservation of biodiversity at any significant scale is unlikely (ten Brink *et al.*, 2009).

Tried and tested strategies exist to address this challenge at the national level. The most common is for governments to secure tax revenue from their citizens and use that revenue either to finance government provision of the public good or pay private actors to provide it. In the case of biodiversity, governments commonly use this mechanism to create and manage national protected area systems. Alternatively, they may offer tax breaks or public funding to encourage private landowners to place their land into conservation easements or set-aside programmes. Where supportive legal structures are in place, private efforts can complement government action by collecting additional 'willingness to pay' for conservation, and channeling funds to provide private landowners with incentives to choose conservation over other land use options (Figure 11.2). This sort of action in effect supplements government redistribution of the cost burden of conservation.

Figure 11.2 **In China, a conservation agreement between a logging company and conservationists protects important panda habitat and municipal water supply. (Photograph by Richard Rice.)**

Constraints to effective conservation in developing countries

Conservation in developed countries largely follows the principles outlined above, with relatively well-funded protected area systems (James *et al.*, 2001) and supportive legal structures that are often widely used by non-governmental organizations (NGOs). For example, the Royal Society for the Protection of Birds and The Nature Conservancy, the largest conservation NGOs in the UK and the US, respectively, dedicate most of their domestic conservation funds to acquiring and managing land for conservation through purchases, easements or related legal options.

In developing countries, conservation differs significantly. This section reviews several of these differences. The subsequent section describes the incentives these differences create for allocating conservation funds, and how these incentives in turn affect the overall effectiveness of that funding in achieving conservation goals.

Inadequate funding

Governments in developing countries typically have extremely limited budgets and face many pressing societal needs that they must weigh against the choice of funding conservation. Even accounting for lower costs and more basic management objectives, the percentage of conservation funding needs that go unmet remains much higher than in developed countries. For example, shortfalls in funding the management of protected area systems in developing countries average perhaps 70% compared with 10–25% in developed countries (James *et al.*, 2001; Wilkie *et al.*, 2001; Bruner *et al.*, 2008).

Furthermore, given poverty levels in developing countries and a heavy reliance by poor people on natural resources (WRI, 2005), there is a strong moral imperative to ensure that both direct and opportunity costs of conservation are not borne locally. To illustrate the potential scope of meeting this objective across all developing countries, the cost of providing compensation to all people in and around existing protected areas has been estimated at approximately US$5 billion per year (James *et al.*, 2001). The scale of this need further increases the gap between the funding required and what is actually available in developing countries, to perhaps 90% just for existing protected areas. With a few notable exceptions such as Costa Rica (Pagiola, 2008), government funding for covering these costs via easements, benefit sharing or related mechanisms is minimal.

Broad distribution of costs and benefits

Another challenge relates to the distribution of costs and benefits. Although conservation is increasingly understood to be economically beneficial at a global level, choosing conservation over resource exploitation frequently implies that significant opportunity costs be borne at the individual, local and national levels (Balmford & Whitten, 2003). From the perspective of addressing the public goods problem described above, transfers of willingness to pay to ensure conservation in developing countries therefore need to cross international borders. This raises a host of complications. For example, such transfers cannot simply rely on national government policy, but must depend largely on voluntary commitments from a range of government and non-governmental agencies in countries with different cultures, languages and values. Institutionalizing such transfers into a transparent and reliable stream of financing for conservation in developing countries remains a daunting challenge.

The degree to which this situation may influence conservation funding decisions should not be understated. For example, the US State of Florida recently agreed to purchase $750\,km^2$ of land for conservation at a price of $1.75 billion (Planet Ark, 2008). By contrast, an equivalent sum could secure more than $100\,000\,km^2$ in the developing world (based on James *et al.*, 2001). In terms of global impact on biodiversity conservation and other ecosystem service values, the latter scenario would achieve incomparably greater benefits. Despite that, international funding of this magnitude has not materialized. The Global Environment Facility (GEF), which represents the contribution of all countries to financing commitments to the Convention on Biological Diversity (CBD), disbursed perhaps $100 million per year to biodiversity projects in the 10-year period from 1992 to 2002 (GEF, 2006).

Lack of legal options for the provision of direct benefits

For various reasons, a significant portion of the financial resources that are transferred internationally to support conservation in developing countries are directed to non-profit organizations. Unfortunately, the legal contexts in most developing countries do not provide any ready means for these organizations to transfer funds to those who bear the costs of conservation in exchange for their commitment to conservation objectives. Easements, for example, are not widely available in developing countries. Where they are, institutional and legal weaknesses often make them a risky proposition, offering the easement holder few guarantees that it will be respected over time. As a result, at the same time that restricted budgets and the diffuse nature of conservation benefits make it difficult to secure adequate revenue, limited legal options and governance issues complicate efforts to channel those limited revenues to the land owners and resource users who ultimately decide whether or not to commit to conservation objectives.

Inadequate monitoring of results

A final issue is that monitoring of on-the-ground conservation impacts is notably poor (Ferraro & Pattanayak, 2006, Kapos *et al.*, this volume, Chapter 5). This is perhaps understandable given the large number of priorities competing for scarce funds. However, despite broad acceptance of the need for adaptive management, the lack of good information deprives conservation practitioners of the basic feedback required to adjust strategies precisely in contexts where the greatest creativity is required.

More perversely, in the absence of data on conservation impacts, other types of information are routinely used to communicate effectiveness and justify continued transfer of funds. These range from indicators of process, such as numbers of workshops held, to stories of possible win–win solutions, to attractive visual materials that may have nothing at all to do with conservation impacts. Even where impact data are used, the individual projects from which they derive may have little relevance to what is possible and cost-effective at a broad scale. Indeed, very few indicators communicate actual efficiency or effectiveness in delivering conservation results on the ground. At best, this situation leaves funding decisions vulnerable to intuition based on incomplete information. At the extreme, it may entirely de-couple success in securing funds from on-the-ground conservation achievement.

Perverse incentives underpinning current conservation practice

Insufficient funding, broad distribution of benefits, and lack of legal mechanisms for direct benefit transfer make it challenging, and perhaps unattractive, for conservation practitioners in developing countries to employ many of the standard responses to the problem of biodiversity as a public good. Therefore, it is not surprising that recent decades have seen numerous innovations that seek to achieve conservation objectives despite these limitations. Among the more enticing ideas are the potential to achieve long-term conservation without incurring long-term financial burdens, and the potential for the market in green products to impact conservation at the landscape scale, without the need to work at the local level.

A complicating factor is that the current culture of fundraising frequently values attractive stories at least as much as demonstrated outcomes. As discussed below, this dynamic appears to have persisted despite evidence of a general lack of effectiveness of many past conservation investments in delivering concrete conservation impact. The resulting incentives to conservation practitioners have been less than conducive to focusing on on-the-ground outcomes. We describe below two large areas of conservation work that respond directly to this dynamic, and explore their implications for maximizing conservation impact with a limited budget.

Over-reliance on attractive but unsubstantiated approaches on the ground

As described earlier, the beneficiaries of conservation must share its costs to make conservation an attractive choice for decision makers at a broad scale. However, few donors or governments are willing to fund direct benefit transfers in developing countries, let alone commit to covering their recurring costs. Taken in combination with severe budget limitations, this situation creates a strong incentive to develop alternative proposals for making conservation economically attractive while minimizing long-term financial burdens.

Recent decades have seen an explosion of strategies that propose to respond to that need by combining conservation with the production of marketable goods. These include the creation of markets for 'green' products, including certified sustainable harvests of timber and ecotourism (IUCN *et al.*, 1991; FSC, 2008). These strategies aim to address the exact limitations described above. Indeed, they hold out the promise of self-financing following an initial investment from external sources, by harnessing the market rather than voluntary transfers to provide benefits to those who choose conservation. Thus, in principle the cost burden is shifted from governments, NGOs and their funding sources to purchasers of the good or service in question. While unquestionably desirable, a closer look suggests that in practice these strategies as implemented to date have some basic limitations.

Some approaches, such as Integrated Conservation and Development Projects (ICDPs), face fundamental business challenges. Frequently, ICDPs aim to start successful businesses in places where transport costs are high, where managers are community members who have little business experience, and where there is often no clear link between income earned and conservation outcome. Predictably, these efforts have not been broadly successful (Figure 11.3). For example, a detailed review in Indonesia found few cases where ICDPs have effectively reconciled local people's development needs with protected area management (Wells *et al.*, 1999). Similar findings are reported across the

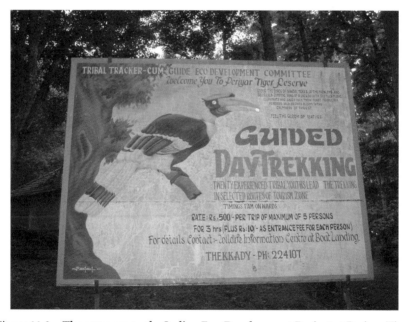

Figure 11.3 **The entrance to the Indian Eco-Development Project at Periyar Tiger Reserve in India. The social legacy of this project was limited apart from for those beneficiaries involved in tourism (see Gubbi *et al.*, 2008). Despite their limited success, ICDPs continue to be funded. (Photograph by kind permission of Sanjay Gubbi.)**

developing world (Robinson & Redford, 2004). Nonetheless, projects that rely on the general principles of the ICDP approach as articulated 20 years ago continue to be launched.

Other approaches rely on optimistic expectations about markets for green production. For example, sustainable forest management (SFM) has been widely touted as a means to blend conservation and production (FSC, 2008). While SFM is typically less profitable than conventional logging as evidenced by an almost complete lack of voluntary adoption of sustainable practice (Rice et al., 2001), the certification of sustainable management practice is widely promoted as a means of ensuring profitability through price premiums and greater market access.

Unfortunately, price premiums from certification have largely failed to materialize (ITTO, 2007). Further, a look at the timber trade suggests that even if such premiums become more common, current and likely future demand for certified timber can be met from only a tiny fraction of forests in developing countries. For instance, all major producers of tropical timber, including Brazil, Indonesia and Malaysia, consume the majority of their timber at home, where there is little or no demand for certified products. Of <20% of total tropical timber production that even enters international trade, the majority is purchased by countries that similarly have little interest in certified products, most notably China, Japan and Korea (ITTO, 2007; Purbawiyatna & Simula, 2008). Finally, even for the fraction of production actually bought by countries where there is demand for certified timber, such demand within those countries is so low (Oliver, 2005) as to make the total potential market for certified logging almost trivial when compared to total timber production. Despite these obvious limitations, hundreds of millions of dollars have been invested in SFM over the years, and funding continues to flow freely to SFM today.

Even where markets are significant, they may not fundamentally alter the distribution of costs and benefits from conservation. For example, the market for ecotourism is large and growing. However, for the rural communities who bear the majority of the costs of living with wildlife, benefits from tourism seldom outweigh costs (Walpole & Thouless, 2005). Even where tourism has served to make conservation attractive, its effectiveness may be fragile and depend heavily on market fluctuations and issues like civil unrest. Other challenges include the danger of misusing the 'eco' label to secure customers regardless of the degree of adherence to ecotourism practices and principles, and the related problem of exceeding resource carrying capacity due to the inherent tension between increasing revenues and limiting environmental footprints (Roe *et al.*, 1997). In other words, while the ecotourism industry may help make conservation broadly valuable, without a much more explicit focus on mechanisms to distribute benefits it will not be likely to make conservation an attractive choice for landowners at a significant scale.

Overemphasis on changing enabling conditions

A second course of action that similarly responds to the perverse incentive described above is to avoid on-the-ground action altogether and instead focus on modifying enabling conditions. At a broad scale, actions such as improved biological priority setting, influencing policy or enhancing planning capacity have the potential to increase the effectiveness of conservation by reducing the number of sites in need of protection, or by increasing overall funding and political support. At a site or regional scale, planning for conservation area creation rather than funding recurrent management or developing participatory conservation frameworks can both be considered strategic means to set the stage for future protection.

The need for these actions and others like them is clear. Conservation practice obviously needs to generate clear priorities for action, build political support at all levels, effectively monitor impacts and focus on innovative ways to address real challenges. However, it is equally clear that these actions do not affect the distribution of costs and benefits of choosing conservation on the ground, and so do not in themselves make conservation happen. In other words, they do not address the problem of public goods that lies at the heart of global biodiversity conservation. From the perspective of optimizing financial resource allocation amongst conservation activities, such approaches should therefore logically form a highly strategic but limited component of a portfolio of investment in conservation that guides and supports a foundation of solid on-the-ground action. If high-level enabling activities *replace* on-the-ground action, then little conservation impact will ultimately be achieved.

Data are not sufficient to quantify the degree to which this may be the case. However, over-investment in high-level activities would be a rational response to the incentive structure we have described. Choosing on-the-ground action requires, among other things, accepting significant limitations to the scale of potential impacts due to funding constraints. These limitations are avoided through focusing instead on 'strategic' or 'catalytic' actions. If indeed this incentive applies broadly, an unjustifiably small percentage of investment is likely to be dedicated to conservation action on the ground and practitioners will allocate an unjustifiably large share of their funding to catalyzing action by others, rather than engaging in conservation action themselves. While observations do not serve as conclusive evidence, every practitioner is aware of a significant and ever-growing number of large international meetings and continuous refinement of priority-setting methodologies. At a local level, 'implementation' all too often consists of participatory plans that lack the funds needed for actual implementation after planning is complete.

In summary, if the reasoning presented in this and the previous section is correct, a disproportionate share of conservation funds may be spent either on continual experimentation with supposed cost-saving approaches, or on changing the enabling environment. The former largely constitute a set of tools that have yet to demonstrate reliable achievements that would justify large-scale continued investment or replication. Meanwhile, in and of themselves, the latter simply do not respond to the problem of global public goods posed by conservation. The likely result, attributable in no small part to the dynamics that have historically governed funding and fundraising, is that a grossly inadequate share of conservation funding may actually be allocated to the on-the-ground actions that truly affect decision makers' choices about biodiversity.

Shifts in the right direction

Although there are strong indications that scarce conservation funds are not being well allocated, there has been a growing awareness of this problem in recent years. Moreover, a number of positive trends are gathering momentum, related both to how the conservation community reacts to the perverse incentive structure that drives funding allocation, and to changing that structure itself. This section describes some of the approaches that appear most promising.

Recognizing diminishing returns to planning and prioritization

From experience, a growing number of colleagues around the world have expressed frustration at the emphasis on planning instead of action. Studies now give strength to this perception by demonstrating the diminishing returns to repeatedly investing in improving planning frameworks prior to investing in action (Wilson *et al.*, this volume, Chapter 2).

Increasing value placed on outcome-based evaluation

Recent studies have stressed the need to evaluate conservation performance against on-the-ground impact, rather than against indicators of process or superficially attractive theories (Kapos *et al.*, this volume, Chapter 5). The value of rigorous evaluation has also been increasingly recognized in theory (Ferraro & Pattanayak, 2006), and a number of such evaluations have been carried out in practice, frequently using deforestation data (e.g. Nepstad *et al.*, 2006). In addition, several initiatives now underway among NGOs and government agencies actively seek to increase focus on rigorous evaluation, for example the Collaboration for Environmental Evidence and the Conservation Measures Partnership.[1] These trends have not yet fundamentally changed the way conservation is evaluated, but they do place welcome emphasis on rigorous evaluation, monitoring and accountability.

More focused implementation

The effectiveness of protected area management in developing countries has begun to receive high-level international attention. The 8th Conference of the Parties to the CBD, for instance, published an official decision urging signatories to ensure that protected areas are adequately financed, and provided guidance on how such finance could be achieved (CBD, 2006). Complementarily, NGOs focused on directly supporting protected area management have begun to appear, including The African Parks Network [2] and Greenvest. [3]

There also has been significant progress in creating direct mechanisms to make conservation attractive to resource owners in developing countries, comparable to easements in the developed countries (Ferraro & Kiss, 2003).

---

[1] http://www.environmentalevidence.org/and http://www.conservationmeasures.org/CMP, respectively.
[2] http://www.african-parks.org.
[3] http://www.greenvest.org.

Growing experience with implementation lends support to the value of these mechanisms as a way to address the problem of public goods posed by conservation and highlights lessons for improving efficiency (Niesten *et al.*, 2008; Wunder *et al.*, 2008). While we freely acknowledge that progress to date could be subject to critiques similar to those presented earlier, we cite as evidence for impact and scalability the work on voluntary agreements by Conservation International and immediate partners that currently protects approximately 10 000 km^2 and involves nearly 100 communities. Moreover, a number of countries, including Costa Rica, Mexico, Brazil and most recently Ecuador are now implementing national programmes based on direct payments for conservation.

Growing appreciation of the distribution of costs and benefits from conservation

Improved knowledge of ecosystem services has drawn a clear link between conservation and human well-being (e.g. Turner *et al.*, 2007; ten Brink *et al.*, 2009; Goldman *et al.*, this volume, Chapter 4). However, there is also growing appreciation that the distribution of costs and benefits from conservation means these links do not in themselves guarantee that resource owners will choose conservation (Balmford & Whitten, 2003). In this context, we believe that the international conservation community is increasingly aware of the need to provide financial support to make foregoing activities like unsustainable fishing or land clearing attractive to decision makers, thereby ensuring the continued provision of significant societal benefit.

Conclusions

We are aware of the irony of having dedicated significant time and resources to arguing on paper that too much money is already spent on discussing conservation strategy. However, that acknowledgment provides a useful basis from which to emphasize several concluding points.

Most basically, we are not advocating an extreme argument that all conservation funding should be dedicated to protected area management or direct payments. There are important roles for high-level actions and for catalyzing innovative approaches at all levels. Nevertheless, the breakdown of overall international conservation investment clearly must be considered far more carefully than has been done to date. For example, while innovation remains vital, the conservation community needs both to control the percentage of resources going toward what is in effect research and development, and be unabashedly realistic about the potential for new approaches to truly meet the objectives of both conservation and human well-being. Similarly, planning, awareness building and other activities that seek to improve the enabling conditions for conservation need to be strategically selected and clearly tied to on-the-ground action. Although practical questions obviously remain, for example those related to finding the right balance between planning and action or who should carry out limited high-level work, we believe that strategic activities in general should attract a smaller share of conservation investment than on-the-ground action.

As noted at the outset, empirical data on the allocation of conservation finance is insufficient to suggest the degree to which such an allocation of funds would represent a marked change of direction. However, our analysis suggests that the necessary shift could be significant, requiring the reorientation of funds, staffing and metrics for success away from strategic planning and into focused implementation. What the international conservation community would look like after that shift is not clear. However, it is likely that protected areas would be better staffed, communities would receive more direct benefits from conservation, and a relatively small number of strategic people and institutions would engage in planning at a large scale.

Such a shift would represent a major undertaking, requiring a willingness to accept real trade-offs and change. However, if the international conservation community can combine that level of honesty and realism with other ideas for increasing funding, we believe that not only will more conservation be achieved, but that conservation will become a far more effective tool for promoting human well-being.

Acknowledgments

While the views expressed here are attributable to the authors, we gratefully acknowledge comments from Fred Boltz, which greatly improved this chapter.

References

Balmford, A. & Whitten, T. (2003) Who should pay for tropical conservation, and how could the costs be met? *Oryx*, 37, 238–250.

Bruner, A., Naidoo, R. & Balmford A. (2008) *Review of the Costs of Conservation and Priorities for Action.* European Commission Contract ENV/070307/2007/486089/ETU/B2, Cambridge.

CBD (Convention on Biological Diversity) (2006) *Decision VIII/24: protected areas.* Eighth Conference of the Parties, 20–31 March, Curitiba, Brazil.

Ferraro, P.J. & Kiss, A. (2003) Will direct payments help biodiversity? *Science*, 299, 1981–1982.

Ferraro, P.J. & Pattanayak, S.K. (2006) Money for nothing? A call for empirical evaluation of biodiversity conservation investments. *PLoS Biology*, 4, 482–488.

FSC (Forest Stewardship Council) (2008) *About FSC: benefits.* Available at: http://www.fsc.org/en/about/about_fsc/benefits.

GEF (Global Environment Facility) (2006) *Biodiversity Matters: GEF's contribution to preserving and sustaining the natural systems that shape our lives.* Available at: http://www.gefweb.org/outreach/outreach-PUblications/GEF_Biodiversity_CRA.pdf.

Gubbi, S., Linkie, M. & Leader-Williams, N. (2008) Evaluating the legacy of an integrated conservation and development project around a tiger reserve in India. *Environmental Conservation*, 35, 331–339.

ITTO (International Tropical Timber Organization) (2007) *Annual Review and Assessment of the World Timber Situation.* ITTO, Japan. Available at: http://www.itto.or.jp/live/PageDisplayHandler?pageId = 199.

IUCN (International Union for the Conservation of Nature), UNEP (United Nations Environment Programme) & WWF (1991) *Caring for the Earth: a strategy for sustainable living.* IUCN, Gland, Switzerland.

James, A.N., Gaston K.J. & Balmford, A. (2001) Can we afford to conserve biodiversity? *BioScience*, 51, 43–52.

Nepstad, D., Schwartzman, S., Bamberger B. *et al.* (2006) Inhibition of Amazon deforestation and fire by parks and indigenous lands. *Conservation Biology*, 20, 65–73.

Niesten, E., Bruner, A., Rice, R. & Zurita, P. (2008) *Conservation Incentive Agreements: an introduction and lessons learned to date.* Conservation International, Washington, DC.

Oliver, R. (2005) *Price Premium for Verified Legal and Sustainable Timber. A study for the UK Timber Trade Federation and Department for International Development.* Department for International Development, London.

Pagiola, S. (2008) Payments for environmental services in Costa Rica. *Ecological Economics*, 65, 712–724.

Planet Ark (2008) *Florida to buy chunk of everglades from sugar firm*. Available at: http://www.planetark.org/dailynewsstory.cfm/newsid/48965/story.htm.

Purbawiyatna, A. & Simula, M. (2008) *Developing Forest Certification: towards increasing the comparability and acceptance of forest certification systems*. ITTO Technical Series No. 29. Available at: http://www.itto.or.jp/live/Live_Server/4092/TS29.pdf.

Rice, R., Sugal, C., Ratay, S. & da Fonseca, G.A.B. (2001) *Sustainable Forest Management: a review of conventional wisdom*. Conservation International, Washington, DC.

Roe, D., Leader-Williams, N. & Dalal-Clayton, B. (1997) *Take Only Photographs, Leave Only Footprints: the environmental impacts of wildlife tourism*. International Institute for Environment and Development, London.

Robinson, J.G. & Redford, K.H. (2004) Jack of all trades, master of none: inherent contradictions among ICD approaches. In *Getting Biodiversity Projects to Work: towards more effective conservation and development*, eds T.O. McShane & M.P. Wells, pp. 10–34. Columbia University Press, New York.

ten Brink, P., Berghöfer, A., Schröter-Schlaack, C. *et al.* (2009) *The Economics of Ecosystems and Biodiversity for National and International Policy Makers: executive summary*. TEEB, Bonn. Available at: www.teebweb.org.

Turner, W.R., Brandon, K., Brooks, T.M. *et al.* (2007) Global conservation of biodiversity and ecosystem services. *Bioscience*, 57, 868–873.

van Schaik, C.P., Terborgh J. & Dugelby, B. (1997) The silent crisis: the state of rainforest nature preserves. In *Last Stand: protected areas and the defense of tropical biodiversity*, eds R. Kramer, C. van Shaik & J. Johnson, pp. 64–89. Oxford University Press, New York.

Walpole, M.J. & Thouless, C.R. (2005) Increasing the value of wildlife through non-consumptive use? Deconstructing the myths of ecotourism and community-based tourism in the tropics. In *People and Wildlife: conflict or co-existence?* eds R. Woodroffe, S. Thirgood & A. Rabinowitz, pp. 122–139. Cambridge University Press, Cambridge.

Wells, M., Guggenheim, S., Khan, A., Wardojo, W. & Jepson, P. (1999) *Investing in Biodiversity: a review of Indonesia's Integrated Conservation and Development Projects*. World Bank, Washington, DC.

Wilkie, D.S., Carpenter, J.F. & Zhang, Q. (2001) The under-financing of protected areas in the Congo Basin: so many parks and so little willingness to pay. *Biodiversity and Conservation*, 10, 691–709.

WRI (World Resources Institute) (2005) *The Wealth of the Poor: managing ecosystems to fight poverty*. World Resources Institute, Washington, DC.

Wunder, S., Engel, S. & Pagiola, S. (2008) Taking stock: a comparative analysis of payments for environmental services programs in developed and developing countries. *Ecological Economics*, 65, 834–852.

Zimmerman, B., Peres, C.A., Malcolm, J.R. & Turner, T. (2001) Conservation and development alliances with the Kayapó of south-eastern Amazonia, a tropical forest indigenous people. *Environmental Conservation*, 28, 10–22.

12

Marketing and Conservation: How to Lose Friends and Influence People

Robert J. Smith, Diogo Veríssimo
and Douglas C. MacMillan

Durrell Institute of Conservation and Ecology, University of Kent,
Canterbury, UK

Introduction

A major challenge in conservation is influencing people's behaviour. Whether encouraging the public to feed garden birds or lobbying governments to tax carbon emissions, conservationists seek to maintain biodiversity by modifying human actions. This work has parallels in the private sector, where companies increase profits by influencing the purchasing behaviour of their customers (Kotler *et al.*, 1999), and this is why many conservation groups use marketing techniques pioneered in the commercial world. One such development is social marketing, which is defined as '*the systematic application of marketing along with other concepts and techniques to achieve specific behavioural goals for a social good*' (French & Blair-Stevens, 2006). However, conservation groups also use marketing in a more traditional sense, and this will be the focus of our chapter. Such marketing campaigns may have little effect on individuals' behaviour, but their impact on fundraising and setting the conservation agenda can be profound (Adams & Hutton, 2007).

When considering such campaigns, it is worth noting that many conservationists are uneasy about relying on those 'dark arts' that are also used to

Trade-offs in Conservation: Deciding What to Save, 1st edition.   Edited by N. Leader-Williams,
W.M. Adams and R.J. Smith.   © 2010 Blackwell Publishing Ltd.

sell cigarettes and soap (Schwartz, 2006). Moreover, some may feel that the conservation ethic is powerful enough without relying on glossy brochures or celebrity-endorsed campaigns. The current extinction crisis suggests otherwise. Therefore, in this chapter we will discuss the role of marketing in the conservation movement, based on the assumption that is vital both for raising funds and for publicizing issues that would otherwise be ignored by a public bombarded with conflicting messages (Foxall *et al.*, 1998). However, we recognize that these marketing campaigns can have negative effects, which partly arise because of the weak links between marketing and conservation success. So, we will also discuss these problems and the trade-offs involved when using marketing in conservation, finishing with some suggestions on how these limitations can be reduced.

Introduction to marketing

Marketing is defined as '*a social and managerial process by which individuals and groups obtain what they want and need through creating, offering and exchanging products of value with others*' (Kotler *et al.*, 1999). This process is an integral part of commerce but its importance grew in the 1960s when demand in developed countries for standardized and undifferentiated products became saturated (Baker, 2008). Companies responded by producing goods and services that were more customer oriented and developed a range of techniques to develop and advertize these products. These techniques vary and so the broad approach is often known as the 'marketing mix', which was originally summarized as the '4 Ps' of product, price, place and promotion. However, this has subsequently been expanded to the '7 Ps' by adding people, process and physical evidence, so that it better covers marketing in service industries. Most of these terms are self-evident, although 'place' refers to the distribution of the product so that it is available to potential customers and 'physical evidence' refers to the physical signs, such as the appearance and behaviour of staff, which customers use to reassure themselves about the quality of the services that they will receive (Drummond & Ensor, 2005).

Marketing and conservation

Many people are keen to conserve biodiversity but lack the time or capacity to get involved directly. Instead, they often provide financial support to conservation non-governmental organizations (NGOs) that then act as their service providers. Sometimes these organizations provide their services directly, for example by buying and managing land or organizing workshops, and sometimes they subcontract projects to local offices or other groups with specific expertise or local knowledge. In this way, they are similar to private companies in the service industry and they generally adopt similar marketing strategies. For example, they adopt conventional promotional techniques based on strong, simple messages that appeal to the target audience. In addition, they focus on developing distinct brand identities (de Chernatony, 2008), as this ensures that they capture most of the benefits of a campaign, rather than incidentally favouring other conservation groups. Such benefits vary but can include attracting new supporters or building influence with donors (Figure 12.1).

Figure 12.1 **Advertizing hoarding outside Maputo International Airport, Mozambique in 2008. The poster uses an English message in a Portuguese-speaking country to stress the financial value of wildlife and highlight the role of the international conservation NGO, the African Wildlife Foundation, in conserving it. (Photograph by Bob Smith.)**

There are, however, key differences between commercial and conservation marketing strategies that stem from the type of service being sold. First, the service cannot be designed based on customer preference alone: the conservation value of a campaign has to be considered. In contrast, commercial campaigns are designed to unlock some existing preference in the consumer, although they can create demand for a previously ignored service. Second, people responding to commercial campaigns receive services that benefit their lives directly. Thus, their purchasing decisions are based both on price and the benefits, whether physical or social, that they expect to gain. In contrast, conservation donors are often inspired by less tangible factors, such as the 'warm glow effect' that derives from moral satisfaction or praise from their peers (Andreoni, 1990). Thus, unless they fund services in their local area, they will generally receive little direct benefit from their donation and this produces two key aspects of conservation campaigns that are discussed further below.

Low cost campaigns

Consumers of commercial services accept that companies will profit from their purchase, so companies include their marketing costs within the price. These businesses may also decide to spend large sums on marketing if it raises sales or allows higher pricing of the product. In contrast, few donors are willing to contribute towards marketing costs, preferring their money to be spent directly on conservation activities. Such costs can be covered through project overheads but these also tend to be kept low because of pressures to reduce bureaucracy. Thus, conservation organizations, and NGOs more generally, spend relatively small amounts on marketing (Pallotta, 2009).

Building trust

Consumers can use a range of approaches to check product value before purchasing and a number of national laws protect them from mis-selling (Drummond & Ensor, 2005). In contrast, it is difficult for people to check whether a marketed conservation project is a genuine priority and most donors have no way of checking whether their money was used wisely. Thus, conservation organizations place a huge emphasis on building trust, as most donors rely on trusted organizations to highlight important projects and make sure their money is well spent. This is part of the reason why international NGOs continue to play such an important role in conservation, even when their involvement is limited to processing and disbursing funds to other organizations, a process that increases bureaucratic costs.

Types of campaign

There are several ways that conservation organizations aim to overcome the constraints described above. The most well known is to base campaigns on flagship species, which are '*popular, charismatic species that serve as symbols and rallying points to stimulate conservation awareness and action*' (Heywood, 1995). In this way, they reduce costs by building on existing awareness of these species and support for their conservation. These species may be threatened, have restricted ranges or fulfil important ecological roles, but they can also be selected for purely strategic reasons to maximize their impact with the target audience (Leader-Williams & Dublin, 2000). The key element is that campaigns must convey a simple message that links positive attitudes towards the flagship species with the desirability of conservation action. Thus, charismatic but potentially dangerous animals, such as elephants, *Loxodonta africana* and *Elephas maximus*, and tigers, *Panthera tigris*, may not be effective flagship species within their range countries, despite their success in raising funds from elsewhere (Kaltenborn *et al.*, 2006). Instead, local campaigns often choose more popular and relatively abundant species as flagships, as these have a higher positive profile with target communities (Bowen-Jones & Entwistle, 2002).

Another important strategy for conservation organizations is to use the news media in the campaigns. Many people are interested in conservation and so it is relatively easy to get such stories publicized (Bradshaw *et al.*, 2007), which has two main advantages. First, it uses the existing infrastructure of the news media and so is a very cheap way of spreading a message widely. Second, it builds trust in the organization by showing that independent news media consider the story important and reliable enough to be broadcast (Ladle *et al.*, 2005). This publicity can be further enhanced by using independent experts, who add authority, or celebrities, who can add credibility, if the public assume that these people would not support causes that could affect their reputation (Brockington, 2008).

All marketing campaigns must also allow for differing levels of donor knowledge and interest. This can vary widely, with some people having little initial knowledge and interest, especially when dealing with projects far from home. Thus, NGOs follow three main strategies: (i) they target a broad audience with a mass appeal campaign; (ii) they establish membership schemes and develop campaigns with an awareness-raising component; and (iii) they target wealthy individuals or organizations and tailor their campaigns accordingly.

Problems with marketing conservation

We have shown above that there are various constraints to designing an effective marketing campaign and these are compounded by conservation-specific limitations. Therefore, it is to be expected that any conservation marketing campaign can produce problems and some of these are reviewed below.

Simplification and audience validation

Most conservation issues are complicated but successful marketing campaigns are simple and appealing. Simplification is not inherently problematic: fundraising around a slogan like 'Save the rainforest!' allows organizations a great deal of freedom in designing their initiatives. However, problems can occur when campaigns simplify the project background and downplay the range of actors and their conflicting demands, aspirations and views (Brockington, 2008). It is tempting to market a project as involving conservation heroes, conservation villains and bystanders (Moore, 2010), without considering whether this portrayal may trivialize the role of some stakeholders and affect decision making (Bradshaw et al., 2007). Perhaps more dangerous is when campaigns focus on how such issues can be resolved, as this leads to the implementation of simple or generic solutions that are appealing to donors but lack input from people with local experience (Brosius, this volume, Chapter 17). Similar issues can occur when producing appealing campaigns, as these must resonate with the wishes or beliefs of the target audience

(Bradshaw *et al.*, 2007). This is particularly problematic when dealing with international audiences, as the cultural norms of the potential donors often conflict with those of the recipient countries (Doherty & Doyle, 2006).

Glamour, novelty and access

Project appeal is not just based on the views of the target audience, it also relates to the type of species and projects involved, leaving less charismatic species and more mundane projects largely ignored (Box 12.1). This has obvious funding implications but it also affects how conservation is represented and perceived. While local campaigns in developed countries focus more on people's relationship with the biodiversity that surrounds them, international campaigns often depict individuals handling, translocating or tracking charismatic species. Mundane fieldwork, like clearing alien vegetation, is ignored and there is little focus on the more quotidian activities, such as meeting with stakeholders or policy development. Thus, international conservation can be perceived by donors as a glamorous activity that has little to do with everyday life. In addition, the rise of the internet means that many people in the recipient countries are more aware of these international campaigns, which may strengthen the perception that conservationists are not interested in their lives.

Box 12.1 **Conservation news stories: what's missing?**

The news media are frequently used by conservation organizations to raise the profile of different issues, so we undertook a preliminary study to investigate the type of information publicized. We used the Google search engine to identify web pages on the BBC website in the international version of the science/nature section, using the keywords 'conservation', 'endangered' and 'threatened' and selecting the 200 web pages with the highest Google ranking. We then described each page based on its content, recording the organizations, conservation issues and taxonomic groups mentioned. We also recorded whether the organization provided the photographs used in the article, as the news media may be more likely to publish stories illustrated with attractive photographs or videos.

We found that 59 species were mentioned in the 200 articles. Twenty-one of these belonged to mammal groups that are traditionally used as flagship species, such as apes, large carnivores, elephants and rhinos. Fifteen species were mentioned more than once and 11 of these belonged to the traditional flagship groups. Some articles focused on groups of species and 27 of these groups were mentioned and they tended to cover a broader range of taxonomic groups (Table 12.1). Thus, the news media seem to discuss a wider range of species than those used in flagship species campaigns. The news stories also mentioned a range of topics, most of which either focused on a call for action or publicizing new results and discoveries (Figure 12.2). Specific conservation issues were mentioned less frequently, although many of these articles highlighted controversial issues such as whaling, international trade and trophy hunting.

Table 12.1   **Species groups mentioned in conservation news articles**

Group name	Frequency	Group name	Frequency	Group name	Frequency
Bears	1	Invertebrates	1	Primates	3
Cedar	1	Magnolias	1	Amphibians	4
Cetaceans	1	Moths	1	Butterflies	4
Chelonians	1	Vultures	1	Corals	4
Cycads	1	Bats	2	Sea turtles	4
Deep sea fish	1	Frogs	2	Birds	7
Dolphins	1	Great apes	2	Plants	8
Equids	1	Rhino	2	Albatrosses	10
Hardwood trees	1	Sharks	2	Whales	13

We found that 66 different institutions were mentioned on the 200 pages, and that 51.6% were international conservation NGOs, 21.6% were multilateral agencies, 17% were universities and 9.8% were government organizations. There were 117 articles illustrated with photographs provided by the organization mentioned in the article

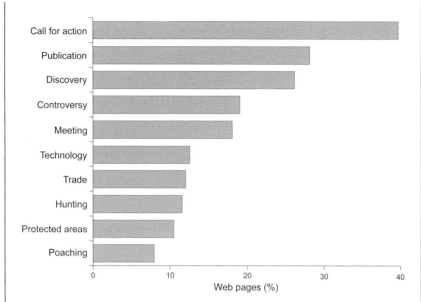

Figure 12.2 **Percentage of conservation news articles mentioning different topics (some articles mentioned more than one topic.)**

and 68% of these were provided by NGOs. The pattern of organizations probably both shows that NGOs are much more interested in publicizing their work and that the stories NGOs produce are more newsworthy. The pattern of articles provided with photographs is probably a stronger reflection of NGOs being better at providing information that will help publicize their story. Both sets of results illustrate the extent to which some groups dominate the conservation news agenda and how groups from developing countries can be excluded from such debates.

Additional problems can occur through the reliance of conservation organizations on the media to publicize their work. First, there are some issues that are much more attractive to the media because they are controversial (Webb & Raffaelli, 2008), so that strategies with near-universal support get less coverage. Second, the media are most interested in novelty (Bradshaw *et al.*, 2007). Thus, new ideas get over-promoted and old ones ignored. This is especially regrettable, given that successful conservation tends to result from the effective implementation of a suite of activities, rather than a succession of hastily conceived and implemented 'win–wins' and 'silver bullets' (McShane, 2003). Problems also occur because the growth in the news media and the fall in cost of content production have made it easy for media-savvy organizations to publicize their work (Davies, 2008). This allows groups with a minority view to have a high media profile, as long as they have sufficient support to cover their running costs. In contrast, many groups lack the capacity or connections to conduct a successful media campaign, leaving them without a voice. Thus, it is possible for small but vocal organizations to dominate the conservation agenda (Norton-Griffiths, 2007).

Distracting with doom, maintaining credibility

Conservation organizations have played an important role by publicizing the current extinction crises, so that many people appreciate the scale of the problem. However, part of this success may relate to the way in which messages about environmental collapse resonate with the general public (Brockington, 2003). Moreover, those organizations that stress the severity of the problem most effectively are likely to receive the most publicity and funding, creating conditions that favour exaggerated pessimism (Ladle & Jepson, 2008). Whilst factual errors are rare because they risk losing trust, conservation campaigns often inaccurately predict the imminent extinction of species and habitats (Ladle *et al.*, 2005; Koh *et al.*, 2010). This could be seen as a precautionary approach or a necessary tactic, given the size of the problem and the power of those supporting the status quo, but it can have three negative consequences: (i) questioning voices tend to be ignored in the face of such apparently serious problems; (ii) inappropriate strategies may be developed based on overly pessimistic assumptions; and (iii) project failure can be blamed on worsening conditions, so that poor planning or implementation are overlooked. Just as importantly, a reliance on marketing discourages organizations from publicizing such problems, discouraging them from learning from their mistakes and sharing solutions (Knight, 2006). Moreover, this status quo is maintained because poor conservation projects can still be successfully marketed, as long as they do not involve cheating the donors by diverting funds to other areas.

Reducing the limitations

There are obvious trade-offs when using marketing in conservation. One solution is to avoid marketing altogether by developing income sources that do not rely on donor funding. The opportunities for developing such schemes through taxation, trust funds, sustainable harvesting or payments for ecosystem services are expanding, but the current conservation funding gap means that donor marketing will continue to play a large role (Balmford & Whitten, 2003). Therefore, we need to develop methods and approaches that recognize the limitations and aim to reduce them. In this section we suggest three broad approaches that would help improve the impacts of conservation marketing.

First do no harm

Conservation marketing campaigns are generally viewed as benign: they raise funds and awareness for good causes. However, we have illustrated above some of the potentially negative effects such campaigns could have. Evidence for such negative effects is very limited, being largely restricted to discussions of the conservation of flagship species. For example, raising the profile of the Zanzibar red colobus monkey *Piliocolobus kirkii* is thought to have increased the blame local farmers gave to this species for crop raiding (Siex & Struhsaker, 1999). There are also suggestions that some protected areas are managed to benefit flagship species to the detriment of other species or habitats (Walpole & Leader-Williams, 2002). Thus, one could imagine similar situations where marketing campaigns have inflamed sensitive issues, creating divisions between different groups and leading to negative conservation outcomes.

There are several ways around this problem: (i) conservation marketing teams need to be more aware that their campaigns could have negative impacts; (ii) conservation groups need to make a clear distinction between what is needed for an effective marketing campaign and what is needed for effective conservation; without such clear thinking there is a danger that marketing ideas start influencing policy; (iii) when discussing marketing strategies, organizations should include input from field staff and stakeholders to develop more appropriate campaigns; and (iv) conservation groups should document the impacts of their marketing process more thoroughly, as little information is currently available. Marketing teams often assess campaigns in terms of their fundraising success but they should also consider changes to donor perception and stakeholder reaction.

Hypothecation and the benefits of being vague

Most conservation campaigns are based on the idea of hypothecation, where money raised though a campaign is used to fund the activities that address the issues mentioned. It might seem tempting to design the most appealing campaign and then spend the money on something more appropriate but this would be unethical and risks breaking the trust between donor and recipient. However, many organizations adopt half-way measures, where they make it clear that if a campaign brings in more than a specified amount then the surplus will be spent on different projects identified by the organization. This helps ensure that other less appealing projects are funded, although it does little to highlight the importance of such projects or change donor opinion in the future. Another way to reduce these impacts is to ensure that marketing campaigns are as unspecific as possible, as these campaigns allow projects to be tailored to local conditions.

Creativity in a creative industry

Many of the marketing campaigns used in conservation rely on the same old litanies about a handful of well-known species. Such a campaigning style will probably always be important because it minimizes costs by building on existing support and because it has a great appeal to some donors. These donors face requests from a range of sectors, so choose the one that appears to be most urgent and appealing. However, we would argue that more creativity is needed to broaden the appeal and impact of conservation issues and that conservationists should first think about what they want to fund and then develop an appropriate campaign. Fortunately, there are already some examples of such creativity being used in conservation. At its simplest, campaigns can use flagships species to raise funds for the broader issues that affect them. Thus, polar bears are used to raise funds for political lobbying to reduce climate change and African elephants to raise money to reduce crop raiding.

Other campaigns have gone further by focusing on species or regions that have been identified for their conservation value rather than their immediate appeal (Box 12.2). These projects are particularly exciting because they were developed with both science and marketing in mind, producing successful campaigns that overcome some traditional limitations. Such campaigns may also help widen the appeal of conservation in general, although it should be noted that some organizations have already successfully broadened their funding base. Thus, projects like Biodiversity Hotspots and the Global 200 Ecoregions have been effective at raising money from wealthy business people who are more interested in projects that stress efficiency, while other campaigns have formed alliances with big business to increase their profile and fundraising opportunities (Goldman *et al.*, this volume, Chapter 4). However, we would suggest more aspirational approaches have been generally neglected, which is perhaps surprising given that most commercial campaigns recognize the importance of this approach. Making people who support conservation feel good about themselves may work in the same way and increase funding levels from the groups who are put off by campaigns that they perceive as supporting hopeless or depressing causes.

Box 12.2 EDGE and Biodiversity Hotspots: beyond traditional flagships

The EDGE project

Flagship species campaigns aimed at international donors have traditionally been based on popular charismatic species, neglecting the many threatened species that are either poorly known and/or less attractive (Sitas *et al.*, 2009). The Zoological Society of London (ZSL) have overcome some of these limitations by launching the Evolutionarily Distinct and Globally Endangered (EDGE) project, identifying 100 mammal and 100 amphibian species that are conservation priorities based on their threat status and unique evolutionary history (Isaac *et al.*, 2007). ZSL have used a number of techniques to make this campaign more attractive to donors. First, they created the EDGE brand and emphasized that this is a novel approach that helps conserve important but neglected species. They also emphasized that the species are selected using a scientifically defensible system, helping to build trust. Second, they have given a higher profile to EDGE species that are highly charismatic, like the red panda *Ailurus fulgens*, or have an appealing but unusual appearance, like the long-eared jerboa *Euchoreutes naso*. In doing so, they help fundraise for the less appealing EDGE species. Third, the project only highlights 10 mammal and 10 amphibian species each year, allowing them to publicize new stories annually and maintain interest in the whole project. The EDGE project also fits within the institutional framework at ZSL, which is a membership organization that focuses on species conservation and scientific research.

Biodiversity Hotspots

Another approach that overcomes even more of the limitations of the traditional campaigns for flagship species is to focus on important regions. The best known example is probably the Biodiversity Hotspot scheme developed by Conservation International (CI), which has identified 34 important regions based on their high levels of plant endemism and habitat loss (Mittermeier *et al.*, 2004; Murdoch *et al.*, this volume, Chapter 3). It could be argued that CI pioneered the approach used by EDGE, in that hotspots are also marketed based on their scientific credentials and their campaigns highlight those appealing species found in hotspots, helping to fundraise for projects for other less charismatic species. The hotspot scheme also allowed for institutional factors as: (i) CI was a relatively new organization, which needed to identify a relatively small number of countries in which to work; and (ii) CI saw the opportunity to fundraise by targeting wealthy philanthropists and organizations like the World Bank, who are interested in projects that emphasize efficiency and maximizing conservation gains.

Creating new types of flagship

When it comes to designing a campaign, conservation NGOs have traditionally marketed themselves as an individual brand. The schemes described above are a radical departure from this approach, as the projects themselves are marketed as brands: each one is described as having independent value, with Biodiversity Hotspots in particular being marketed as the best system for conserving biodiversity, so that other individuals and organizations are encouraged to help conserve these species and regions (Myers *et al.*, 2000). This approach is important because it allows these schemes to be marketed in the same way as traditional species campaigns, with the main differences being that each flagship is a group of species or regions, rather than an individual species, and their appeal is based on their objectively measured conservation importance, rather than their popularity or charisma. Thus, we would argue that these schemes could be seen as a new type of flagship, as they also fulfil the criteria of serving as 'symbols and rallying points to stimulate conservation awareness and action'. Donors may choose to spend money on particular EDGE species or hotspots but the key aspect is that the marketing campaigns focus on the value of conserving the group.

Such an approach is initially more difficult because it involves building brand awareness from scratch. However, creating new flagships also has significant benefits for the organizations that develop them. First, marketing these schemes helps raise the profile of the associated organization and portrays them as being objective and efficient. Second, these flagships are linked with the organization, so they can then act as gatekeepers for dispersing funds: donors interested in tigers send money to a range of organizations; donors interested in EDGE species or hotspots generally send money to ZSL and CI (Ellison, 2008). This is obviously beneficial for the organizations involved but it also creates tensions, especially in the case of Biodiversity Hotspots that have been marketed as the most effective way of conserving biodiversity. The criticisms of hotspots have also been more vocal because the scheme was based on research published in high-profile scientific publications, leading to a lively debate in a number of conservation journals (Smith *et al.*, 2009). In particular, authors have questioned whether hotspots are as scientifically valid as claimed by their developers (Whittaker *et al.*, 2005) and expressing unease over the role of marketing (McShane, 2003).

Conclusions

Biodiversity conservation is seen by many people as a luxury, an irrelevance or a threat, despite the many benefits that it provides mankind. This has led to calls for the mainstreaming of conservation, so that different groups from all countries and sectors combine to promote conservation activities (Balmford & Cowling, 2006). Unfortunately, marketing campaigns often work against this trend because they identify the groups within society that would provide the most benefits and target their actions accordingly, often alienating other stakeholders. In this chapter, we have described the often tenuous relationship between marketing and conservation success and suggested some ways to reduce the negative aspects of marketing in conservation. These aspects are little discussed in the literature and we would argue that there needs to be greater debate about the impacts of marketing in driving funding patterns and policy development. Moreover, we think that conservation organizations need to think carefully when designing their activities, so that they explicitly consider these problems. Marketing campaigns play a key role in conservation and have the potential to play an even more important funding role in the future. However, this will depend on recognizing the trade-offs involved and developing new approaches that broaden both involvement and appeal.

References

Adams, W.M. & Hutton, J. (2007) People, parks and poverty: political ecology and biodiversity conservation. *Conservation and Society*, 5, 147–183.

Andreoni, J. (1990) Impure altruism and donations to public-goods: a theory of warm-glow giving. *Economic Journal*, 100, 464–477.

Baker, M. (2008) One more time: what is marketing? In *The Marketing Book*, eds M. Baker & S. Hart, pp. 3–18. Butterworth-Heinemann, Oxford.

Balmford, A. & Cowling, R. M. (2006) Fusion or failure? The future of conservation biology. *Conservation Biology*, 20, 692–695.

Balmford, A. & Whitten, T. (2003) Who should pay for tropical conservation, and how could the costs be met? *Oryx*, 37, 238–250.

Bowen-Jones, E. & Entwistle, A. (2002) Identifying appropriate flagship species: the importance of culture and local contexts. *Oryx*, 36, 189–195.

Bradshaw, C.J.A., Brook, B.W. & McMahon, C.R. (2007) Dangers of sensationalizing conservation biology. *Conservation Biology*, 21, 570–571.

Brockington, D. (2003) Myths of skeptical environmentalism. *Environmental Science and Policy*, 6, 543–546.

Brockington, D. (2008) Powerful environmentalisms: conservation, celebrity and capitalism. *Media Culture and Society*, 30, 551–568.

Davies, N. (2008) *Flat Earth News: an award-winning reporter exposes falsehood, distortion and propaganda in the global media*. Vintage, London.

de Chernatony, L. (2008) Brand building. In *The Marketing Book*, eds M. Baker & S. Hart, pp. 306–326. Butterworth-Heinemann, Oxford.

Doherty, B. & Doyle, T. (2006) Beyond borders: transnational politics, social movements and modern environmentalisms. *Environmental Politics*, 15, 697–712.

Drummond, G. & Ensor, J. (2005) *Introduction to Marketing Concepts*. Elsevier Butterworth-Heinemann, Oxford.

Ellison, K. (2008) Business, as usual. *Frontiers in Ecology and the Environment*, 6, 512–512.

Foxall, G.R., Goldsmith, R.E. & Brown, S. (1998). *Consumer Psychology for Marketing*, 2nd edn. International Thomson Business Press, London.

French, J. & Blair-Stevens, C. (2006) From snake oil salesmen to trusted policy advisors: the development of a strategic approach to the application of social marketing in England. *Social Marketing Quarterly*, 12, 29–40.

Heywood, V.H. (ed.) (1995) *Global Biodiversity Assessment*. Cambridge University Press, Cambridge.

Isaac, N.J.B., Turvey, S.T., Collen, B., Waterman, C. & Baillie, J.E.M. (2007) Mammals on the EDGE: conservation priorities based on threat and phylogeny. *PLoS One*, 2, e296.

Kaltenborn, B.P., Bjerke, T., Nyahongo, J.W. & Williams, D.R. (2006) Animal preferences and acceptability of wildlife management actions around Serengeti National Park, Tanzania. *Biodiversity and Conservation*, 15, 4633–4649.

Knight, A.T. (2006) Failing but learning: writing the wrongs after Redford and Taber. *Conservation Biology*, 20, 1312–1314.

Koh, L.P., Ghazoul, J., Butler, R.A. *et al.* (2010) Wash and spin cycle threats to tropical biodiversity. *Biotropica*, 42, 67–71.

Kotler, P., Armstrong, G., Saunders, J. & Wong, V. (1999) *Principles of Marketing*, 2nd European edn. Prentice Hall, Harlow, UK.

Ladle, R. & Jepson, P. (2008) Towards a biocultural theory of avoided extinction. *Conservation Letters*, 1, 111–118.

Ladle, R.J., Jepson, P. & Whittaker, R.J. (2005) Scientists and the media: the struggle for legitimacy in climate change and conservation science. *Interdisciplinary Science Reviews*, 30, 231–240.

Leader-Williams, N. & Dublin, H. (2000) Charismatic megafauna as 'flagship species'. In *Priorities for the Conservation of Mammalian Diversity: has the panda had its day?*, eds A. Entwistle & N. Dunstone, pp. 53–81. Cambridge University Press, Cambridge.

McShane, T.O. (2003) The devil in the detail of biodiversity conservation. *Conservation Biology*, 17, 1–3.

Mittermeier, R.A., Robles-Gil, P., Hoffmann, M. *et al.* (2004) *Hotspots Revisited: Earth's biologically richest and most endangered ecoregions.* Cemex, Mexico City.

Moore, L.E. (2010) Conservation heroes versus environmental villains: perceiving elephants in Caprivi, Namibia. *Human Ecology*, 38, 19–29.

Myers, N., Mittermeier, R. A., Mittermeier, C. G., da Fonseca, G.A.B. & Kent, J. (2000) Biodiversity hotspots for conservation priorities. *Nature*, 403, 853–858.

Norton-Griffiths, M. (2007) How many wildebeest do you need? *World Economics*, 8, 41–64.

Pallotta, D. (2009) *Uncharitable: how restraints on nonprofits undermine their potential.* Tufts University Press, Medford, MA.

Schwartz, M. W. (2006) How conservation scientists can help develop social capital for biodiversity. *Conservation Biology*, 2, 1550–1552.

Siex, K.S. & Struhsaker, T.T. (1999) Colobus monkeys and coconuts: a study of perceived human–wildlife conflicts. *Journal of Applied Ecology*, 36, 1009–1020.

Sitas, N., Baillie, J.E.M. & Isaac, N.J.B. (2009) What are we saving? Developing a standardized approach for conservation action. *Animal Conservation*, 12, 231–237.

Smith, R.J., Veríssímo, D., Leader-Williams, N., Cowling, R.M. & Knight, A.T. (2009). Let the locals lead. *Nature*, 462, 280–281.

Walpole, M.J. & Leader-Williams, N. (2002) Tourism and flagship species in conservation. *Biodiversity and Conservation*, 11, 543–547.

Webb, T.J. & Raffaelli, D. (2008) Conversations in conservation: revealing and dealing with language differences in environmental conflicts. *Journal of Applied Ecology*, 45, 1198–1204.

Whittaker, R.J., Araújo, M.B., Paul, J. *et al.* (2005) Conservation biogeography: assessment and prospect. *Diversity and Distributions*, 11, 3–23.

(13)

Trade-offs between Conservation and Extractive Industries

*Manuel Pulgar-Vidal[1], Bruno Monteferri[1]
and Juan Luis Dammert[1,2]*

[1]Peruvian Society for Environmental Law, Lima, Peru
[2]Department of Social Sciences, Catholic University of Peru,
Lima, Peru

Introduction

Conservation can prove a difficult issue for politicians and decision makers in many developing countries. Conservation is often only perceived as providing nebulous long-term benefits, while extractive industries such as mining and hydrocarbons are seen to contribute to short-term economic growth, for which conservation initiatives may further serve to impede progress. However, are the different goals of conservation and development really incompatible, and should trade-offs be expected between these two agendas? If so, who should represent conservation interests in a developing country? Should it be international conservation organizations, local non-governmental organizations (NGOs), public authorities, local settlers or indigenous people?

This chapter seeks to examine some of the tensions between conservation and development. With an emphasis on Latin America, and a special emphasis on Peru, we aim to show how complex it can be to face trade-offs between conservation and development. First, we outline some the tensions perceived between conservation and development in developing countries, using the

Trade-offs in Conservation: Deciding What to Save, 1st edition.   Edited by N. Leader-Williams, W.M. Adams and R.J. Smith.   © 2010 Blackwell Publishing Ltd.

example of extractive industries focused on non-renewable natural resources. Second, we set the interplay between conservation and extractive industries in its regional context within Latin America. Third, we discuss some of these issues based on specific case studies from Peru. Finally, we consider some of the challenges that conservation currently faces from extractive industries.

The challenge of sustainability for developing countries

Globalization adds complexity to the coexistence of developed and developing countries, and dramatically illustrates the challenges over considering the most appropriate spatial and temporal scales for sustainable development. Development makes an implicit assumption that developing countries wish to emulate and attain a similar development status to developed countries. However, if all developing countries followed the same growth trajectories as developed countries, and consumed resources and produced pollution at the same levels as the United States or Europe, such 'development' would quickly exceed Earth's ecological 'carrying capacity'. Indeed, human demand may well have already exceeded the biosphere's regenerative capacity (Wackernagel *et al.*, 2002). Hence, developing countries face the challenges, not only of development, but of sustainable development.

This development challenge must be achieved in a context where economic considerations appear to play the most influential role in mainstream sustainable development (Adams, 2009). The strong influence of economics in conceptualizing debates over sustainability and development arises from the application and extension of the notion of capital beyond the spheres of economics, business and finance. Consequently, it is necessary to make distinctions between human-made capital, natural capital, critical natural capital, social capital and cultural capital (Blewitt, 2008).

'Natural capital' is created by biogeophysical processes rather than by human action, and represents the ability of the environment to meet human needs, whether though directly providing raw materials, such as fish or timber, or what are called by the functionalist term 'indirect services' (Adams, 2009).[1] Natural resources and ecosystem services may in turn be subdivided into 'renewable resource capital' and 'non-renewable resource capital'. The further concept

---

[1] According to Adams (2009), the concept of natural capital has been key for the engagement of economics and environmentalist critiques to development that gave rise to ideas about sustainable development.

of 'critical natural capital' refers to those aspects of the global ecosystem upon which human lives and cultures ultimately depend, and that should be non-negotiable in discussions about development objectives.

These concepts are all related to the ideas of 'strong sustainability' and 'weak sustainability'. The idea of strong sustainability insists that there should not be any decline in natural capital over time, and that future generations should inherit the same stocks of natural resources as did their forebears (Blewitt, 2008: 5). As with so much else, policy makers, academics, sustainability practitioners and others throughout the world rarely seem to fully agree in debates over development options. Consequently, alternative sustainability conditions have been conceptualized, namely 'weak sustainability' that involves no reduction in critical natural capital, and 'very weak sustainability' in which the loss of natural capital must not be more than the increase in human capital and man-made capital (Blewitt, 2008: 5).

Strong sustainability is rarely achieved in developing countries, where it is often common to find development models that are based around the increase of human-made capital. However, such increases in human-made capital are usually achieved by depleting natural capital, for example through export of primary, non-renewable resources. In the setting of developing countries, the classic and traditional triangle of sustainability is constantly challenged by uneven confrontations, in the context of complex social issues and tensions between micro- and macroeconomic factors. Consequently, by placing special emphasis on natural capital and requiring that the stocks of both natural and human-made capital be maintained (Blewitt, 2008), strong sustainability is unlikely to appeal to grassroots environmentalists in the South facing the daily human tragedy of poverty (Adams, 2009).

The notion of weak sustainability that allows trade-offs between natural and human-made capital appears more realistic in the context of developing countries. Thus, debates tend to arise over when, how and which trade-offs must be negotiated. In other words, conservationists are constantly challenged to define what is negotiable and what is not. Furthermore, hard choices (McShane *et al.*, 2010) have to be made in a context where extractive industries are now reaching some of the most remote and biodiversity-rich ecosystems, driven by growing global demand for minerals and rapidly changing technologies and economics. Until recently, many such areas were closed to foreign investment and remained largely unexplored and undeveloped for minerals and other natural resources, thus allowing them to remain ecologically pristine. Now, economic liberalization, privatization of resource extraction and other incentives for investment in developing countries have opened up these areas to an unprecedented scale of industrial development (Rosenfold Sweeting & Clark, 2000: 7).

Developing countries rich in non-renewable resources face the enormous challenge of simultaneously addressing the demands of three different, but strong, drivers: (i) increasing global demand for non-renewable natural resources; (ii) the imperative of sustainability and wider environmental concerns; and, (iii) poverty reduction in their national contexts. Theoretically, these challenges could be addressed simultaneously by using the revenue from sustainable resource exploitation for development and poverty reduction. In practice, these challenges are really difficult to accomplish, and several trade-offs are involved.

In developing countries, development agendas are usually determined by the imperative of economic growth and poverty reduction, while conservation agendas are of much lower priority. The urgency of poverty alleviation has contributed to consolidating the hegemony of the economic growth discourse or to the 'dictatorship of economics' (Badiou, 2000: 7). The dynamics of extractive industries, conservation and development visions in Latin American countries provide an interesting perspective from which to explore these complex issues.

Development visions, extractive industries and conservation in Latin America

During the last decade, Latin America has been the scene of struggles between different development models, struggles in which extractive industries have played a crucial role. The growing profile of the Venezuelan president in regional politics has heightened tensions in relations between two blocks of countries within the Andean region. These blocks comprise the anti-imperialist and pro-socialists in Venezuela, Bolivia and Ecuador on one side, and the neo-liberal allies of Washington in Peru and Colombia on the other. The more powerful countries of Brazil, Chile and Argentina act to stabilize regional Latin American politics, influenced by different variants of development models at both ends of the spectrum.

In Ecuador, Bolivia and Venezuela, government intervention in economic affairs has increased and public oil and gas companies have been strengthened. In Venezuela, the giant oil company Petroleos de Venezuela Sociedad Anónima

(PDVSA) earned US$126 000 million in 2008,[2] and is 100% owned by the Venezuelan state. According to its website,[3] the goals of PDVSA are to promote the harmonious development of the country, to consolidate the sovereign use of resources, strengthen endogenous development and to bring about a decent and worthy existence for Venezuelan people.

The nationalization of hydrocarbons and the enormous oil profits provided by PDVSA have allowed the Venezuelan president to conduct an ambitious political project that promotes socialist goals, not only in Venezuela but across the whole region. Latin American integration is a key objective of this project, and energy is a crucial means to this end. Bolivia has been a partner in the project since the Bolivian government nationalized hydrocarbon natural resources in 2006.[4] Bolivia established that 82% of the hydrocarbon production would go to the state, and that private companies would have to accept the new law or leave the country. Since the start of the process, the public hydrocarbons company Yacimientos Petroleros Fiscales Bolivianos (YPFB) has signed several co-operation agreements with PDVSA. In the case of Ecuador, the government established in 2006 that oil companies had to pay the state 50% of their extraordinary incomes, a percentage that was raised to 99% in 2007.[5] As in the case of Bolivia and Venezuela, many conflicts arose between hydrocarbon companies and the Ecuadorian government.

Extractive industries are also important for the more liberal countries in Latin America. Although more supportive of transnational private investments, these countries have large public hydrocarbons companies. Historically in Chile, the most important source of national income has been the world's biggest copper company (Corporación Nacional del Cobre, CODELCO), which was nationalized under the socialist government in 1971. In Brazil, the state company PETROBRAS, has a market value of US$173 000 million and has achieved its goal of supplying Brazilian domestic oil needs, and is planning to invest $174 000 million between 2009 and 2013.[6]

The huge profits that the pro-socialist countries have generated by nationalizing their mining and hydrocarbon operations have financed more pro-poor development strategies (CIDSE, 2009) and political projects directed at

---

[2] http://www.aporrea.org/energia/n137011.html.
[3] http://www.pdvsa.com/.
[4] Bolivian Supreme Decree 28701 (2006).
[5] A new legal proposal that will regulate hydrocarbon activities is under debate. More information is available at www.minasypetroleos.gov.ec.
[6] http://www.petrobras.com.

implementing socialist policies in the region. However, this grand project design has met resistance at different scales in various countries, and been critiqued for a number of reasons, including: (i) supporting rentier states; (ii) through constant denouncements that follow a strongly authoritarian approach; (iii) through reducing the rights of the public to freely express their opinion; and (iv) a lack of transparency. In contrast, liberal and often right-wing governments have been accused of giving away natural resources on behalf of global private interests, thereby favouring small national elites to the detriment of the rights of the poor in the name of a false democracy and freedom.

This chapter does not seek to enter a geopolitical discussion on different economic models and political systems, but rather to offer a panoramic view of the ground across which conservation agendas are set with reference to extractive industries. Within their different national contexts, environmental concerns are shaped by political issues. In some cases, the objective could be to protect ecosystems and local settlers from careless transnational companies. In other cases, it could be to advance exploitation in pristine areas in the name of national sovereignty. Furthermore, it is difficult to strongly criticize oil companies without being accused of attacking the model or defending socialist penetration. Anyway, whether trying to advance national participation in the benefits of resource extraction, or in the values of liberal democracy and free markets, the imperative of economic growth is the key issue for whatever model is followed in the region, and so pressure on ecosystems will continue to increase (see Figure 13.1).

In this context, public policy debates generally pay only marginal attention to conservation objectives. On the other hand, the extraction of mineral and hydrocarbon resources is of central policy and economic importance. Any recent increases in concern for nature and biodiversity are still not as strong as the traditional, longer lasting emphasis placed on extractive industries in Latin America.

The exploitation of natural resources is considered the quickest and easiest path to attain economic growth. For this reason, temporal criteria acquire special relevance when navigating trade-offs between conservation and extractive industries, given that short-term criteria in favour of economic decisions are the rule rather than the exception. Electoral interests are also an important factor to consider. Thus, hunger for quick profits diminishes the possibilities

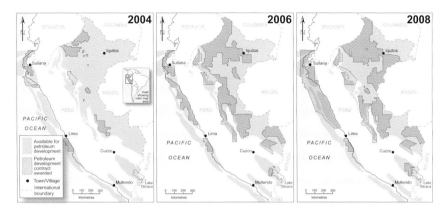

Figure 13.1 **Advance of hydrocarbon operations in Peru. (Source: www.perupetro. com.pe, accessed under Public Information Access and Transparency Law, Law 27806 (Articles 3, 5, 7 and 15).)**

of a responsible design of the extractive process in terms of environmental and social care. According to CIDSE:

> 'The current development model in Latin America promotes and prioritizes eco- nomic growth above other dimensions of development, notably environmental sustainability, equity, social justice and respect for human rights. The push for accel- erated world economic growth has led to increasing demand for and pressure on nat- ural resources such as minerals and other primary materials. As a result, companies have increasingly exerted pressure on States to open up territory to feed the expansion of the world economy. This has put pressure on what are often fragile environments and vulnerable people who share their land with minerals and energy sources'
>
> (CIDSE, 2009: 2).

There are cases, however, where alliances are established between conservation organizations and extractive companies. For example, Conservation Interna- tional in Suriname (CI-S) has enjoyed some notable success in engaging large mining companies to participate in environmental initiatives that go beyond their commercial operations, and to reinforce landscape conservation strate- gies. CI-S has trained staff from mining companies to use camera traps for assessing and monitoring the impact of their exploration and mining activities on large mammals.

> 'Suralco and BHP Billiton also funded a rapid assessment expedition in one of their concessions as part of their assessment process. They decided not to mine that area, in part because of the endemic species discovered there. The companies have showed their willingness to exclude areas from their activities and will propose to the government to give those areas a protection status. With the support of Suralco, Brownsberg Natural Park was created by the government within their mining concession'
>
> (Nelson, 2009: 282).

Despite the cultural, legal, political, economic and social differences among Latin American countries, we believe that making explicit trade-offs between extractive industries and conservation can lead to better decisions and strategies. The trade-offs that require to be navigated include those between short-term economic growth and conservation, but also between energy supply and conservation, as well as several other issues. In any case, economic considerations are not the only factor that should be taken into account, and decision models and tools that include economical, ecological, social and political factors are required.[7] However, the problem seems to be that ethical values and moral discourses are rarely included as part of the tools used for decision-making processes.

In the next section we discuss some of these issues based on Peruvian experiences. Peru is firmly engaged with a free trade approach, and its economic growth is highly dependant on extractive industries. Since the government and business sectors are not particularly concerned about conservation of biodiversity relative to commercial concerns, civil society and international pressures are of key importance for achieving conservation goals.

Conservation and extractive industries in Peru

Peru is one of Earth's 17 mega-diverse countries, and is very rich in biodiversity. Peru has a surface area of $1\,285\,216\,km^2$, making it the 20th largest country in the world. It supports a human population of $\sim$28 million inhabitants, of which 74% live in urban areas and 26% in rural areas. Peru is currently

---

[7] For more information regarding Advancing Conservation in a Social Context (ACSC) visit www.tradeoffs.org.

experiencing an economic boom that has only slowed as a result of the recent global financial crisis. Thus the growth in annual GDP in Peru dropped from 9.84% in 2008[8] to an estimated 1.8% in 2009 (BCR, 2009). For over a decade, Peru has been committed to a neo-liberal development model, even though its constitution promotes a social market economy. In turn, this suggests that the national government believes that only private investment will allow the country to advance, and so places little emphasis on statecraft and long-term planning. Hence, any participatory consensus over conditions and strategies followed to promote environmental conservation can be undermined by the conditions and requisites imposed by investors.

Recent economic growth has not significantly contributed to reducing social inequalities within Peru. Since Peru's economy is based upon primary export commodities, its growth merely reflects the high prices of minerals on the international market.[9] However, extractive industries do not usually generate productive linkages, nor do they require huge amounts of manual labour. Throughout its history, Peru has experienced successive periods of boom-and-bust, thanks to the extraction and primary export of resources such as guano, niter, copper, rubber, zinc and, most recently, gold (Bebbington & Bury, 2009) (Table 13.1). Furthermore, economic booms do not last long and generally result in little re-investment in the future of the exporting country (Castro de la Mata, 2005). Due to unsustainable use of its resources, and low commodity prices, the periods of prosperity have never lasted long, and advantage was not taken of them while they did last. As a result, the steady loss of natural capital has not resulted in a significant increase in human-made capital, especially in rural areas, where ~75% of people are poor, and ~40% of people are extremely poor.[10] Hence, tensions between conservation and development must be analyzed through the lens of redistributive conflict (SPDA, 2008).

---

[8] www.inei.gob.pe.

[9] The increasing prices of minerals and oil have had several consequences, including: (i) increased pressure to explore and exploit oil and minerals within any area of Peru, whether natural protected areas, conservation concessions, lands of indigenous communities, timber concessions or ecotourism concessions; (ii) more public policies that promote extractive investment through weakening environmental and social considerations; (iii) increasing conflicts over pollution, social and economic benefits, or political decision making; (iv) more conflicts over exploring and exploiting oil and minerals in conservations areas; and (v) improved environmental and social practices among extraction companies, agreed through direct 'negotiation' with local settlers, while leaving the government as an outside observer.

[10] Source: National Statistics and Informatics Institute (INEI, 2004), *National Homes Survey 2004*. The measurements of poverty in Peru can be problematic since the criteria used vary according to the year and the responsible institution. In this case, we state the official amount reported by the Peruvian government, handed to us by INEI.

Table 13.1 **Peru as a world leader in mining exports, based on data for 2008. (Source: Banco Central de Reserva del Peru and Sociedad Nacional de Minería Petróleo y Energía. Available at: www.snmpe.org.pe)**

Mineral	Annual exports for 2008 (US$ million dollars)
Cooper	7662
Gold	5529
Zinc	1465
Silver	594
Iron	377
Lead	1131

Decision making and conflict resolution over the use of natural resources are generally achieved on the basis of negotiations, pressure and resistance between the state, local and foreign investors, conservationists, civil society and local stakeholders. The diversity of interests and visions, the weak presence of the state in areas where natural resources are most rich, and the institutional weakness in establishing and enforcing coherent policies based on a long-term vision are key issues (Bebbington & Bury, 2009). In many cases, this simply generates scenarios in which the text of the law is just referred to in negotiations, while law enforcement itself is seriously lacking. Thus, the agendas for conservation and extractive industries, which many see as equivalent to development, run along separate and often antagonistic paths. In the case of hydrocarbons, for example, as the international price of oil rises, and the danger of a global energy crisis looms closer, the number of petroleum lots in Peruvian territory has increased exponentially. Indeed, as a direct consequence of a policy to promote investment in hydrocarbons, the coverage of oil fields across the Peruvian Amazon has greatly increased since 2004 (Figure 13.1). Likewise, there has been an exponential increase in the numbers of mining concessions granted in recent years (Bebbington & Bury, 2009). Inevitably, oil fields and mining concessions overlap with natural protected areas, ecotourism concessions, conservation concessions and the lands of indigenous peoples, which can result in serious and explicit conflicts (Bebbington & Bury, 2009). The irony is that different sectors of government grant the same area to investors to develop incompatible activities. Unless, that is, birdwatchers start enjoying seeing oil infrastructure and pollution (Figure 13.2) in the background to their pictures!

Figure 13.2 **Aerial view of petroleum pollution flowing from Quebrada Piedra Negra into River Tigre in Intuto, Loreto Region, Peru in 1995. The source of the pollution was an oil field managed by the Occidental Petroleum Company that left Peru under claims of Peruvian citizens and international pressure. Despite improving environmental standards for oil exploration and exploitation in the Amazon, pollution and conflicts still tend to occur. (Photograph by kind permission of Thomas Mueller.)**

The government's position on natural resources

The Peruvian government has promoted investment in the mining and hydrocarbon sectors as a political priority (see also Bebbington & Bury, 2009). Between late 2007 and early 2008, the president of Peru published a set of three articles known as 'The dog in the manger' articles.[11] These sought to provide arguments in support of the investment policy over the

---

[11] 'The dog in the manger' is a fable concerning a selfish dog that cannot eat but will not let others eat. The expression in Spanish, *El Perro del Hortelano, no come y no deja comer*, is very commonly used to express this idea in diverse contexts. The articles were published in *El Comercio*, Peru's most prestigious newspaper, on 28 October 2007 and are available at: http://www.elcomercio.com.pe/edicionimpresa/Html/2007-10-28/el_sindrome_del_perro_del_hort.html.

sustainable management of natural resources, minimizing rather than recognizing environmental and social considerations, and calling those who promote such considerations as 'dogs in the manger'. The articles mostly featured forests and issues of land, mining and hydrocarbons in the Amazon, and the seas.

It is important to recognize the president's initiative to open a public debate over the use of natural resources, and to express the government's viewpoint over how these should be used. However, when referring to environmentalists as communists in disguise, the president contributed to polarizing debates between the positions of the actors involved. Equally, it is common in countries of the global South, such as Peru, to promote co-operation between high echelons of the government and private foreign companies over the exploitation of natural resources within national territory (Martínez Alier, 2009).

In the first article[12] the president criticized the obstacles for private investment and concluded that there were several resources in Peru that were not used, thus attracting no value (García, 2007). In this sense, the president promoted the granting of large areas of the Amazon rainforest to big companies in order to promote investment and productivity, as well as to construct dams in the Amazon, and to establish aquaculture in the seas.

Several ideas and proposals contained in the president's articles were later included in legislative decrees passed by the Executive Power as part of the implementation process of the Free Trade Agreement with the United States. However, some of these decrees mobilized Amazonian native communities, who claimed that the laws passed affected their rights, including the right to prior informed consent considered in Convention 169 of the International Labour Organization. As a result of the tensions between indigenous people and the government, more than 30 persons died in Bagua during the most violent moment of the conflict and four of the legal decrees passed by the government are no longer in force. This situation was a potential turning point for the role of indigenous people within the Peruvian political arena. We now discuss three cases where there have been conflicts between extractive industries and conservation in Peru.

---

[12] http://www.elcomercio.com.pe/edicionimpresa/html/2007-10-28/el_sindrome_del_perro_del_hort.html.

Majaz mining project

The environmental constraints that Peru faces in imposing good governance is well illustrated by the Rio Blanco mining project, more familiarly known under the name of the mine. Majaz lies in the north highlands of Peru, in the buffer zone of the Tabaconas Namballe National Sanctuary, which was established to conserve spectacled bears *Tremarctos ornatus*, mountain tapirs *Tapirus pinchaque* and *Podocarpus* forests. The mining project has faced strong opposition from local people since mining operations began. Opposition has included a local referendum, in which approximately 93% of the people expressed their opposition to the mine because of its potentially negative environmental impacts. However, the referendum was not binding and the national government continued to support the mining project.

After the Majaz case had appeared across all the front pages of national newspapers, a process of dialogue was announced by the government. However, it is not clear what solution will be negotiated, since the local population sees no solution other than an end to the mining operation, while the mine and the government see the dialogue process as a way of convincing the local population of the benefits of the mine, even though the local population is unwilling to change its position.

In 2007, the Chinese mining company Zijin bought the mine from the British company Monterrico Metals. The commercialization of the mine stocks on the financial market has added to the large profit expected from the mining operation, which in turn represents serious pressures on the government to continue supporting the mine. As a result, the government refuses to accept the local referendum result and has decided to continue with the mining project. Indeed, the government has promoted a legal instrument to declare 20 mining projects as being in the national interest, including Majaz, which clearly illustrates its concept of democracy.

The case took an even more dramatic twist when cases of torture of local settlers by mine personnel came to light in 2008. It is unfortunate that the government of Peru accuses people concerned about environmental impacts of being 'against investment' and 'opposed to development'. However, it was left to the UK, as the country of origin of the former mine owner, Monterrico Metals, to take action over the cases of torture. A London court ordered the freezing of more than UK£5 million from the accounts of Monterrico Metals in 2009, in order to guarantee the reparation payments to the 29 torture victims in the mining camp in 2005. This action of British justice contrasts with that of the Peruvian justice system, unable so far to hold to account any mine officer or their security staff responsible for these events.[13]

Following Sklair (2003), García Llorens (2009) has identified in the Majaz case the existence of a new transnational capitalist class, based on a triple alliance between the local bourgeoisie, transnational corporations (Zijin mining company) and the pro-capitalism governmental bureaucracy (the president and the agencies with executive power). According to Sklair (2003), this transnational capitalist class is more evident *'particularly in those societies marked by foreign investment oriented to exportation'* as in Peru (Sklair, 2003: 171). In the future, a complex negotiation process can be expected as mistrust has become widespread. In this case, as in most cases related to extractive industries, conservation is only a matter of concern for the local population and some NGOs, while extractive industries receive support from the private sector, most of the press and the government.

Hydrocarbons within natural protected areas

The Sierra del Divisor is part of a complex of mountains that rise spectacularly from the middle of the Amazonian plain, a long way from where the easternmost Andes end, on the Peruvian border with Brazil. Native communities know the area as Siná Jonibaon Manán, the 'Land where fierce men live' (Vriesendorp *et al.*, 2006). This unique and isolated geological formation supports different climates across its altitude range, which in turn have support a wide range of species including endemics. A rapid biological inventory Vriesendorp *et al.*, 2006) found 18 species of primate, including the bald uakari *Cacajao calvus* and Goeldi's marmoset *Callimico goeldi*, both considered vulnerable according to the list of Threatened Species of Wild Fauna in Peru.[14] Moreover, 300 species of birds were documented, many of which are extremely rare and little known, including the acre antshrike *Thamnophilus divisorius*, recorded for the first time in Peru.

For these reasons, the Sierra del Divisor was declared a Reserved Zone, which provides a transitory status of natural protected area until the information

---

[13] For more information visit http://www.guardian.co.uk/world/2009/oct/18/peru-monterrico-metals-mining-protest and http://www.todosobremajaz.com/.

[14] Approved by Supreme Decree No. 034-2004-AG, published 22 September 2004.

required to define its final category and extension is to hand. However, the Sierra del Divisor Reserved Zone faces trade-offs between the different visions of conservation and development as it is superimposed over the indigenous Isconahua Territorial Reserve and also contains hydrocarbon blocks and mining concessions. The categorization process for the Sierra del Divisor began in April 2006, but the outcome is currently unknown. Among the interests that come together in the area are: indigenous and conservation organizations; oil, mining and timber companies; infrastructure projects; and local non-indigenous settlers. Although the technical process determined the area should be categorized as a national park, this has not happened due to the reluctance of the energy sector.

The different positions can be simplified as follows. Conservation organizations promoted its categorization as a national park, a category of natural protected area that does not allow the extraction of natural resources. Meanwhile, indigenous organizations manifested their interest in establishing two indigenous reserves within the area to protect isolated indigenous people. In contrast, the government's mining and energy sectors proposed the category of national reserve, which allows hydrocarbon activities. Even though most of the actors in the categorization commission supported the national park option based on technical information, the area was not categorized because the mining and energy ministry blocked this alternative.

This case illustrates the chasms that exist between the interests of conservation and indigenous organizations, and the importance of promoting alliances between these actors when more powerful stakeholders sit at the negotiation table. Moreover, Sierra del Divisor shows the shortcomings of participation in the categorization processes when explicit power issues are involved. So far, the categorization process only appears to be a participatory performance, while in parallel and unilaterally the outcome is preordained no matter what, given the presence of the extractive industries in the area (Monteferri et al., 2009).

In contrast, the Bahuaja Sonene National Park (BSNP) is a flagship of the Federally Protected National System of Natural Areas (SINANPE) in Peru. The national park was established for the conservation of threatened species such as the maned wolf Chrysocyon brachyurus and marsh deer Blastocerus dichotomus, and of unique ecosystems such as the hidden valley of Candamo and the wet savannas of Pampas del Heath. In 2002, BSNP was declared by the National Geographic Society as one of the most emblematic natural sanctuaries in the world. During 2007, a legal initiative that proposed the reduction of the core zone of BSNP by some 2098 km^2 was discussed before the Council of Ministers. The proposal was then sent to parliament for debate where it was likely to win approval, given its objective was to facilitate the use of hydrocarbons found in the area, at a time when the price of oil was higher than US$100 per barrel. Approval was likely, despite the technical opinion of the national authority of protected areas responsible for SINANPE, which was against the proposal. However, the Peruvian government withdrew the proposal because the US threatened not to ratify the Amendment Protocol of the Free Trade Agreement. Nevertheless, the proposal could re-surface in future if oil prices rise.[15] At the outset, the highest echelons of the Peruvian government have expressed their support for reducing the size of BSNP in order to facilitate oil exploitation.[16]

As a part of the strategy to defend BSNP from being reduced in size, an exercise was conducted based on the value of the ecosystem services in the national park and the benefits it provides. The objective was to make explicit the trade-offs between immediate and long-term benefits. The valuation exercise showed the practical problems of fully capturing the value of ecosystem services in commercial markets, as well as of adequately quantifying the value of the services provided (see Costanza *et al.*, 1997, 1998). According to the exercise, the total economic valuation of BSNP lay in the range of US$254 million to 2399 billion dollars (Barrantes & Cardenas, 2008). The lack of detailed information on the biodiversity of the area was a major limitation of the valuation exercise, and a limitation that is commonly faced when undertaking valuation exercises in developing countries where research on biodiversity is not a priority (León, 2007). Nevertheless, using various techniques, the exercise recognized the direct and indirect use values, and non-use values, of the ecosystems and natural resources in BSNP. This work stimulated debate regarding how strategic it was to value environmental services in this specific case, and how the valuation exercise was done. Similar debates have taken

---

[15] Civil society responded immediately against the proposal and many people raised their voices in protest. The debate remained in the public eye for over 2 weeks. Finally, a US Congressman sent a letter to the Peruvian Ambassador to the US, which established that the proposal to reduce BSNP was against the Amendment Protocol of the Free Trade Agreement because it reduced environmental legal standards in order to promote investment.

[16] The removal of the Protected Natural Areas Council Officer who opposed the reduction in size of BSNP is a disturbingly clear sign of this interpretation of government intentions. Before taking the proposal to reduce BSNP to the Ministers Council, the Agriculture Minister first sent it to the Protected Natural Areas Council Officer, who rightly observed that it threatened biological diversity, violated national legislation and jeopardized ongoing negotiations of the Free Trade Agreement with the USA.

place elsewhere, since the ecosystem service value of the entire biosphere was first estimated to be worth US$33 trillion per year (Costanza *et al.*, 1997).

So what can be learned about how decisions are made concerning extractive industries and conservation in Peru? In its broader context, there is strong opposition towards protected areas, conservationists and the indigenous communities who inhabit them. All these constituencies are seen as obstacles to investment in extractive industries. In ensuing offensives, the government maligns those who oppose specific projects promoting mining and hydrocarbon interests, as terrorists, communists, antisystemic and poverty negotiators. Ironically, in defending national resources, the patriotism of the critics is also questioned and the accusations include betraying Peru's aspirations for development and, consequently, defending Chilean interests.[17] Consequently, the powerful *anti-ecologist* lobby has to be faced down by a united conservation lobby primarily mobilized from the global South (Martínez Alier, 2009).

Conclusions

Most biodiversity-rich countries are also developing countries, and so face particular concerns about alleviating poverty and integrating themselves with the global economy. The priority accorded by developing countries to economic growth places disproportionate power in the hands of economists, who in turn reinforce the paradigm of economic growth in development (Adams, 2009). However, while the economy is the sector that shapes the mainstream of sustainability, not all economic approaches promoted in developing countries follow a sustainable approach. The urgency of alleviating poverty allows easy hegemony that the most important idea is economic growth – which is the only, or fastest, way of accomplishing this goal – through comparative advantages, in many cases through exploitation of natural resources provided by nature.

The greatest challenge for conservation in developing countries is to build legitimacy and consensus for the importance of conservation initiatives within development policies, based on ethical, economic, social and ecological concerns for the future of wild nature. New policies and actions are required to avoid polarized debates over deciding how to use or misuse natural resources.

---

[17] Chile lies at the southern border of Peru and has long been considered its most serious security threat. A war between the countries at the end of the 19th century ended with a Chilean invasion of Peruvian territory. The pro-Chilean argument was mentioned by the current president himself.

To accomplish this challenge, our experience has stressed the importance of explicitly acknowledging the benefits, risks and costs of the trade-offs between conservation and extractive industries. Hard choices often need to be made and these must be openly and honestly negotiated (McShane *et al.*, 2010). Not to do so leads to unrealized expectations and, ultimately, to unresolved conflict.

Power relationships can involve people and organizations from different cultures and world views, and from different professions and geographical zones (Brosius, this volume, Chapter 17). People can hold different priorities, interests and visions over what to accomplish in a specific area, how and by whom. In this context, conservation initiatives represent one among various alternatives. However, in cases where there are very asymmetric power relations, the only way for conservation organizations to move forward appears to be through political struggle and activism, rather than by acknowledging and negotiating trade-offs.

Conservation will not be always the winner. Negotiations will need to take place and sacrifices made. The overall goal will be to distinguish between those situations that are negotiable and non-negotiable, to obtain the information to sustain these positions and to be more efficient when implementing conservation actions. This last point will require important modifications to the design of conservation projects, especially of those aspects related to social issues and the dissemination of project outcomes. All projects must be systematically assessed, including those that fail, and the results disseminated in order to improve the design of future projects (Kapos *et al.*, this volume, Chapter 5).

Thus, the relationships between funding agencies, intermediate organizations, public agencies and local settlers are driven to evolve in order for conservation to adapt to the new context. Legitimacy is crucial. As new actors and interests become involved, new mechanisms and alliances must be developed to face the challenges in the form of opportunities instead of conflict situations.

References

Adams, W.M. (2009) *Green Development: environment and sustainability in a developing world*, 3rd edn. Routledge, London.

Badiou, A. (2000) *Movimiento Social y Representación Política*. Acontecimiento No. 19-20. Available at: www.grupoacontecimiento.com.ar.

Barrantes, R. & Cardenas, M.K. (2008) *Valoración Económica del Parque Nacional Bahuaja Sonene*. Sociedad Peruana de Derecho Ambiental, Lima, Peru.

Banco Central de Reserva del Peru (2009) *Reporte de Inflación 2009. Panorama actual y proyecciones macroeconómicas 2009–2011*. Available at: http://www. bcrp.gob.pe/docs/Publicaciones/Reporte-Inflacion/Reporte-Inflacion-23-Setiembre-2009/Reporte.pdf.

Bebbington, A.J. & Bury, J.T. (2009) Institutional challenges for mining and sustainability in Peru. *Proceedings of the National Academy of Sciences of the USA*, 106, 17296–17301.

Blewitt, J. (2008) *Understanding Sustainable Development*. Earthscan, London.

Castro de la Mata, G. (2005). *Un Mendigo Sentado en un Banco de Oro: reflexiones sobre desarrollo y medio ambiente en el Peru*. PROFONANPE and WWF, Lima, Peru.

CIDSE (2009) *Impacts of Extractive Industries in Latin America: analysis and guidelines for future work*. Available at: http://www.cidse.org/uploadedFiles/Regions/Latin_America/EPLA%20analysis%20final%20ENG.pdf.

Costanza, R., d'Arge, R., de Groot, R. *et al.* (1997) The value of the world's ecosystem services and natural capital. *Nature*, 387, 253–260.

Costanza, R., d'Arge, R., de Groot, R. *et al.* (1998) The value of the world's ecosystem services: putting the issues in perspective. *Ecological Economics*, 25, 67–72.

García, A. (2007) El síndrome del Perro del Hortelano. *Diario El Comercio*, 28 October.

García Llorens, M. (2009) *La democracia extractiva: estado, corporaciones y comunidades en el caso Majaz*. Manuscript, Consejo Latinoamericano de Ciencias Sociales, Buenos Aires, Argentina.

INEI (Instituto Nacional de Estadística e Informática) (2004) *Encuesta Nacional de Hogares 2004*. Available at: www.inei.gob.pe.

León, F. (2007) *Aporte de las Áreas Naturales Protegidas a la Economía Nacional*. Instituto Nacional de Recursos Naturales, Lima, Peru.

Martínez Alier, J. (2009) *El Ecologismo de los Pobres: conflictos ambientales y lenguajes de valores*. Icaria, Barcelona.

McShane, T.O., Tran Chi, T., Songorwa, A.N. *et al.* (2010) *Hard Choices: making trade-offs between biodiversity conservation and human well-being*. Biological Conservation, in press.

Monteferri, B., Canziani, E., Dammert, J.L. & Silva, J.C. (2009) *Conservación, Industrias Extractivas y Reservas Indígenas: el proceso de categorización de la Zona Reservada Sierra del Divisor*. Cuaderno de Investigación No. 2. Sociedad Peruana de Derecho Ambiental, Lima, Peru.

Nelson, R. (2009) Private and community-based conservation in Suriname. In *La Conservación Privada y Comunitaria en Los Países Amazónicos*, eds B. Monteferri & D. Coll, pp. 257–286. Sociedad Peruana de Derecho Ambiental, Lima, Peru.

Rosenfeld Sweeting, A. & Clark, A.P (2000) *Lightening the Lode: a guide to responsible large-scale mining*. Conservation International, Washington, DC.

Sklair, L. (2003) *Sociología del Sistema Global: el impacto socioeconómico y político de las corporaciones transnacionales.* Gedisa, Barcelona.

SPDA (Sociedad Peruana de Derecho Ambiental) (2008) *Cuestión de Perspectiva No. 1.* Sociedad Peruana de Derecho Ambiental, Lima, Peru. Available at: http://www.spda.org.pe.

Vriesendorp, C., Schulenberg, T.S., Alverson, W.S., Moskovits, D.K. & Rojas Moscoso, J-I. (eds) (2006) *Perú: Sierra del Divisor.* Rapid Biological Inventories Report No. 17. The Field Museum, Chicago.

Wackernagel, M., Schulz, N.B., Deumling, D. *et al.* (2002) Tracking the ecological overshoot of the human economy. *Proceedings of the National Academy of Science of the USA*, 99, 9266–9271.

(14)

A Fighting Chance: can Conservation Create a Platform for Peace within Cycles of Human Conflict?

Rosalind Aveling[1], Helen Anthem[1] and Annette Lanjouw[2]

[1]Fauna & Flora International, Cambridge, UK
[2]Arcus Foundation, Cambridge, UK

Introduction

Conflict occurs as a result of actual or perceived differences in needs, values and interests (Matthew *et al.*, 2009), and can lead to a struggle for power, including claims over resources. Consequently, human conflict over land and natural resources will increase as an almost inevitable consequence of global population growth and declining economic performance, overlaid by the effects of a rapidly changing climate. Pressure points are emerging over land values, over forest, marine and water resources, as well as over extraction sites for minerals and fossil fuels (Pulgar-Vidal *et al.*, this volume, Chapter 13). Any insights gleaned on achieving conservation in conflict scenarios could be critical to retaining Earth's biodiversity and ecological balance over the long term.

Wider reviews (Dudley *et al.*, 2002) and more detailed case studies, for example from former Zaire (Hart & Hall, 1996) and Cambodia (Loucks *et al.*, 2009), document the outcomes for wildlife populations and habitats experiencing armed conflict. Furthermore, Hanson *et al.* (2009) show that over 80% of major armed conflicts between 1950 and 2000 occurred within

Trade-offs in Conservation: Deciding What to Save, 1st edition.   Edited by N. Leader-Williams, W.M. Adams and R.J. Smith.   © 2010 Blackwell Publishing Ltd.

Biodiversity Hotspots (Murdoch *et al.*, this volume, Chapter 3). Therefore, conservationists will need to increasingly confront the trade-offs of operating in conflict situations. Using insights drawn from experience gained over many years in Africa and Asia, this chapter examines characteristics of armed conflict, its interaction with natural resources and the potentially positive role of conservation in some conflict scenarios. First, we describe the impact of prolonged armed conflict on the conservation of natural resources. Second, with reference to two case studies, both of which are in areas of high biodiversity value, both of which have charismatic flagship species around which conservation action can rally, we draw out some of the trade-offs that have to be negotiated in conflict situations. Third, we use both cases to demonstrate how a focus on persistence of the natural environment and its resources can help achieve stability through changes of governance and even contribute to maintaining peace.

Characteristics of armed conflict

The armed conflicts that we consider here have a political element, resulting in the disruption of the rule of law and civil authority. This does not include the generally more localized conflicts where traditional or judiciary authority persists and can be applied to resolve the conflict. An example of this latter type of conflict would be between communities over water use at a water-hole, or between communities, government and corporations over the construction of a hydroelectric dam. It is also important to consider what is meant by the widely used term 'post-conflict'. This term suggests a linear progression, from pre-conflict to conflict to post-conflict, with different challenges and different types of assistance required for each stage of progression (UNDP, 2004). However, in reality, the nature of modern conflicts means that they are not so easily compartmentalized. Consequently, unambiguous post-conflict situations are rare. Indeed, some of the constraints faced in conventional development settings – including corruption, weak civil society and shortages of skilled professionals – cannot be isolated as challenges peculiar to post-conflict situations, but are common to many other situations (UNDP, 2004) in which conservationists seek to implement projects.

Conflicts often set a downward spiral in motion, resulting in increasing resource degradation (Figure 14.1). Conflict disrupts peoples' lives, preventing them from working their fields, accessing clinics and hospitals, and going

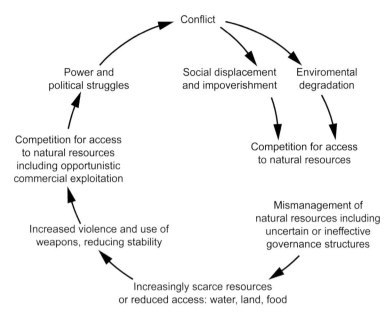

Figure 14.1 **Conflict and resource degradation: a downward spiral. (After Lanjouw, 2003.)**

to work. This can further exacerbate trade-offs between conservation and poverty (Roe & Walpole, this volume, Chapter 9) and lead to problems that will persist long after hostilities cease. In the Democratic Republic of Congo (DRC), 0.4 million deaths since 1998 can be directly attributed to violence. The majority of the countrywide total of 5.4 million 'excess deaths' (i.e. those in excess of regional norms) reported since 1998 have been attributed to infectious diseases, malnutrition and neonatal and pregnancy-related conditions (Coghlan *et al.*, 2007), all of which are severely exacerbated as a consequence of existing conflicts. Indeed, famine frequently kills more people than violence during armed crises. However, in many cases, famine occurs as a direct result of the war (Sen & Dreze, 1989), so subsequent deaths are an indirect consequence of the war.

A key distinguishing feature of modern conflicts is the involvement of civilians (Figure 14.2), both as combatants and as targets. In World War I, approximately 5% of casualties were civilians. In World War II, the figure approached 50%. However, in contemporary conflicts, up to 80% of casualties

Figure 14.2   **Mines laid around protected areas continue to pose a danger to civilians and an impediment to conservation or development. As here in Cambodia, and along the boundary of the Parc Nacionales de Virunga in Rwanda, mines have been cleared by United Nations and military munitions disposal teams, who work closely with protected area staff. (Photograph by kind permission of FFI.)**

are civilians, many of whom are women and children (Ingram, 1994, in Lanjouw, 2003). There are around 640 million small arms and light weapons in the world today and 8 million more are produced every year.[1] The increasing availability of arms, and the ease with which they can be obtained, contributes to increasing civilian casualties in modern conflicts. Two other key features of

---

[1] Control Arms (2005), at http://www.controlarms.org/find-out-more/the-issues.

modern conflict are that the majority are national or internal, although they may not remain within the confines of one country. They are also driven by a wide-ranging combination of factors, including access to resources, ethnicity, ideology, weak states and poor leadership (Shambaugh *et al.*, 2001).

Conflict situations differ in type, cause, speed of onset, scale and impact. While the contexts vary, some common characteristics can be identified around economic, political and social features. In economic terms the following can be expected: high transaction costs; increased illicit activities; high unemployment; high national debt; and a weak human resource base. Politically, features of conflict include: often limited national government resources and capacity; cases of warlordism; armed insurrectionist groups; weak legitimacy of police or judiciary; and a restructuring political system. Prevalent social features include population displacement, weak social services, erosion of trust and armed criminality (UNDP, 2004).

The United Nations Development Programme (UNDP) differentiates two general categories of issue in the aftermath of conflict – sudden or rapid changes, and endemic challenges. Sudden or rapid changes include: an influx of returnees; renewed armed conflict; national economic crisis; collapse of the government; change of government and government policy; military occupation or withdrawal; and change or collapse of local interlocutors or partners. The endemic challenges include: weak baseline data; high staff turnover; limited public sector capacity; incomplete peace processes and divided communities; lawlessness and criminality; and erosion of trust and social capital (UNDP, 2004).

Links between conflict and natural resources

More than 90% of 150 wars fought between the end of World War II and the mid-1990s were in the developing world (Amnesty International and Oxfam, 2003, in Brookstein, undated). In developing countries, many people depend directly on natural resources, and these can provide both the spark and the fuel for conflict. Therefore, competitive access to natural resources is an important contributory factor to many conflicts, and the prevalence of natural resource-driven conflicts may well increase as those resources become more scarce (Lanjouw, 2003). Many conflicts are fought along national or provincial borders that were often drawn along natural divisions such as mountain ranges, rivers or lakes. These same features were also the justification for the siting of many protected areas (Leader-Williams *et al.*, 1990), so many armed conflicts occur in or near protected areas (Lanjouw, 2003). Equally, it must be acknowledged that natural resources are rarely the sole cause of conflict. Other factors such as ethnicity and identity, or political differences, are often the underlying drivers of conflict. However, research indicates that at least 40% of all intrastate conflicts over the last 60 years are linked to natural resources, and that these natural resource-linked conflicts are twice as likely to relapse into conflict from apparent 'post-conflict' situations (Matthew *et al.*, 2009).

Conflict has three main impacts on natural resources (Shambaugh *et al.*, 2001): (i) habitat destruction and loss of wildlife; (ii) over-exploitation and degradation of natural resources; and (iii) pollution, this latter being a particular problem in conflicts over water resources. The breakdown of law enforcement and traditional controls often intensifies pre-existing pressure on natural resources to unsustainable levels. Habitat destruction and loss of wildlife can be as a result of subsistence needs, for strategic reasons or for commercial gain. Large numbers of refugees and internally displaced persons (IDPs) often subsist in marginal areas, adding to the degradation pressure through clearing of land for farming or fuel. Communities also turn to over-exploitation of bushmeat and wild food plants for survival. Social disruption can be used strategically by armed forces, controlling food supplies and using hunger or disease as weapons. For the strategic reason of improving mobility and visibility, cutting and burning vegetation may also accompany insurgent and military activity (Shambaugh *et al.*, 2001). Conflict can also be a cover for the illicit commercial exploitation of natural resources, and such exploitation can be a primary motivating factor behind conflict (Lanjouw, 2003). Humanitarian agencies can exacerbate this downward pressure, for example by using unsustainably logged wood for reconstruction.

The cycle of conflict, environmental degradation and social impoverishment indicates that conflict resolution and peace building efforts could benefit from a significant environmental component (Lanjouw, 2003). Clearly, investment in environmental conservation and the more sustainable and equitable management and use of natural resources can address one of the root causes of conflict and insecurity. When closely linked to livelihood strategies (Roe & Walpole, this volume, Chapter 9), conservation can be particularly successful at reducing the likelihood of conflict and the vulnerability of communities to natural disasters (Matthew *et al.*, 2002).

The two case studies below both represent sites of high biodiversity value that have been subject to prolonged and repeated cycles of conflict. Both can also demonstrate the impact of conservation initiatives involving a range of partners locally as well as international conservation agencies working closely with national and provincial management authorities. Work in both areas is ongoing as new threats to wildlife and natural habitats emerge, and as new opportunities present themselves to secure sustainability for these habitats, their neighbours and humanity, through investment in natural ecosystem services values.[2]

Guerillas and gorillas

Our first case involves the mountain gorilla *Gorilla beringei beringei* in the Virunga volcanoes of Central Africa, in an international border area that straddles the national boundaries of Rwanda, Uganda and eastern DRC (Figure 14.3). This is an area where different political maps have frequently overlain the current borders in the recent past.

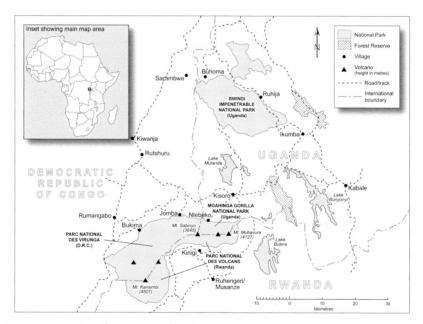

Figure 14.3 **Map of Virunga and Bwindi, home to mountain gorillas, and a zone of conflict lying on the borders of Rwanda, Uganda and DRC.**

[2] Updates on work at these sites, and in other conflict sites around the world, are available at www.fauna-flora.org.

Context and conservation issues

The Great Lakes Region of central Africa includes eastern DRC, formerly Zaire, Rwanda, Burundi, western Uganda and northern Tanzania. The area is mountainous and fertile, and control over its resources has oscillated over the past century between groups divided along political and economic, rather than ethnic or racial, lines (Chretien & Triaud, 1999). Clashes between groups have led to cross-border tensions and movement of people over the whole region, whilst wars throughout the region have allowed access to small arms and light weapons (Boutwell & Klare, 2000). The region around the Virunga volcanoes has seen veritable waves of refugees, militias and rebel groups around and across the forests. The 1994 genocide in Rwanda killed up to 1 million people within 100 days, while up to 2 million people fled to neighbouring countries (Joint Evaluation of Emergency Assistance to Rwanda, 1996). The refugees spent more than 2 years in refugee camps and during that time the former members of the Rwandan army (FAR) and the extremist rebels (Interahamwe) formed political and military groupings to recapture control of Rwanda (Jongmans, 1999). The insurgency that followed caused cycles of conflict within and around the Virunga volcanoes, with its beleaguered mountain gorillas, and the situation remains dynamic. The complex emergency in the Great Lakes Region has taken an enormous toll on the civilian populations (International Crisis Group, 2000) with up to 4.5 million people perishing since 1998 as a result of conflict and associated impacts in eastern DRC (Coghlan *et al.*, 2007). The reasons for the conflicts along the border regions between DRC, Rwanda and Uganda have not been fully resolved (Duly, 2000), but some element of control over land and its natural resources is likely.

Environmental destruction in DRC from the refugee crisis in 1994–1996 included heavy deforestation, depletion of freshwater sources, soil erosion and problems with the disposal of waste and corpses. The toll on wildlife, biodiversity and protected areas in Rwanda and DRC has been described in numerous reports and articles (Biswas & Tortajada-Quiroz, 1996; Henquin & Blondel, 1996; Kalpers & Lanjouw, 1999). Indirect threats to the environment included: economic collapse and loss of livelihood opportunities; social collapse and loss of social support structures; increased dependence on natural resources by displaced people and military groups; armed combatants and landmines within protected areas; and political opportunism in an insecure context. Direct threats included: agricultural encroachment and deforestation for firewood and construction; illegal harvest of plant resources; and poaching that included killing of 18 mountain gorillas from 1995 to 1998. Park staff and non-governmental organization (NGO) colleagues also had to deal with other factors, including: control of human diseases that affect wildlife; their vulnerability to armed groups; and a perceived loss of their independent mandate when working alongside military residents in the park. Nevertheless, despite these problems, we now examine how conservation has served as a platform for peace in Central Africa.

A fragile peace and a fragile resource

The International Gorilla Conservation Programme (IGCP) is a coalition of the African Wildlife Foundation, Fauna & Flora International (FFI) and WWF that has been working in Rwanda, DRC and Uganda since 1991. IGCP has focused on strengthening the park management capacity for the afromontane forests of the Virunga volcanoes and Bwindi Impenetrable Forest, to enhance regional collaboration over their shared species and ecosystem, and to improve capacity (technical support and policy framework) for effective conservation. Emphasis was placed on involving a range of interest groups in order to mitigate the impact of regional conflict on the mountain gorillas and their forest habitat, and included joint initiatives with the United Nations High Commission for Refugees (UNHCR) and the United Nations Environment and Development Programmes (UNEP and UNDP), as well as a group of bilateral and multilateral development agencies.

Conservation NGOs worked with the park authorities towards transboundary collaboration between national parks. In 2001, Rwanda, DRC and Uganda signed a Transboundary Agreement and Declaration on Collaboration. In 2006, the three countries signed a Memorandum of Understanding for Transboundary Collaboration, of which the IGCP was official facilitator, and agreed on a Transboundary Strategic Plan to be jointly implemented according to a Funding Accord (February 2008).

Despite losses of gorillas during more than a decade of conflict, surveys (Kalpers *et al.*, 2003) have shown that a co-ordinated but adaptable effort to protect the world's remaining mountain gorillas has been astonishingly successful. Key to this is an integrated approach, forging links between humanitarian, developmental, environmental, political and military sectors. This includes building relationships between local and external actors, such as state structures, external organizations and civil society institutions. Investing in people, and local structures, can sustain impact. Bringing together the interests of warring parties can act as a strategy for building peace. Flexibility to respond to a changing situation is crucial, as is perceived neutrality of external agencies, and co-ordination between them.

Based on evidence of the past two decades, can the security of mountain gorillas and their habitat be maintained if the region remains under cycles of conflict? The fragility arises because Virunga and Bwindi are the only habitats left for mountain gorillas (see Figure 14.3). Furthermore, it is difficult to establish sufficient security in this cross-border habitat to allow for a more stable pattern of secure livelihoods and local economies to develop and persist. Ecotourism remains a part of the financial underpinning of gorilla conservation and local livelihoods in Uganda and Rwanda, and to a certain extent in DRC. However, the protected area authorities are starting to look to sustainable support through linkages between conservation and ecosystem services (Goldman *et al.*, this volume, Chapter 4). Given the relatively small size of the protected areas within Virunga, this may not focus on carbon, but may rather follow a different model around water and biodiversity services to surrounding agriculture.

People of the forest

Our second case involves the orang utan *Pongo abelli* and valuing ecosystem services in the province of Aceh in the Indonesian island of Sumatra (Figure 14.4). Increasingly promoted as 'Green Aceh', this case is intranational rather than international, where the conflict was between the province and the state.

Context and conservation issues

Aceh is one of Indonesia's poorest provinces, where 26.5% of the human population lives below the poverty line (World Bank, 2008). Nevertheless, Aceh is rich in natural resources including oil, natural gas, timber and minerals, but these resources have previously provided national rather than local benefits. Over recent decades the province has seen waves of centralization followed by waves of decentralization of political authority. Consequently, Aceh has suffered from weak local institutions and issues of corruption,

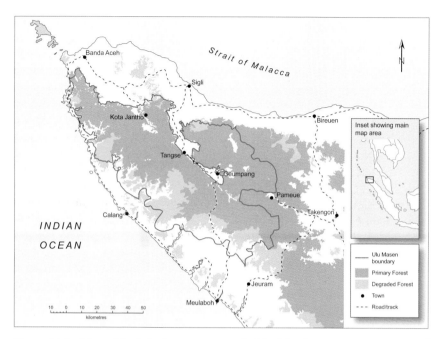

Figure 14.4 **Map of Aceh in Sumatra, stronghold of the Sumatran orang utan and scene of a zone of internal conflict.**

military repression and human rights violations. During more than 30 years of conflict it is estimated that up to 15 000 people have been killed and 1.5 million displaced. The roots of Aceh's conflict lie in the 1940s, but the Free Aceh Movement (GAM) emerged in 1976, and counter-insurgency operations were mounted by the national government (Renner, 2006).

The early years of the 21st century saw martial law, civil emergency and negotiations that eventually led to a peace agreement being signed between the government and GAM in August 2005, in the aftermath of the tsunami in December 2004. The tsunami killed an estimated 150 000 people and displaced upwards of another 1 million. Societal infrastructure was damaged or destroyed and the civil administration decimated, with loss of both people and records. Direct environmental damage was extensive, and the tsunami also hugely increased the scale of pressure on local resources.

The years of conflict acted as a veil for natural resource destruction and illegal activities. In the aftermath of the tsunami the high demand for timber for reconstruction of the thousands of houses, and other buildings, that were destroyed or damaged was felt within the Ulu Masen ecosystem and across the province. Despite an attempt to provide legal supplies of timber from elsewhere, there was an increase in illegal logging along with road building and other infrastructure projects. The former rebel leader was elected Governor of Aceh in 2006 and promoted the concept of 'Green Aceh', outlining a 'green future' for the province. This included an exploration of the prospects for monetizing ecosystem services. In 2007, an examination by the international conservation NGO FFI of how effective partnerships can be forged for conservation in the wake of conflict or natural disaster included an analysis of the perspectives of Acehnese stakeholders. This characterized the shift in governance that accompanied resolution of the conflict, summarized in Table 14.1. However, these shifts in governance are still not complete and depend on continuing security underpinned by a socially acceptable economic framework.

Conservation as a platform for peace

Despite the ongoing conflict, local groups and international organizations such as FFI have worked in Aceh since 1998 on conservation of Asian elephant *Elephas maximus* landscapes and the institutional framework to manage the Ulu Masen forest landscape for conservation. Their hope is to trade conflict and degradation for peace, development and conservation.

From 2001 to 2003, access to the project areas was restricted due to instability. However, relationships and legitimacy were still developed with local stakeholders, including the network of Mukims and the Mukim Council.

Table 14.1 **Governance shifts in Aceh prior to the formalization of partially autonomous provincial administration through elections in 2009.**

Factor	Before 2005 peace agreement	After 2005 peace agreement
Governance	Centralized and autocratic	With an 'Aceh perspective'
Provincial governor	Politically appointed	Locally elected
Local accountabilty	Lacking	Improved
Local institutions	Weak	Strengthened

These established relationships enabled an immediate response to the 2004 tsunami, linking humanitarian agencies with local communities and promoting sustainable reconstruction, livelihoods and forest protection, all of which strengthened local institutions. Local entities were linked to a multi-stakeholder trust fund and the potential for monetization of ecosystem services was explored while building the social frameworks to share benefits from conservation of forest diversity. However, this is a classic case of easier said than done. As multilateral negotiation establishes a regulatory framework for carbon in intact ecosystems, and investors anticipate a market in carbon credits, the groundwork is being laid in Aceh and other pilot projects, where the complexities are being worked through. It is a long-term strategy, but sustainable finance requires sustaining the ecosystem, which in turn requires stability and a prolonged break in the cycle of conflict.

Ambition versus pragmatism

These case studies highlight trade-offs between the desired and the possible when trying to achieve conservation in a conflict scenario. The persistence of both habitat and flagship species in the Virunga volcanoes and Bwindi Impenetrable Forest of Africa, and in the Ulu Masen forest landscape of Aceh, show just what strong frameworks, and conservation success, can be achieved within conflict scenarios if trusting, non-partisan local relationships can be developed and goals aligned. Of course, there are blemishes in this otherwise optimistic picture, and they revolve around issues of gender, short-term imperatives, ethics, impartiality, investment choices, capacity, governance, development and economics, which we now consider further.

Gender

In most violent conflicts, men are involved as combatants and perpetrators of violence, as well as political leaders in the conflict. In contrast, many of the victims of conflict are women, often as specific targets. Equally, many women become household leaders within otherwise patriarchal societies as a direct consequence of conflict. Conservationists need to have knowledge and awareness of local social relations and dynamics, and be able to meaningfully include all segments of society. Attendees at conservation workshops around the world will be aware that women remain in the minority in the forestry and wildlife sectors generally, and particularly in the often lucrative forestry sector, where this disparity can be extreme. Fortunately, some notable female role models are emerging, including in all three countries that share the Virunga volcanoes. Furthermore, despite their male-dominated nature, the Mukims around the Ulu Masen forests have recognized and welcomed the technical support of women working locally for an international NGO, especially where credibility has been forged by working together through both conflict and natural disaster.

Short-term imperatives

Violent conflict leads to human casualties and suffering, and immediate humanitarian perspectives naturally predominate in this context. During and after conflict, donors often reallocate funding from long-term needs, including sustainable management of the environment, to short-term, humanitarian needs. As conflict can disrupt local economic activity, people often turn to food they can grow or obtain outside the marketplace. This leads to a higher dependence on natural resources from wild lands and protected areas, such as hunting, fishing and collection of plants, grazing or cultivation. Such dependence can lead to habitat destruction and reduced options for post-war reconstruction and sustainability.

Ethics and impartiality

A pragmatic and non-partisan approach is critical during cycles of conflict, but achieving this can be challenging. During conflict, control over natural resources and habitats can often shift from one side to the other. Working in different forests, it is sometimes only possible to tell which side has been in the area by the type of munitions they have left behind (Figure 14.5). Ethical constraints arise when those who control the resources are engaged in atrocities, or prevent vulnerable communities from accessing resources for their livelihood needs. The challenge is to work together without supporting one particular side in the conflict. A clear focus on conservation goals that can be understood by all parties can provide a temporary route out of trouble and, often, a long-term framework for peace. However, impartiality demands

Figure 14.5   **Among the detritus left behind by warring parties moving through the forest, protected area managers have had to clear serious quantities of munitions in the wake of conflict, such as this haul from the Cardomom Mountains of Cambodia. (Photograph by kind permission of FFI.)**

acute awareness, and has to be seen as impartial by combatants. Knowledge of divisions arising from ethnic, tribal, religious or political allegiances can be used to ensure that local personnel are not put in positions where it is impossible to retain credible impartiality – if it is lost, they should be removed from the conflict. However, there are many examples where ex-combatants are reintegrated into the societies that emerge from conflict and work alongside each other. As with ex-poachers, those who retreated into forests during conflict can prove the most effective guardians of its diversity, as conservationists can demonstrate from Cambodia to Liberia, and from Indonesia to the DRC (Hammill *et al.*, 2008).

Investment choices

Working in conflict areas for long-term resilience of natural habitats also requires resilient conservationists. With an eye to the long-term integrity of the habitat, and in support of those with a mandate to protect it, conservation groups have often had to re-equip ranger stations and provide other physical infrastructure multiple times, and under difficult circumstances, for example in DRC and Liberia. This increases the costs of maintaining a presence just as the conflict decreases available funding, as development funding is either suspended or shifted to humanitarian goals. However, it has proved critical to retaining both habitats and the infrastructure of relationships that will sustain them – again as demonstrated by the remaining forest landscapes in both DRC and Liberia, which are of increasing conservation importance globally.

Capacity

Over 100 rangers have died working for the protection of mountain gorillas in the DRC portion of the Virunga National Park during the years of conflict (Eckhart & Lanjouw, 2008) (Figure 14.6). Investing in people can be expensive and risky, but can be critical to post-conflict success in achieving conservation objectives (Lanjouw, 2003). Financial as well as technical support is vital to keep people going during the crisis, and to be able to restore the environmental management sector once the conflict is over. In the 1960s, Frankfurt Zoological Society stepped in to pay salaries of national park staff in DRC when the local economy collapsed during a period of conflict, and returned to help rebuild conservation infrastructure subsequently as a trusted partner. The Rapid Response Facility is currently providing such critical, short-term support to sustain conservation actions for areas of global biodiversity importance in emergency situations.[3]

Governance

Sustainability can be built by a combination of local people, communities and local authorities, working together as decision makers, participants and beneficiaries. This is enhanced where governance over natural resources and local environments is clear and where there is a local mandate or stake in stewardship of those resources. However, in situations of conflict, authority

---

[3] The Rapid Response Facility is a collaboration between UNESCO, UNF and FFI, and has been supported by USAID and the Arcadia Land Trust. Details available at http://www.rapid-response.org/.

Figure 14.6   **Rangers working in the Virunga volcanoes protect wildlife and tourists, but have often been caught in crossfire themselves. Their weapons are usually older and less reliable than those available to military or insurgent factions. (Photograph by kind permission of J.P. Moreiras/FFI.)**

becomes centralized and local authority relationships are superseded by military command that may be less driven by concern for a sustainable local environment. Decisions and interventions by central authorities under these circumstances can prove less sustainable, through lack of local participation and interests. Re-establishing or securing a local mandate is important in post-conflict scenarios. FFI has found that this can be stimulated by a comprehensive process of local reassessment and zonation planning, as in Cambodia, or through clearer local connections into emerging national policies, as in Liberia.

Development and economics

The need to rebuild national economies after prolonged conflict can lead to hard choices over speed and scale. Large-scale investors have more substantive budgets and can start up rapidly, with faster returns on investment, but may not prioritize either sustainability or local capacity. Smaller scale initiatives may take longer to become successful, but can often generate more value over time. Unless beneficiaries of either type of investment include those whose livelihoods have been damaged by the conflict but are linked to a sustainable natural environment, further loss of the natural resource base can be assumed.

In some cases, conflict may actually be beneficial to biodiversity, either through the creation of 'no go' zones between opposing sides, or as people migrate to towns and settlements, so reducing resource use in certain areas (Hanson *et al.*, 2009). This has been observed in Aceh and also around Sapo National Park in southeastern Liberia.

Conclusions

We have drawn on experience and analysis of conservation in conflict zones to outline some of the critical factors that need to be taken into account to successfully maintain environmental integrity. We recognize that there are also examples where armed conflicts have prevented exploitation of remote areas, with both people and commercial activities avoiding violence and risk. This is the case in parts of southern Sudan, where commercial extraction was prevented as large areas of the Sudd were rendered too volatile by separatist activities over many years, and is also the case in Aceh, where many people migrated to coastal settlements during periods of heightened conflict. However, once such conflicts are over, no local structures have been built for sustainable use and effective environmental governance, increasing the risk that exploitation and over-harvesting can rapidly resume. We argue that building these local frameworks can be achieved through careful engagement on behalf of natural resources during cycles of conflict, and that it will also help safeguard natural areas as they come under increased threat from human societies adapting to the impacts of a changing climate. Although the direct benefits of national parks and natural resources to building and rebuilding shattered economies and livelihoods may be relatively small, the contribution can be highly significant, as in the case of Rwanda. In addition, the protection and maintenance of basic ecological functions can be the critical foundation for development for (low-income) human populations.

Acknowledgments

We thank our colleagues Mark Infield, Helene Barnes and James Murray for telling us of their personal observations in Aceh and Liberia.

References

Biswas, A.K. & Tortajada-Quiroz, H.C. (1996) Environmental impact of the Rwandan refugees on Zaire. *Ambio*, 25, 403–408.

Boutwell, J. & Klare, M.T. (2000) Waging a new kind of war: a scourge of small arms. *Scientific American*, 282, 48–53.

Brookstein, A (undated) *The Rough Guide to Conflict and Peace Policy*. CAFOD briefing paper, London.

Chretien, J-P. & Triaud, J-L. (1999) *Histoire d'Afrique: les enjeux de memoire*. Editions Karthala, Paris.

Coghlan, B., Ngoy, P., Mulumba, F. *et al.* (2007) *Mortality in the Democratic Republic of Congo: an ongoing crisis*. International Rescue Committee, New York.

Dudley, J.P., Ginsberg, J.R., Plumptre, A.J., Hart, J.A. & Campos, L.C. (2002) Effects of war and civil strife on wildlife and wildlife habitats. *Conservation Biology*, 16, 319–329.

Duly, G. (2000) Creating a violence-free society: the case of Rwanda. *Journal of Humanitarian Assistance*. Available at: web.archive.org/web/20021231082638/http:/www.jha.ac/greatlakes/b002.htm.

Eckhart, G. & Lanjouw, A. (2008) *Mountain Gorillas: biology, conservation and coexistence*. Johns Hopkins University Press, Baltimore.

Hammill, A., Crawford, A. & Bescanson, C. (2008) *Gorillas in the Midst: assessing the peace and conflict impacts of the International Gorilla Conservation Program activities. Final report*. International Institute for Sustainable Development, Winnipeg, Canada.

Hanson, T., Brooks, T.M., da Fonseca, G.A. B. *et al.* (2009) Warfare in Biodiversity Hotspots. *Conservation Biology*, 23, 578–587.

Hart, J.A. & Hall, J.S. (1996). Status of eastern Zaire's forest parks and reserves. *Conservation Biology*, 10, 316–324.

Henquin, B. & Blondel, N. (1996) *Etude par télédétection sur l'évolution récente de la couverture boisée du Parc National des Virunga*. EEC/UNHCR/IZCN, Goma, Democratic Republic of Congo.

International Crisis Group (2000) *Uganda and Rwanda: friends or enemies?* International Crisis Group, Nairobi and Brussels.

Joint Evaluation of Emergency Assistance to Rwanda (1996) *The International Response to Conflict and Genocide: lessons from the Rwanda experience*. Overseas Development Institute, London, and Danida, Copenhagen.

Jongmans, B. (1999) Rwanda. In *Searching for Peace in Africa: an overview of conflict prevention and management activities*, eds M. Mekenkamp, P. van Tongeren & H. van de Veen, pp. 247–248. European Centre for Conflict Prevention, Utrecht, the Netherlands.

Kalpers, J. & Lanjouw, A. (1999) *Protection of ecologically sensitive areas and community mobilisation*. Report, Environmental Management Training Workshop for UNHCR.

Kalpers, J., Williamson, E., Robbins, M. *et al.* (2003) Gorillas in the crossfire: population dynamics of the Virunga mountain gorillas over the past three decades. *Oryx*, 37, 326–337.

Lanjouw, A. (2003) *Women and the role of conservation in conflict*. Paper presented at the International Bar Association Conference, June 2003.

Leader-Williams, N., Harrison, J. & Green, M.J. B. (1990). Designing protected areas to conserve natural resources. *Science Progress*, 74, 189–204.

Loucks, C., Mascia, M.B., Maxwell, A. *et al.* (2009) Wildlife decline in Cambodia, 1953–2005: exploring the legacy of armed conflict. *Conservation Letters*, 2, 82–92.

Matthew, R., Brown, O. & Jensen, D. (2009) *From Conflict to Peacebuilding: the role of natural resources and the environment*. United Nations Environment Programme, Nairobi.

Matthew, R., Halle, M. & Switzer, J. (2002) *Conserving the Peace: resources, livelihoods and security*. International Institute for Sustainable Development, Winnipeg, Canada.

Renner, M. (2006) *Aceh: peacemaking after the tsunami*. Worldwatch Institute, Washington, DC.

Sen, A. & Dreze, J. (1989) *Hunger and Public Action*. Oxford University Press, Oxford.

Shambaugh, J., Oglethorpe, J., Ham, R. & Tognetti, S. (2001) *The Trampled Grass: mitigating the impacts of armed conflict on the environment*. Biodiversity Support Program Publication No. 139. WWF-US, Washington, DC.

UNDP (United Nations Development Programme) (2004) *Institutional Flexibility in Crises and Post-Conflict Situations: best practices from the field*. UNDP, New York.

World Bank (2008) *Aceh Poverty Assessment 2008: the impact of the conflict, the tsunami and reconstruction on poverty in Aceh*. World Bank, Washington, DC.

Part IV
Social and Institutional Constraints

15

Trading-off 'Knowing' versus 'Doing' for Effective Conservation Planning

Andrew T. Knight and Richard M. Cowling

Department of Botany, Nelson Mandela Metropolitan University, Port Elizabeth, South Africa

Introduction

Implementing effective conservation planning initiatives is difficult, challenging work. Read through any forthright case study that has attempted to translate maps of priority areas generated from a spatial prioritization into action and it becomes quickly apparent that realizing success is typically a real struggle (e.g. Noss *et al.*, 1997; Cowling & Pressey, 2003; Beier, 2007). This struggle stems, in part, from a complex suite of inter-related trade-offs that must be navigated if effective conservation action is to be implemented. These trade-offs may require resolution by an individual conservation planner, or, more broadly, throughout an entire conservation planning initiative.

In this chapter, we outline the role of spatial prioritization analyses in broader conservation planning processes. We do this in the context of the trade-offs faced by conservation planners attempting to bridge the 'research implementation gap' by translating spatial prioritizations into real-world action. We then detail the challenges these trade-offs present. Finally, we outline six hallmarks defining conservation planning initiatives that actively attempt to navigate these trade-offs in pursuit of translating spatial prioritizations into effective conservation action.

Trade-offs in Conservation: Deciding What to Save, 1st edition.   Edited by N. Leader-Williams, W.M. Adams and R.J. Smith.   © 2010 Blackwell Publishing Ltd.

'Knowing' and 'doing' as trade-offs for effective conservation planning

The resources available for conservation planning are woefully inadequate (Balmford *et al.*, 2002). Therefore, conservation planning initiatives should strategically decide where to invest their limited resources (Wilson *et al.*, this volume, Chapter 2; Murdoch *et al.*, this volume, Chapter 3) before setting about attempting to implement effective conservation actions at identified important sites (Aveling *et al.*, this volume, Chapter 14). Spatial prioritizations are analyses that identify areas for conservation action to achieve explicit targets (Wilson *et al.*, this volume, Chapter 2; Murdoch *et al.*, this volume, Chapter 3), and are but one of many integrated activities simultaneously conducted by effective conservation planning initiatives (e.g. Knight *et al.*, 2006a) (Figure 15.1). Recently, the number of spatial prioritizations published in peer-reviewed journals has grown exponentially (Pressey, 2002).

Translating maps of important areas for conservation into effective conservation is challenging for at least four reasons. Firstly, the world is a complex place. Species, ecosystems, people, livelihoods, relationships, development, business, climate and more, all interact. In turn, this creates a multitude of possible futures across multiple geographic and temporal scales. Secondly, our world is ever-changing. Natural changes in ecological systems are often unpredictable and heterogeneous in both time and space. Some changes are irreversible, and many are non-linear. Human-induced changes have increased significantly since the Industrial Revolution, impacting the Earth in unprecedented ways. Thirdly, people have widely differing attitudes and values regarding nature, which effectively means that Earth, in a natural resource management context, comprises multiple realities (Sayer & Campbell, 2004). Often these multiple realities manifest as divergent goals. Finally, by and large, people do not place appropriate value upon natural resources. As a result, the implementation of conservation plans often directly competes with development (Roe & Walpole, this volume, Chapter 9). Given that these four reasons often operate synergistically, it is little wonder that translating spatial prioritizations into effective conservation action is so challenging.

Translating spatial prioritization outputs into effective conservation action requires trading off a diverse variety of entities and issues. We define a conservation trade-off as *a situation in which a balance must be struck by a project, programme or individual between two or more alternative courses of*

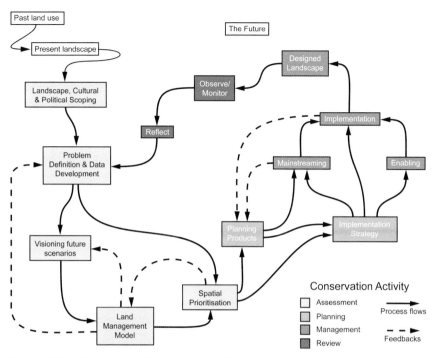

Figure 15.1   **Spatial prioritization as an element of systematic assessment in an operational model for implementing conservation action. (After Knight *et al.*, 2006a and reproduced by kind permission of ©2006 Society for Conservation Biology.)**

action that ultimately influence the delivery of effective conservation action. In this situation, compromises must be made in order to achieve conservation goals. Indeed, sometimes all desired goals may not be achievable at the same time (Leader-Williams *et al.*, this volume, Chapter 1). Examples include balancing the investment in conducting research versus implementing conservation action (Possingham *et al.*, 2007; Wilson *et al.*, this volume, Chapter 2), or balancing the theoretical, optimal achievement of conservation targets in a protected area network against the constraints of land available for purchase (Knight *et al.*, in review) (Box 15.1). Many of these trade-offs are rarely identified and explored by conservation planners. Nevertheless, explicitly addressing these trade-offs will assist conservation planners to navigate them more effectively.

Box 15.1 **Priority versus opportunity**

Formally protected areas are regarded as the foundation instrument for securing nature conservation goals. Spatial prioritization analyses are increasingly used for designing protected area networks, and are increasingly sophisticated, but still typically apply ecological data alone. Recently there have been calls for the inclusion of human and social data in spatial prioritizations. Most spatial prioritizations commonly assume that all land throughout a planning region is available for acquisition. However, land managers' willingness to sell can be highly variable. We mapped the willingness of land managers to sell their land in a part of the Makana Municipality of the Eastern Cape, South Africa. This planning region forms part of the southwestern portion of the Maputaland–Pondoland–Albany hotspot, a global conservation priority. Specifically, this planning region forms part of a priority conservation corridor in the subtropical thicket biome, which has a high level (20%) of plant endemism. Current rates of habitat destruction are very low, although were historically high, but very few major conservation initiatives are currently operational. We examined: (i) the degree to which habitat-type targets are achieved across a landscape; (ii) the area and cost efficiency; (iii) the spatial configuration; and (iv) the cost effectiveness of areas identified as important for achieving conservation targets. It was found that only 10 out of 48 land managers were willing to sell their land to conservation interests. Only seven, five and one of the 19 vegetation types in the planning region, respectively, achieved conservation targets of 10%, 30% and 50% when unwilling land managers were removed from the analysis. Assuming land managers who were unwilling to sell could be convinced to sell if offered a price substantially above market value, the cost of acquiring all lands ranged between 6.2% and 30.7% more expensive than the estimated 2006 land price. Accounting for implementation opportunities and constraints, such as land managers' willingness to sell, is therefore of fundamental importance for undertaking spatial prioritizations that can be effectively translated into conservation action.

Trade-offs for effective conservation planning

In order to better prepare to navigate trade-offs, conservation professionals require a sound understanding of the social-ecological system of their planning region. Planners require a systematic appreciation of, firstly, stakeholders' needs and behaviours (Cowling & Wilhelm-Rechmann, 2007) and, secondly, points of intervention where social systems link with, and influence, ecological systems (Briggs, 2001; Brunckhorst, 2002). In our experience, a suite of trade-offs that we detail below need to be navigated by conservation planners, and often the stakeholders working with them, to ensure spatial prioritizations are translated into effective conservation action. We propose ways of approaching these trade-offs in real-world situations, and identify both successes and failures of our approach (see Cowling & Pressey, 2003; Pierce *et al.*, 2005), in the hope that other conservation professionals can learn and benefit from our experiences. We now discuss five trade-offs in turn.

Publishing research versus 'doing' conservation

Many conservation professionals agree that conservation decisions are best informed by scientific information. Application of the scientific method allows the limits of knowledge to be understood, and provides consensus through peer review. For these reasons, scientific knowledge is regarded as providing information for decision making that is more defensible than other knowledge traditions, for example local or traditional ecological knowledge, which is particularly important in litigious contexts (Noss *et al.*, 1997).

Science has significant limitations, however, in the context of implementing effective conservation action. Firstly, science is limited in the questions to which it can provide answers (Cowling *et al.*, 2004; Cabin, 2007). Many scientists are reticent to involve themselves in conservation decision making as they often desire complete certainty before committing to a management action (Knight *et al.*, 2006b). Unfortunately, planners will never have all the scientific information they would wish for to support all the decisions that must be made. Secondly, many scientists appear to regard practical applications of science as unprofitable (Diamond, 1986; Starbuck, 2006), and so focus primarily upon producing publications rather than saving species and habitats. Career progression advances through where and how many publications a scientist produces, not on how many species he or she saves. Thirdly, science is not infallible and can provide wrong conservation 'solutions' (e.g. Gross, 2005; Noon & Blakesley, 2006). Fourthly, science can be poorly targeted and so provide few results of practical use. For example, few of the highly sophisticated spatial prioritization techniques pioneered in the peer-reviewed literature have been integrated into the core business of implementing organizations (Prendergast *et al.*, 1999; Hopkinson *et al.*, 2000). Finally, the return on investment in scientific information, such as species surveys, has been overestimated, meaning it makes little sense to postpone conservation action whilst conducting more research (Grantham *et al.*, 2008). A trade-off needs to be made between gathering, analyzing and publishing data (Figure 15.2) versus implementing conservation action (Cabin, 2007). What

Figure 15.2   **Members of the technical team of South Africa's Working for Woodlands Project assess survival of spekboom (*Portulacaria afra*) cuttings planted to restore degraded subtropical thicket in the Baviaanskloof World Heritage Site. This restoration project, which is funded by the Department of Water and the Environment, seeks to create jobs and skills by nurturing a green economy. Annual Thicket Forum meetings provide an opportunity for people from different backgrounds to discuss progress and problems in implementing this project. (Photograph by kind permission of Mike Powell.)**

we now require is a new approach to information for decision making that balances science against action through the application of critical thinking.

Funding and capacity versus effectiveness

Spatial prioritization techniques were developed in response to the limited availability of resources and funding for conservation. Data on the economic costs of implementing conservation action are increasingly included in spatial prioritizations (Naidoo *et al.*, 2006). However, data on the availability of human resources required to effectively spend conservation funds are rarely included (Rodríguez *et al.*, 2006). Spatial prioritizations detail where we have the potential to save money, but not whether we can actually undertake implementation effectively (Knight & Cowling, 2007).

A lack of capacity almost guarantees that funding will be spent poorly, if at all (Rodríguez *et al.*, 2006). Conservation activities on private land often require behaviour and technology change, and so may impose a variety of costs upon a land manager. Inadequate financial capital can rate as a far larger constraint to changing land use practices than a lack of knowledge, skills or access to information (Curtis *et al.*, 2001). These implementation costs imposed on land managers should be assessed when implementing private land conservation initiatives, with a view towards establishing cost-sharing arrangements so as to provide land managers the financial capacity to implement.

That said, a lack of human capital, namely the stock of health, nutrition, education, skills and knowledge, along with access to the services providing these, and the leadership quality of individuals, must also be strong to ensure effective conservation action. Commitment and skills within a programme to establish processes that build trust and competency amongst citizens and agencies is essential (Curtis *et al.*, 2002). For this reason, education and training should be strategically targeted and conducted prior to implementation (e.g. Pierce *et al.*, 2005). Strong social capital is also critical (Pretty & Ward, 2001; Folke *et al.*, 2005), as a 'capacity to concert' is typically required to mobilize collective action for conservation planning initiatives (Daubon & Saunders, 2002) as well as to establish effective governance (Brunckhorst, 2002) and social learning institutions (Folke *et al.*, 2005; Knight *et al.*, 2006a), which are prerequisites for adaptive management.

Understanding the trade-off between the financial, human and social capacity for implementation and effective action is critical to the efficient and effective spending of funding, and so should be included in spatial prioritizations (Knight & Cowling, 2007). Given that an absence of capacity requires mainstreaming and capacity building to be conducted prior to implementation (Knight *et al.*, 2006a), and that the most heterogeneous data have the greatest influence on a spatial prioritization (Perhans *et al.*, 2008), it is essential that a trade-off be made between the relative investments in ecological, vulnerability, cost and human and social capital data.

'Optimal' protected area networks versus land availability

Most spatial prioritizations assume that candidate areas identified by a spatial prioritization will be available for potential action. However, substantial proportions of land in many regions of the world are privately owned, meaning the availability of land for conservation is a function of the choices made by individual land managers (Cowling & Pressey, 2003). Typically, these people are not obliged to engage in conservation. Therefore, securing important areas is contingent upon people's willingness to be involved, both as individuals and as a collective, for effective implementation (Knight & Cowling, 2007).

Individual people make idiosyncratic, sometimes irrational, choices over the use of natural resources and how they manage land (Hardin, 1968; Ostrom, 2008). The conservation instruments available for securing important areas are diverse. For example, protected areas span a continuum of security levels from strict formal protection (i.e. IUCN, the International Union for the Conservation of Nature categories Ia and Ib) to multiple-use areas (i.e. IUCN category VI), whilst the range of economic instruments is continually expanding to include direct payments, subsidies, reverse auctions, payments for ecosystem services, and more (for a review see Young *et al.* 1996; Bruner *et al.*, this volume, Chapter 11). When investigating people's willingness to sell their land in the Makana Municipality of the Eastern Cape province of South Africa, we found that the vast majority were not interested in selling their land and moving away (Knight *et al.*, 2010). However, the majority (90%) of people were 'possibly interested' in some form of voluntary conservation agreement, with a reassuring proportion (41%) 'possibly interested' in negotiating a legally binding conservation agreement (Knight *et al.*, 2010).

The Makana case highlights the importance of understanding the trade-off between the 'optimal' protected area configuration and the type of conservation activity in which land managers are prepared to engage. Unfortunately, many spatial prioritizations published in the peer-reviewed literature do not identify specific conservation instruments for securing important areas, and instead simply identify places of biological importance. In many regions of the world there will be little choice in the areas available, as willingness may be low, meaning conservation planners must build genuine, trusting, long-term relationships with land managers, and negotiate mutually satisfactory outcomes. Failure to account for land manager willingness to participate guarantees that spatial prioritizations will have to be re-done (Margules & Pressey, 2000; Knight *et al.*, in review), and hence waste more resources, funding and, perhaps most importantly, time.

Spatial prioritization outputs versus conservation planning products

The rapid advancement of spatial prioritization techniques has led to a massive diversification of useful tools for allocating conservation resources. Numerous off-the-shelf (e.g. CPLEX, LINDO) and specialist softwares (e.g. C-Plan, Marxan) are available, as are geographic information systems (GIS). Whilst substantial effort and expense has been committed towards developing these tools, little effort has been directed towards developing useful products for practitioners.

Spatial prioritization tools have almost universally been developed and utilized by researchers for general application, and not by practitioners. Accordingly, the raw outputs produced by spatial prioritization tools typically attempt to deliver a panacea solution, which is rarely entirely effective (Ostrom *et al.*, 2007), and so fails to meet the exact needs of practitioners. This means spatial prioritization outputs may complicate the very transition to action they are supposed to facilitate (e.g. Cowling & Pressey, 2003). Conservation planners rarely invest significant time in refining spatial prioritization outputs into conservation planning products, and almost no specific research exists into conservation planning product (e.g. maps or decision support systems) development (but see Theobald *et al.*, 2000; Pierce *et al.*, 2005; Reyers *et al.*, 2007).

Cartography, a discipline that has existed for thousands of years, represents an extensive body of research into the development of map products, and highlights the importance of an interdisciplinary approach to spatial prioritization by linking spatial representation to psychology and visual arts.

The development of conservation planning products appears to slow the rapid deployment of conservation information, but spending time developing conservation planning products probably improves buy-in by, and utility for, practitioners. For example, Pierce *et al.* (2005) spent a year refining the raw outputs of Rouget *et al.* (2006) into four distinct products: (i) an A3 hardcopy mapbook for decision makers lacking GIS skills; (ii) digital maps for GIS-literate decision makers; (iii) a handbook explaining the maps and how to apply them to the development approval process; and (iv) training for decision makers. This year was equivalent to the time spent undertaking the spatial prioritization itself.

A trade-off exists between rapidly producing a spatial prioritization output and designing a conservation planning product that is both user-useful and user-friendly for practitioners (Pierce *et al.*, 2005). Researchers may argue that threats are often degrading and destroying valued nature so rapidly that information on important areas should be developed and distributed as quickly as possible. However, a poorly designed product is likely to be of only limited use to practitioners (e.g. Cowling & Pressey, 2003; Knight *et al.*, 2006b), and so may ultimately lead to practitioners becoming disillusioned with the conservation agenda because the product performs poorly, and so exit from conservation planning initiatives.

Scientists as impartial observers versus activists

Historically, most spatial prioritizations have been conducted by scientists rather than by practitioners, and the outputs are used to guide decision making. This potentially places the scientist in a position of considerable influence, and in so doing raises questions regarding the role he or she should play. Does the scientist act as the impartial observer, providing information as requested, or does he or she become the activist facilitating the conservation planning process? The traditional scientist's role of impartial observer is problematic for translating spatial prioritizations into effective action in two ways.

Firstly, although applying a scientific approach is important for establishing defensibility, it is critically important that spatial prioritizations should manifest societal values, not those of conservation professionals (Theobald *et al.*, 2000). Scientists should not offer solutions based upon their personal values, as they have historically done. Rather, they should engage society, pressing citizens for their vision of future landscape management (i.e. collaboratively develop a landscape management model *sensu* Knight *et al.* (2006a)), so that they can assist society towards achieving these goals. The scientist must intimately engage various segments of society so as to understand their values and goals.

Secondly, conservation planning is fundamentally a social, not a scientific, process. The translation of a spatial prioritization into effective action requires committed, genuine collaboration. This necessitates a long-term relationship between scientists and practitioners, with regular, ongoing, trusting interaction (Theobald *et al.*, 2000). Ideally, the scientist is housed within an implementing organization, or, at least, teams conducting spatial prioritizations should include staff from implementing organizations (Knight *et al.*, 2006b). In this way, spatial prioritizations are user driven, which is critically important, as useful spatial prioritizations should answer specific practitioner needs (Knight *et al.*, 2006b).

Scientists are most effective where they adopt the role of facilitator (Sayer & Campbell, 2004), a role located mid-way between impartial observer and virulent activist. This requires that they lead by serving (Beier, 2007), and advocate the benefits of nature conservation and spatial prioritization, whilst promoting a broader conservation planning operational model as a mechanism for achieving conservation goals (Knight *et al.*, 2008) – specifically those of social learning and adaptive management, which are required to ensure long-term effectiveness of conservation planning initiatives (Knight *et al.*, 2006a).

Navigating trade-offs for effective conservation planning

Here we propose actions or processes for resolving the trade-offs that potentially hinder the translation of spatial prioritizations into effective action. These are structured as integrated hallmarks of an effective conservation planning operational model (e.g. Knight *et al.*, 2006a), not as single solutions matched to individual trade-offs. This is essential for two reasons. Firstly, conservation planning initiatives are processes whose complexity should match that of the social-ecological system they are attempting to influence (Knight *et al.*, 2006a). Secondly, trade-offs interact in complex, non-linear ways, meaning, in some cases, a single action can be an essential component of multiple solutions. We now discuss six hallmarks in turn.

Plan for implementation: adopt an operational model

Trading off research and action is helped by situating spatial prioritization within a broader operational model for conservation planning (e.g. Knight et al., 2006a). This promotes linking the spatial prioritization with other activities essential for translating the spatial prioritization process into action. Some operational models are focused primarily upon spatial prioritization (e.g. Margules & Pressey, 2000; Groves et al. 2002). These models give the impression that spatial prioritization is the dominant activity in a conservation planning process in all of time, resourcing and importance. It is not. In South Africa, we believe that spatial prioritization is 5% of the process. An operational model assists conservation planners to plan and integrate their activities, to ensure all necessary activities are completed, to monitor their progress, and to assist in keeping them directed towards implementation even when they lose their focus.

First conduct a social assessment

A prerequisite for developing spatial prioritizations that can be effectively translated into action is to develop an understanding of the social-ecological system of a planning context, particularly the links between social and ecological systems (Cowling & Pressey, 2003; Cowling & Wilhelm-Rechmann, 2007). The social assessment should be undertaken prior to the spatial ecological assessment for the following reasons. Firstly, it provides an understanding of the existing interactions between people and natural ecosystems, including probable future land use pressures. Secondly, it reveals opportunities for linking conservation actions synergistically with initiatives in other sectors thereby enabling mainstreaming of conservation interests. Thirdly, it provides an assessment of institutional capacity to deliver effective action. In the context of spatial prioritization, an initial social assessment promotes the assessment of conservation opportunity (i.e. what it is possible to achieve), not simply areas of important valued nature.

Involve practitioners

If spatial prioritizations are to be designed and undertaken to deliver effective conservation action they must involve practitioners. Practitioners, ultimately, must define conservation problems because they are the people positioned to solve them, and so conservation planners should collaborate with practitioners to define problems (Knight *et al.*, 2008). The involvement of practitioners should be encouraged both in the design and undertaking of the spatial prioritization (Knight *et al.*, 2006b) and to provide data for spatial prioritizations (Pearce *et al.*, 2001; Cowling *et al.*, 2003; Beier, 2007). Without the input of practitioners, we almost guarantee the perpetuation of the research implementation gap in conservation planning.

Map opportunity, not simply priority

Applying biological data alone to spatial prioritization tells us what areas have valued plants and animals, but little on the practical necessities of implementation – what it is going to cost, who is willing and has the capacity to do it, and where relationships and institutions exist that can be synergistically encouraged to achieve conservation goals. To this end, mapping conservation opportunity (Knight & Cowling, 2007) offers significant advantages over simply mapping conservation priority (defined as a measure of an area's conservation value coupled with its vulnerability). This allows proactive conservation actions to be scheduled, as well as reactive activities such as decision making for land use development applications. It also better ensures that spatial prioritizations will not have to be repeated when it is discovered that some actions (whilst biologically appropriate) are either socially unacceptable, inappropriate or not feasible, and facilitates linking a spatial prioritization to an implementation strategy (Knight *et al.*, 2006a). This alleviates the likelihood of practitioners becoming disillusioned with spatial prioritization approaches that provide them with little information relevant for their activities.

Collaboratively develop an implementation strategy

Spatial prioritizations provide guidance on 'where' conservation resources could be most efficiently allocated to increase the probability they can be translated into effective conservation action through the achievement of meaningful targets for valued nature. This is useful information. However, the act of proclaiming a protected area involves many other tasks, none of which are trivial. These tasks include: arranging funding, conducting biological and cultural surveys, negotiating with land managers, purchasing the land, gazetting the new protected area and fencing it, recruiting staff, and more. These activities are also typically the responsibility of a diverse range of people. Spatial prioritizations do not offer (nor should they) instructions as to how to undertake these activities, or who should be responsible (Knight *et al.*, 2006a). It is therefore essential to complement a spatial prioritization with an implementation strategy that details the tasks required for effective implementation, along with allocating responsibility for specific tasks. The process of developing an implementation strategy provides an opportunity for proactive collaboration between stakeholders, which produces a clear vision of future land management activities, 'buy-in', a benchmark for future monitoring, interdisciplinary decision making and accountability.

Support people and strategies with social learning institutions

Ultimately, effective conservation planning hinges upon mobilizing adaptive collective action because social-ecological systems are highly dynamic, and it is rare that valued nature and the processes that sustain it are managed by any single individual. Getting people to collaborate and work effectively together over long periods of time is a highly challenging task. Maintaining motivation and group harmony, avoiding burn-out (Byron *et al.*, 2001) and operationalizing democratic group decision making that improves over time, all takes significant technical, emotional and financial input. Establishing social learning institutions provides a structure for guiding these activities, and making the most of the human capital available, whilst building the necessary social capital.

Conclusions

There are trade-offs to be made between the 'knowing' provided by a spatial prioritization on the theoretically best areas to implement conservation, and the 'doing' of effective conservation action at these locations. These trade-offs represent an essential compromise between the theory and practice of conservation. Bridging these trade-offs typically proves highly challenging for both individual conservation planners and conservation organizations. Building this bridge requires consilience – a transdisciplinary approach that integrates the natural sciences, the social sciences and humanities (Wilson,

1998). This will better ensure that a focus on 'planning for implementation' is maintained (Knight *et al.*, 2006a), and will provide the impetus for the social learning that is a prerequisite for adaptive management.

References

Balmford, A., Bruner, A., Cooper, P. *et al.* (2002) Economic reasons for conserving wild nature. *Science*, 297, 950–953.

Beier, P. (2007) Learning like a mountain. *Wildlife Professional*, Winter 2007, 26–29.

Briggs, S.V. (2001) Linking ecological scales and institutional frameworks for landscape rehabilitation. *Ecological Management and Restoration*, 2, 28–35.

Brunckhorst, D.J. (2002) Institutions to sustain ecological and social systems. *Ecological Management and Restoration*, 3, 108–116.

Byron, I., Curtis, A.L. & Lockwood, M. (2001) Exploring burnout in Australia's Landcare Program: a case study in the Shepparton region. *Society and Natural Resources*, 14, 901–910.

Cabin, R.J. (2007) Science-driven restoration: a square grid on a round earth? *Restoration Ecology*, 15, 1–7.

Cowling, R.M. & Pressey, R.L. (2003) Introduction to systematic conservation planning in the Cape Floristic Region. *Biological Conservation*, 112, 1–14.

Cowling, R.M. & Wilhelm-Rechmann, A. (2007) Social assessment as a key to conservation success. *Oryx*, 41, 135.

Cowling, R.M., Knight, A.T., Faith, D.P. *et al.* (2004) Nature conservation requires more than a passion for species. *Conservation Biology*, 18, 1674–1677.

Cowling, R.M., Pressey, R.L., Sims-Castley, R. *et al.* (2003) The expert or the algorithm? Comparison of priority conservation areas in the Cape Floristic Region identified by park managers and reserve selection software. *Biological Conservation*, 112, 147–167.

Curtis, A.L., Lockwood, M. & MacKay, J. (2001) Exploring landholder willingness and capacity to manage dryland salinity in the Goulburn Broken catchment. *Australian Journal of Environmental Management*, 8, 79–90.

Curtis, A.L., Shindler, B. & Wright, A. (2002) Sustaining local watershed initiatives: lessons from Landcare and watershed councils. *Journal of the American Water Resources Association*, 38, 1207–1216.

Daubon, R.E. & Saunders, H.H. (2002) Operationalizing social capital: a strategy to enhance a communities' "capacity to concert". *International Studies Perspectives*, 3, 176–191.

Diamond, J.M. (1986) The design of a nature reserve system for Indonesian New Guinea. In *Conservation Biology: the science of scarcity and diversity*, ed. M.E. Soulé, pp. 485–503. Sinauer Associates, Sunderland, MA.

Folke, C., Hahn, T., Olsson, P. & Norberg, J. (2005) Adaptive governance of social–ecological systems. *Annual Review of Environment and Resources*, 30, 441–473.

Grantham, H., Moilanen, A., Wilson, K.A., Pressey, R.L. & Rebelo, A.G. (2008) Diminishing return on investment for biodiversity data in conservation planning. *Conservation Letters*, 1, 190–198.

Gross, L. (2005) Why not the best? How science failed the Florida panther. *PLoS Biology*, 3, 1525–1531.

Groves, C.R., Jensen, D.B., Valutis, L.L. *et al.* (2002) Planning for biodiversity conservation: putting conservation science into practice. *BioScience*, 52, 499–512.

Hardin, G. (1968) The tragedy of the commons. *Science*, 162, 1243–1248.

Hopkinson, P., Evans, J. & Gregory, R.D. (2000) National-scale conservation assessments at an appropriate resolution. *Diversity and Distributions*, 6, 195–204.

Knight, A.T. & Cowling, R.M. (2007) Embracing opportunism in the selection of priority conservation areas. *Conservation Biology*, 21, 1124–1126.

Knight, A.T., Cowling, R.M. & Campbell, B.M. (2006a) An operational model for implementing conservation action. *Conservation Biology*, 20, 408–419.

Knight, A.T., Cowling, R.M., Difford, M. & Campbell, B.M. (2010) Mapping human and social dimensions of conservation opportunity for the scheduling of conservation action on private land. *Conservation Biology*, 24, in press.

Knight, A.T., Cowling, R.M., Rouget, M. *et al.* (2008) Knowing but not doing: selecting priority conservation areas and the research-implementation gap. *Conservation Biology*, 22, 610–617.

Knight, A.T., Driver, A., Cowling, R.M. *et al.* (2006b) Designing systematic conservation assessments that promote effective implementation: best practice from South Africa. *Conservation Biology*, 20, 739–750.

Knight, A.T., Grantham, H., Smith, R.J. *et al.* (in review) Landowner willingness to sell defines conservation opportunity for protected area expansion. *Biological Conservation*.

Margules, C.R. & Pressey, R.L. (2000) Systematic conservation planning. *Nature*, 405, 43–53.

Naidoo, R., Balmford, A., Ferraro, P.J. *et al.* (2006) Integrating economic costs into conservation planning. *Trends in Ecology and Evolution*, 21, 681–687.

Noon, B.R. & Blakesley, J.A. (2006) Conservation of the northern spotted owl under the Northwest Forest Plan. *Conservation Biology*, 20, 288–296.

Noss, R.F., O'Connell, M.A. & Murphy, D.D. (1997) *The Science of Conservation Planning: habitat conservation under the Endangered Species Act*. Island Press, Washington, DC.

Ostrom, E. (2008) The challenge of common pool resources. *Environment*, 50, 8–20.

Ostrom, E., Janssen, M.A. & Anderies, J.M. (2007) Going beyond panaceas. *Proceedings of the National Academy of Sciences of the USA*, 104, 15176–15178.

Pearce, J.L., Cherry, K., Drielsma, M., Ferrier, S. & Whish, G. (2001) Incorporating expert opinion and fine-scale vegetation mapping into statistical models of faunal distribution. *Journal of Applied Ecology*, 38, 412–424.

Perhans, K., Kindstrand, C., Boman, M. *et al.* (2008) Conservation goals and the relative importance of costs and benefits in reserve selection. *Conservation Biology*, 22, 1331–1339.

Pierce, S.M., Cowling, R.M., Knight, A.T. *et al.* (2005) Systematic conservation planning products for land-use planning: interpretation for implementation. *Biological Conservation*, 125, 441–458.

Possingham, H.P., Grantham, H. & Rondinini, C. (2007) How can you conserve species that haven't been found? *Journal of Biogeography*, 34, 758–759.

Prendergast, J.R., Quinn, R.M. & Lawton, J.H. (1999) The gaps between theory and practice in selecting nature reserves. *Conservation Biology*, 13, 484–492.

Pressey, R.L. (2002) The first reserve selection algorithm: a retrospective on Jamie Kirkpatrick's 1983 paper. *Progress in Physical Geography*, 26, 434–441.

Pretty, J. & Ward, H. (2001) Social capital and the environment. *World Development*, 29, 209–227.

Reyers, B., Rouget, M., Jonas, Z. *et al.* (2007) Developing products for conservation decision-making: lessons from a spatial biodiversity assessment for South Africa. *Diversity and Distributions*, 13, 608–619.

Rodriguez, J.P., Rodriguez-Clark, K.M., Oliveira-Miranda, M.A. *et al.* (2006) Professional capacity building: the missing agenda in conservation priority setting. *Conservation Biology*, 20, 1340.

Rouget, M., Cowling, R.M., Lombard, A.T., Knight, A.T. & Kerley, G.I.H. (2006) Designing large-scale conservation corridors for pattern and process. *Conservation Biology*, 20, 549–561.

Sayer, J.A. & Campbell, B.M. (2004) *The Science of Sustainable Development: local livelihoods and the global environment*. Cambridge University Press, Cambridge.

Starbuck, W.H. (2006) *The Production of Knowledge: the challenge of social science research*. Oxford University Press, Oxford.

Theobald, D.M., Hobbs, N.T., Bearly, T. *et al.* (2000) Incorporating biological information in local land-use decision making: designing a system for conservation planning. *Landscape Ecology*, 15, 35–45.

Wilson, E.O. (1998) *Consilience: the unity of knowledge*. Abacus, London.

Young, M.D., Gunningham, N., Elix, J. *et al.* (1996) *Reimbursing the Future: an evaluation of motivational voluntary, price-based, property-right, and regulatory incentives for the conservation of biodiversity, Parts 1 and 2*. Department of the Environment, Sport and Territories, Canberra.

$$\textbf{16}$$

Path Dependence in Conservation

William M. Adams

Department of Geography, University of Cambridge,
Cambridge, UK

Introduction

The painter George Catlin, travelling in the American west in 1932, wrote about the conservation of the threatened American bison *Bison bison* in the second volume of his letters, '*What a beautiful and thrilling specimen for America to preserve and hold up to the view of her refined citizens and the world, in future ages. A nation's Park, containing man and beast, in all the wild and freshness of their nature's beauty!*' (Catlin, 1841). This vision of 'a nation's park' was ahead of its time, although protected areas of various kinds, especially private hunting reserves and forests, had long been recognized. But the idea of national parks, eventually realized in 1872 at Yellowstone (Figure 16.1), began a tradition of conservation that not only persevered in the USA, but that spread throughout the world. Once established, the idea of national parks ran like a marked path through the byways of conservation policy and practice.

Ideas in conservation, such as that of national parks, tend to become so strongly established that they start to frame the way people think about conservation challenges and how they respond to them. When this happens, policy evolution starts to become path-dependent. Ideas do change over time, sometimes slowly and organically and sometimes rapidly, when long dominant or commonly held ideas are abandoned. Sometimes, change does

Trade-offs in Conservation: Deciding What to Save, 1st edition.   Edited by N. Leader-Williams, W.M. Adams and R.J. Smith.   © 2010 Blackwell Publishing Ltd.

Figure 16.1   **American bison *Bison bison* in Yellowstone National Park. Declared in 1872, Yellowstone became the model for many national parks worldwide, and still epitomizes the category II protected area. (Photograph by kind permission of Kent Redford.)**

not happen when it should. Then ideas get stuck in a rut, and conservation action gets locked into outmoded and dysfunctional ways of working. Path dependence has huge significance for conservation action, and especially for the trade-offs made among conservation goals, and between conservation and other goals. This chapter discusses how conservation ideas change and how they can become fixed. The first section discusses path dependence in conservation policy, the second how we should understand the process of change in ideas, and the third uses the example of protected areas to illustrate change and stasis. The chapter then explores what factors encourage path dependence, and finally how conservationists might respond to the issue.

Path dependence in conservation policy

The critical link between ideas and action in conservation is the arena of policy. Policy can be defined as the decisions taken by those with responsibility for a particular area of public life (Keeley & Scoones, 2003). Policy can describe a broad course of action (or inaction), and it needs to be conceived of as an inherently political process, a web of inter-related decisions.

Environmental policy making inevitably involves dealing with uncertainty, for example about the nature of environmental change or the behaviour of people and their economies and societies. Conservation scientists tend to hope that research will reduce such uncertainties to a negligible level, but certainty is elusive. Conservation decisions are characterized by uncertainty about facts and the associated factual claims of diverse actors (Woodward & Bishop, 1997). Science cannot supply policy makers in environmental decision making with certainty. Indeed it is a misunderstanding of the nature of scientific enquiry to expect certainty (Dovers *et al.*, 1996; Bradshaw & Borchers, 2000).

It has been argued that, in order to make coherent decisions in the face of uncertainty, policy makers create self-referencing stories, convincing each other that their understanding of problems is correct and their choice of solutions appropriate (Roe, 1991, 2004). These stories or scenarios simplify ambiguities, control (or appear to control) uncertainty and provide a secure basis for debate and action. Roe (1991, 2004) calls such stories 'narratives'. In development and environment, narratives are typically devised by researchers and consultants, and taken up by international organizations and their clients in governments and national non-governmental organizations (NGOs), through training, institution building and patterns of investment. They tell stories about what will happen if events develop as they describe, and they enable (or persuade) their listeners to respond to the predefined standard problem in particular kinds of ways. In effect they offer 'off the shelf' analysis and prescription, circumventing the need for lengthy processes of scientific enquiry or other data gathering. Narratives enable action to start at once, by borrowing generalized explanations of what is wrong and what needs to be done. In this, narratives are extremely useful, for without simplifications of both problem and context, policy making and action would be paralyzed.

The case of desertification in the Sahel is a classic example of the power of environmental narratives (Swift, 1996; Sullivan, 2000). Faced with the Sahel drought of 1972–1974 and a policy context that emphasized neo-Malthusian concerns about the disastrous consequences of population growth, the desertification narrative met the needs of specific groups of powerful policy actors: national governments in Africa; international aid bureaucracies, especially United Nations (UN) agencies; and scientists (Swift, 1996). In the 1970s, recently independent African governments were restructuring their bureaucracies and strengthening central control over natural resources. Drought, and the assumptions about human-induced environmental degradation linked to them, legitimated such claims and made centralized top-down environmental planning seem a logical strategy. Pastoralism could be portrayed as doomed and self-destructive, and its replacement by sedentary agriculture made to seem necessary and beneficial (Sinclair & Fryxell, 1985). Aid donors, meanwhile, found in desertification a problem that seemed to transcend politics and legitimated 'large, technology-driven international programmes' (Swift, 1996: 88). To scientists developing new fields such as remote sensing, desertification offered fertile terrain for expansion–satellite imagery offered scientific answers without the need for lengthy and tedious fieldwork, and desertification became a source of funding and legitimacy for new cadres of technicians and researchers (Adams, 2009).

Entrenched narratives that define widely recognized policy problems give rise to formulaic 'blueprint' approaches to planning, oblivious to context, resistant to new data or observations that challenge established ways of thinking, and impenetrable to local actors with different understandings of processes or the consequences of policy action. Thus in the context of African rangelands, the narrative of desertification, backed by assumptions about overgrazing and carrying capacity, has led to stereotyped pastoral policies aimed at confining, controlling and often settling nomadic pastoralists (Horowitz & Little, 1987).

Policy change

It might seem obvious that policies change in response to new evidence, new ideas or new conservation challenges. In fact, the process of policy change is complex and sometimes unpredictable or perverse. The most common way in which conservationists think of policy is as an essentially linear process by which problems are identified, studied and understood, and policy responses are formulated that change outcome (Figure 16.2). Thus, policy formation is portrayed as a top-down decision-oriented process based on rational

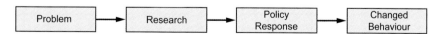

Figure 16.2 **Policy making as a linear process.**

instrumental behaviour, and policy analysis focuses on decisions taken and their impacts (Keeley & Scoones, 2003). In conservation, calls are often made for science-driven policy, implying an assumption that science is a source of expert knowledge, and policy will be improved if science and scientists have a higher profile in decisions. This simple and simplistic conceptualization of policy hides a number of important political and social dimensions of policy. These include the questions 'Who is in a position to undertake research and translate it into policy? On what are claims to expertise based, and who frames or defines "problems" and "solutions"?'

In fact policy making is a multiple, incremental and complex process (Keeley & Scoones, 2003). Policy change arises from interactions and negotiations between different groups with different interests over time. Such engagements include those between state, civil society and business organizations, for example debates over new development projects, or between different tiers of government, for example when State and Federal governments in the USA dispute the regulation of vehicle pollution, or between different interest groups, for example between animal welfare and conservation NGOs over questions such as the culling of elephants (Dickson & Adams, 2009; Harrop, this volume, Chapter 7).

Policy does not, however, result in any simple way from the outcomes of argument or bargaining between different actors: policy outcomes cannot be predicted by weighing the strength of different protagonists, as if it were an exercise in mathematical mechanics. Policy outcomes also reflect the knowledge and power embedded in expertise and decision making (Hajer, 1995). Some sorts of knowledge or information are typically ignored in conservation assessments. In particular, lay or indigenous knowledge is often disregarded in favour of formal scientific knowledge (Brosius, this volume, Chapter 17), even where the latter is manifestly lacking in basic data, as for example in debates over wild cougars in northern Ontario (Lemelin, 2008). Science and policy interact in a process of mutual construction, where science and society are co-produced through mundane daily activities such as the conducting of fieldwork, the compilation of reports, or workshops and conferences (Latour, 1999).

Ideas, concepts and categorizations are expressions of knowledge and power (Hajer, 1995). Policy can be understood as what Foucault (1975) calls a 'political technology', because of the way it categorizes the world through processes of ordering and labelling, and hence leads to social regulation or rule making. The way problems are expressed involves embedded assumptions, and these frame and constrain the way people understand the need for and possibility of action. Language and ideas are therefore a currency of power in the transaction between science and policy, and between different interests and actors.

In the language of policy narratives, what is important is the way the ideas they represent become entrenched in the minds of key actors. Policy narratives are remarkably persistent, and cannot be overturned by simply showing that they are untrue in a particular instance, but only by providing a better and more convincing story (Roe, 1991, 2004; Leach & Mearns, 1996).

Path dependence emerges where narratives are supported by elite interest groups, for example researchers, scientists and recognized policy 'experts', and when they are supported by resource flows, for example bilateral aid or other funding through powerful international organizations. Often narratives are established globally, for example through northern-dominated conservation science conferences or international meetings (Brosius, this volume, Chapter 17), and spread vertically down to national scale through funded programmes, institutional strengthening projects and (more slowly) elite education such as masters courses or PhDs for students from developing countries. Within developing countries, they are disseminated by urban-based media and specialized 'conservation education' materials, including increasingly film, video and web-based materials. Dominant narratives are often linked to celebrities and political figures (Brockington, 2008). They are disseminated locally as community leaders learn what to say to get assistance (Brockington, 2006).

The important aspect of the path dependence induced by environmental narratives is that they do not endure because they lead to successful policy outcomes. Indeed, their influence and durability are often not related at all to their actual economic, social or environmental consequences! Nor do they endure because they are continuously endorsed by empirical evidence or experience. Indeed, narratives often persist in the face of strong empirical evidence against their storyline. Instead, path dependence results from the way in which narratives become institutionalized as part of 'received wisdom', becoming culturally, institutionally and politically embedded in the way key actors, especially technical experts within government and non-governmental organizations, think.

Rapid and permanent change from one set of ideas to another can be understood using the concept of paradigm shift by analogy with work on ideas in science (Kuhn, 1962). A paradigm includes both current theory and the entire world view that supports it. At any time, scientists tend to share common understandings of how things work. As research is done, significant anomalies tend to be revealed that destabilize the current paradigm. Eventually there is a crisis, and new ideas eventually form a new paradigm, which is then once again disputed between followers of old and new. A classic example of such a 'paradigm shift' was acceptance of the theory of continental drift, many decades after Wegener first proposed it: nothing else explained the tectonic behaviour of Earth's surface.

The idea of a paradigm shift has been extended to social sciences, although not by Kuhn (1962), and is also used loosely to describe rapid changes in thought in other contexts, such as policy. As in natural science, there is huge resistance to new ways of thinking about society and social processes. The persistence of environmental policy narratives means that they are extremely resistant to new data. Roe (1991) observes that they cannot be undermined by case-by-case demonstration that they are wrong. He suggests that an established environmental narrative can only be replaced by a plausible counter-narrative that 'tells a better story' (Roe, 1991: 290). Moreover, these counter-narratives have to be as parsimonious, plausible and comprehensive as the original. Point-by-point rebuttals do not break the power of entrenched narratives: they are not stories, for they do not have a beginning, middle and end. Nor do case studies, even when based on extensive and careful research, do much to undermine policy certainties entrenched by strong narratives, although they can be used to form the basis for new narratives.

Parks and narrative change

The tensions between long-established narratives and new ideas that challenge them is well illustrated by that most constant of concerns, the establishment and management of national parks (Figure 16.3). Protected areas have formed the dominant narrative of terrestrial conservation since the end of the 19th century. Thus national parks were created in Australia in 1879, Canada in 1887, New Zealand in 1894, Switzerland in 1914, the Belgian Congo in 1925 and in South Africa in 1926 (Adams, 2004). International meetings on the protection of nature were held in Paris in 1931 and in London in 1933, and in 1936 an international convention was signed in Paris, which included a commitment to national parks (Adams, 2004). Following World War II, African colonies underwent a 'conservation boom', as numerous national parks were established (Neumann, 2002). In 1958, IUCN, the International

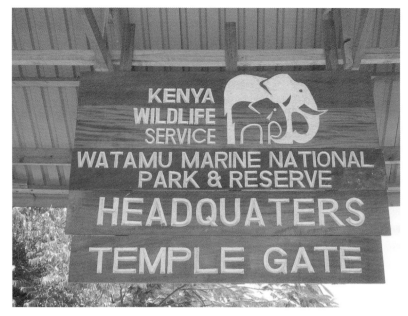

Figure 16.3   **The entrance to the Watamu Marine National Park and Reserve in Kenya, established in 1968. Worldwide, the extension of protected areas to marine environments has lagged terrestrial parks, but the model is receiving increasing attention. (Photograph by Bill Adams.)**

Union for the Conservation of Nature established a Provisional Committee on National Parks, which developed into the World Commission on Protected Areas. In 1962, the UN General Assembly adopted a 'World list of national parks and equivalent reserves', and the IUCN developed a classification that defined different kinds of protected areas (Ravenel & Redford, 2005). The area protected globally continued to grow. By the start of the 21st century, protected areas had become the central strategy for conservation. In 2005, over 100 000 protected areas covered more than 2 million km^2, or 12% of Earth's land surface (Naughton-Treves et al., 2005).

Although national parks have been a dominant theme in conservation for many decades, the concept has not been unchanging. Organic and gradual change reflects the resilience of the narrative and its capacity to incorporate new thinking. In this category might be included: the changing role of big game hunting in the way people thought about national parks, from parks as a reservoir of game to a refuge; from parks as places of natural importance in their own right, to the central place given to tourists; from parks as inviolate places, to places that serve a national purpose and places of recreation and economic development. The recognition of different categories of protected areas, embracing resource extraction (Rosser & Leader-Williams, this volume, Chapter 8) and eventually cultural heritage, also reflects an evolution of the central narrative without significant challenge to its core structure.

There have also been, however, more explicit attempts to create counter-narratives for protected areas in the last 30 years. The most established was an offshoot of the wider narrative of a 'community-based conservation' (Western *et al.*, 1994; Hulme & Murphree, 2001). This narrative recognized both a moral imperative to take account of the interests and needs of local people, and the pragmatic problem that the imposition of conservation against their wishes in a 'fortress conservation' strategy was likely to be expensive, difficult to implement and consequently unlikely to succeed. The new narrative was articulated at the Third and Fourth World Congresses on National Parks and Protected Areas in 1982 and 1992, and in the 'zoning' approach of a strictly protected core and a surrounding buffer zone in the UNESCO (United Nations Educational, Scientific and Cultural Organization) Biosphere Reserves. This approach led to the adoption at the Fifth World Parks Congress in 2003 of a new paradigm for protected areas, that they should be integrated with the interests of 'all affected people' to provide benefits *'beyond their boundaries on a map, beyond the boundaries of nation states, across societies, genders and generations'* (IUCN, 2005: 220).

This emphasis on the need to ensure the participation of local people in protected areas deliberately emphasized continuity with what had gone before. However, it was also a deliberate attempt to reorientate and restructure the way conservationists internationally thought about protected areas. In Roe's (1991, 2004) terms, this was a counter-narrative that told a better story than the thinking it sought to replace.

The new narrative's storyline drew on long-standing romantic ideas about the 'community' (Agrawal & Gibson, 1999). Its success was partly a matter of timing (Adams & Hulme, 2001). It tied conservation to the upwelling of political and policy commitment to sustainable development arising from the Brundtland Report in 1987 and the UN Conference on Environment and Development at Rio in 1992. It also caught the shift in development fashion

Table 16.1　**New park narratives.**

Narrative	Key references
Strictly protected parks	Brechin *et al.*, 2003; Hutton *et al.*, 2005
Private sector involvement in parks	Carter *et al.*, 2008
Community conserved areas	Lockwood *et al.*, 2006
Direct payments for conservation	Ferraro & Kiss, 2002
Pro-poor conservation	Roe & Elliott, 2004; Kaimowitz & Sheil, 2007
Ecosystem services	Wunder, 2006; Egoh *et al.*, 2007
Carbon and standing forests (REDD)	Miles & Kapos, 2008

from 'top-down', 'technocratic' approaches to 'bottom-up' and 'participatory planning', and it reflected the rampant neo-liberal mantra of the 1980s with its faith in the market (and consumers) to deliver policy change. The narrative of community conservation had a significant and lasting effect on the way protected areas were understood within international conservation policy. The national park narrative was not overthrown, but it was transformed.

Since the 1990s, a number of new attempts have been made to swing protected area thinking in novel directions (Table 16.1). These include renewed arguments for the strict protection of parks, interest in private sector involvement in protected areas, and in community conserved areas. Related counter-narrative attempts involve calls for 'pro-poor conservation', debates about the economic value of ecosystems, including ecosystem services, and the idea of payments for ecosystem services, as well as debates about the value of carbon in standing forests (Table 16.1). The fragmentation of the formerly rather monolithic narrative about protected areas has ushered in an era of 'policy churn', in which novel ideas are tried, found not to work, and abandoned in favour of yet more novelty.

What factors encourage path dependence?

Path dependency in conservation policy develops where ideas about what needs to be done, and how, emerge from formulaic and standardized narratives, particularly where these are believed to derive from scientific consensus, or are advocated by those who claim scientific credentials. Of course a careful scientist approaching a conservation problem would be extremely careful before pulling a recipe for action 'off the shelf' without the benefit of detailed observations. However, scientists undertaking scientific research rarely define conservation policy, but rather planners and project managers, perhaps trained as scientists, take decisions with limited information and limited time acquire it. They fall back on narratives with a superficial scientific credibility (Roe, 1991).

Several factors exacerbate this tendency. The first is what Chambers (1983) calls 'project bias': where certain key projects are referred to so often in the literature for their success (or failure) that people stake the stories about them as reflecting general truths about what works or does not. Premature overenthusiastic dissemination of positive outcomes leads to highly misleading narratives of 'success', as in the case of the experimentation with community-based natural resource management (CBNRM) particularly in southern Africa (Hutton *et al.*, 2005). Premature evaluation of programmes such as CAMPFIRE, often by people engaged with them professionally, created a powerful narrative of success (Box 16.1). Subsequent appraisals suggest a more complex balance of success and failure, and clear trade-offs between different interests (Box 16.1).

Box 16.1 **The CAMPFIRE programme in Zimbabwe**

In the 1980s, CBNRM programmes were developed in southern Africa, notably as CAMPFIRE in Zimbabwe, that allowed local authorities, and by implication local people, to benefit from the profits of the private professional safari hunting industry. In dry districts it was argued that hunting would produce higher and more tangible economic benefits than other land uses such as livestock or farming (Hutton *et al.*, 2005). The success of this approach was widely reported. An explanatory storyline was developed that communities were efficient and low-cost managers of natural resources, that wildlife revenue could make a significant difference to rural livelihoods, contribute to poverty reduction and provide economic incentives for conservation, leading to greater tolerance of wildlife and positive conservation outcomes (Hutton *et al.*, 2005).

Stories of the success were reported at international conferences, published in academic journals and books, press-released by participating organizations (not least during debate on the banning of international trade in ivory in the 1990s under the Convention on International Trade in Endangered Species of Wild Fauna and Flora (CITES)), and repeated by academic teachers to attentive students. In the 1990s, in-depth studies of CAMPFIRE began to be published that questioned the early optimism. It was successful in a few districts, but elsewhere success was elusive (Murphree, 1994, 2000; Murombedzi, 1999, 2001; Rihoy *et al.*, 2010). The reasons for success or failure were complex, demanding an in-depth understanding of social, economic and political processes and the volatile nature of Zimbabwean politics and economy. CAMPFIRE was a novel and important programme, but not the runaway success that was claimed for it.

Yet this re-appraisal of CAMPFIRE made almost no difference to the 'lessons' disseminated about it internationally. The programme was elevated to mythical status, with a symbolic power out of all proportion to the depth of empirical or theoretical understanding of social, economic and political change. Particular locations, such as Masoka in the Zambezi Valley, or Mahenye in the southeast Lowveld, were cited as evidence of success by people claiming expertise in CBNRM, but who had never been to these places. Even today, it is not uncommon to hear references by international 'experts' to 'the CAMPFIRE project in Zimbabwe' (*sic*) used to endorse community-involved approaches to conservation of all kinds, often by people who have read little original research about the region.

A second factor that promotes path dependence in conservation policy is the intensive mobilization of public campaigning for fundraising and to generate support for conservation. This can work in several ways. Campaigns can reinforce biases towards charismatic species that act as flagships for campaigns, or are used in the corporate branding of organizations that lead campaigns (Leader-Williams & Dublin, 2000). It is also possible for campaigns based on particular narratives to be undermined when those narratives change (Smith *et al.*, this volume, Chapter 12). One example is the use of compassionate or welfare arguments in campaigns about species and habitat conservation, for example the cruelty of hunting in a campaign about rainforest conservation, or the cruelty of poaching in a generalized appeal for funds for savanna conservation. In the case of whaling, for example, the conservation movement initially treated whaling as an issue of sustainable harvesting. However, campaigns portraying whales as sentient creatures brought about a shift in public consciousness towards animal welfare and rights that demanded a different position: whaling was widely see to be morally wrong, not just badly managed (Harrop, 2003; Harrop, this volume, Chapter 7).

A third factor promoting path dependence is the 'tyranny of success', or the need of conservation actors of all kinds to present success and suppress failure. This task is made easier by the lack of agreed measures of conservation 'success', and the cost and difficulty of such measurement. If projects do not work, there is every incentive for the individuals and organizations involved to move on without full analysis of why, and especially for there to be no public debate. NGOs in particular understandably wish to present projects in a reasonably favourable light to donors and partners. As a result, even where project experiences are reported in the public literature, they are often described in glowing terms, and even if the appraisal is scrupulously even-handed, such analyses are typically undertaken as the project closes, while residual funds are still flowing and institutions set up are still functioning. Such studies often report short-term success, but are unable to demonstrate if achievements endure. Where projects are analyzed later, it is often by social scientists working with local communities, and their analyses are often missed by conservation planners (because they are published in theses and abstruse journals) or dismissed as biased. Individuals will discuss failure when speaking off the record, but the literature on the selfsame projects often tells a very different tale. There are few safe places for conservationists to talk about mistakes, and therefore little is learned from them. Path dependence is promoted, but success is not.

Addressing path dependence in conservation policy

There are several ways in which the perverse effects of path dependence in conservation policy can be countered. The first requirement is to understand policy making not as a linear process (see Figure 16.2), but as a recursive or cyclic process (Figure 16.4). To avoid narratives that lead to perverse

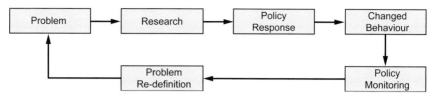

Figure 16.4 **Reflexive policy making.**

outcomes, policy implementation must feed back into the way problems are understood, and hence into the next stage of work. Policy making becomes a spiral of change and adaptation (Lee, 1993). This explicitly builds in the expectation of learning from mistakes (and successes).

A second requirement is to understand the role of science and other research. Science does not offer the policy maker certainty, although it may help reduce uncertainty. Science should help identify important parameters and processes, and it may help guide decision making. But 'good science' demands careful empirical research and the testing of hypotheses. Premature synthesis of inadequate data, or conclusions based on theorization or analogy from other contexts, should not be treated as gospel to be obeyed. Science-based policy making must be just that: drawing on scientific understanding where it exists, and to the extent that it is trustworthy, not scientist-driven guesswork.

A third, and related, requirement is to take account of the full diversity of knowledge and interpretation. It is often those who work on the ground in communities affected by projects who understand how they do or do not work. Sometimes these people will be lowly field conservation staff. More often they will be local people: lacking formal education, certainly lacking high-level scientific qualifications, lacking the means to be heard, let alone taken seriously by project planners (Brosius, this volume, Chapter 17). Such people are often the surest guides to questions of fit between policies and local realities (Figure 16.5). They see the effect of the narratives taken for granted by project planners, and if there are faults they will most readily spot them. If project planners listened to them, and Jones *et al.* (2008) suggest that even the hardened scientific sceptic may have little reason not to, it might be possible to avoid the perverse effects of path-dependent but ultimately unsuccessful policy.

A fourth requirement is to understand the role of narratives in policy path dependence and change. 'Policy making' is not an abstract process. All policy processes are located in space and time: people, events and place all matter. Narratives are created and changed in particular places, whether

Figure 16.5   **A farmer speaking at a meeting to discuss crop raiding by elephants in Laikipia, Kenya. Local residents tend to have clear and logical views about wildlife that can often differ from those of conservation planners, although in this case the problems the farmer is highlighting are the subject of specific human–elephant conflict management efforts. Further details are available at: http://www.geog.cam.ac.uk/research/projects/heccapacity/. (Photograph by Bill Adams.)**

NGO boardrooms, international meetings, web-based networks of interest and activism, coffee bars, student seminar rooms, or the pages of learned conservation journals.

Who, therefore, can change narratives? First, and most obviously, organizations can, through their policies and strategies, their planning of programmes, and their work to determine the structure and agendas of international meetings. Second, conservation scientists have a huge power in determining and changing narratives by their research and (more particularly) their debates about the implications of research and their interpretation of policy experience. Third, narratives are changed by critical engagement from outside conservation organizations and their circle, for example by human rights

NGOs campaigning on issues such as population displacement from protected areas, or the rights and needs of indigenous and mobile people.

Lastly, there is the potential for narratives to be changed by citizens, by ordinary people affected by biodiversity loss, and demanding more conservation, or by conservation programmes that change their life for the worse, demanding less conservation. But dysfunctional narratives can only be overturned in this way if conservation planners open up their thinking to the people that their policies affect. And despite the success of the language of participation and community-based conservation, this is not currently how most conservation planning is done. Mostly we are on a different path. Changing it would be difficult, but if this chapter is right, not impossible.

References

Adams, W.M. (2004) *Against Extinction: the story of conservation*. Earthscan, London.

Adams, W.M. (2009) *Green Development: environment and sustainability in a developing world*, 3rd edn. Routledge, London.

Adams, W.M. & Hulme, D. (2001) Conservation and communities: changing narratives, policies and practices in African conservation. In *African Wildlife and Livelihoods: the promise and performance of community conservation*, eds D. Hulme & M. Murphree, pp. 9–23. James Currey, London.

Agrawal, A. & Gibson, C. (1999) Enchantment and disenchantment: the role of community in natural resource management. *World Development*, 27, 629–649.

Bradshaw, G. A. & Borchers, J.G. (2000) Uncertainty as information: narrowing the science policy gap. *Conservation Ecology*, 4(1), 7. Available at: http://www.consecol.org/vol4/iss1/art7/.

Brechin, S.R., Wilhusen, P.R., Fortwangler, C.L. and West, P.C. (eds) (2003) *Contested Nature: promoting international biodiversity with social justice in the twenty-first century*. State University of New York Press, New York.

Brockington, D. (2006) The politics and ethnography of environmentalisms in Tanzania. *African Affairs*, 105, 97–116.

Brockington, D. (2008) Powerful environmentalisms: conservation, celebrity and capitalism. *Media, Culture and Society*, 30, 551–568.

Carter, E., Adams, W.M. & Hutton, J. (2008) Private protected areas: management regimes, tenure arrangements and protected area categorization in East Africa. *Oryx*, 42, 177–186.

Catlin, G. (1841) *Letters and Notes on the Manners, Customs and Condition of North American Indians*. Published by the author at the Egyptian Hall, London.

Chambers, R. (1983) *Rural Development: putting the last first*. Longman, London.

Dickson, B. & Adams, W.M. (2009) Science and uncertainty in South Africa's elephant management debate. *Environment and Planning D: Government and Policy*, 27, 110–123.

Dovers, S.R., Norton, T.W. & Handmer, J.W. (1996) Uncertainty, ecology, sustainability and policy. *Biodiversity and Conservation*, 5, 1143–1167.

Egoh, B., Rouget, M., Reyers, B. *et al.* (2007) Integrating ecosystem services into conservation assessments: a review. *Ecological Economics*, 63, 714–721.

Ferraro, P.J. & Kiss, A. (2002) Direct payments to conserve biodiversity. *Science*, 298, 1718–1719.

Foucault, M. (1975) *Discipline and Punish: the birth of the prison*. Gallimard, Paris (in translation: Allen Lane, London, 1997).

Hajer, M.A. (1995) *The Politics of Environmental Discourse: ecological modernization and the policy process*. Oxford University Press, Oxford.

Harrop, S.R. (2003) From cartel to conservation and on to compassion: animal welfare and the International Whaling Commission. *Journal of International Wildlife Law and Policy*, 6, 79–104.

Horowitz, M.M. & Little, P.D. (1987) African pastoralism and poverty: some implications for drought and famine. In *Drought and Hunger in Africa: denying famine a future*, ed. M. Glantz, pp. 59–82. Cambridge University Press, Cambridge.

Hulme, D. & Murphree, M. (eds) (2001) *African Wildlife and Livelihoods: the promise and performance of community conservation*. James Currey, London.

Hutton, J., Adams, W.M. & Murombedzi, J.C. (2005) Back to the barriers? Changing narratives in biodiversity conservation. *Forum for Development Studies*, 32, 341–370.

IUCN (International Union for the Conservation of Nature) (2005) *Benefits Beyond Boundaries. Proceedings of the Vth World Parks Congress*. IUCN, Cambridge.

Jones, J.P.G., Andriamarovololona, M.M., Hockley, N., Gibbons, J.M. & Milner-Gulland, E.J. (2008) Testing the use of interviews as a tool for monitoring trends in the harvesting of wild species. *Journal of Applied Ecology*, 45, 1205–1212.

Kaimowitz, D. & Sheil, D. (2007) Conserving what and for whom? Why conservation should help meet basic human needs in the tropics. *Biotropica*, 39, 567–574.

Keeley, J. & Scoones, I. (2003) *Understanding Environmental Policy Processes: cases from Africa*. Earthscan, London.

Kuhn, T. (1962) *The Structure of Scientific Revolutions*. University of Chicago Press, Chicago.

Latour, B. (1999) *Pandora's Hope. Essays on the reality of science studies*. Harvard University Press, Cambridge, MA.

Leach, M. and Mearns, R. (eds) (1996) *The Lie of the Land: challenging received wisdom on the African environment*. James Currey, London.

Leader-Williams, N. & Dublin, H.T. (2000) Charismatic megafauna as 'flagship species'. In *Priorities for the Conservation of Mammalian Diversity: has the panda had its day?*, eds A. Entwistle & N. Dunstone, pp. 53–81. Cambridge University Press, Cambridge.

Lee, K. (1993) *Compass and Gyroscope: integrating science and politics for the environment*. Island Press, Washington, DC.

Lemelin, R.H. (2008) Doubting Thomases and the cougar: the perceptions of puma management in northern Ontario, Canada. *Sociologia Ruralis*, 49, 56–69.

Lockwood, M., Worboys, G. & Kothari, A. (eds) (2006) *Managing Protected Areas: a global guide*. Earthscan, London.

Miles, L. & Kapos, V. (2008) Reducing greenhouse gas emissions from deforestation and forest degradation: global land-use implications. *Science*, 320, 1454–1455.

Murombedzi, J.S. (1999) Devolution and stewardship in Zimbabwe's CAMPFIRE Programme. *Journal of International Development*, 11, 287–93.

Murombedzi, J.S. (2001) Why wildlife conservation has not economically benefited communities in Africa. In *African Wildlife and Livelihoods*, eds D. Hulme & M. Murphree, pp. 208–226. James Currey, Oxford.

Murphree, M. (1994) The role of institutions in community-based conservation. In *Natural Connections: perspectives in community-based conservation*, eds D. Western, R.M. White & S.C. Strum, pp. 403–427. Island Press, Washington, DC.

Murphree, M. (2000) *Community-based conservation: the new myth?* Unpublished paper presented at the Conference on African Wildlife Management in the New Millennium, Mweka, December.

Naughton-Treves, L., Holland, M.B. & Brandon, K. (2005) The role of protected areas in conserving biodiversity and sustaining local livelihoods. *Annual Review of Environmental Resources*, 30, 219–252.

Neumann, R.P. (2002) The postwar conservation boom in British colonial Africa. *Environmental History*, 7, 22–47.

Ravenel, R.M. & Redford, K.H. (2005) Understanding IUCN Protected Area categories. *Natural Areas Journal*, 25, 381–389.

Rihoy, L., Chirozva, C. & Anstey, S. (2010) "People are not happy": crisis, adaptation and resilience in Zimbabwe's CAMPFIRE Programme. In *Community Rights, Conservation and Contested Land: the politics of natural resource governance in Africa*, ed. F. Nelson, pp. 174–201. Earthscan, London.

Roe, D. & Elliott, J. (2004) Poverty reduction and biodiversity conservation: rebuilding the bridges. *Oryx*, 38, 137–139.

Roe, E. (1991) Development narratives, or making the best of blueprint development. *World Development*, 19, 287–300.

Roe, E. (2004) *Narrative Policy Analysis: theory and practice*. Duke University Press, Durham, NC.

Sinclair, A.R.E & Fryxell, J.M. (1985) The Sahel of Africa: ecology of a disaster. *Canadian Journal of Zoology*, 63, 987–994.

Sullivan, S. (2000) Getting the science right, or introducing science in the first place? Local 'facts', global discourse – 'desertification' in north-west Namibia. In *Political Ecology: science, myth and power*, eds P. Stott & S. Sullivan, pp. 15–44. Edward Arnold, London.

Swift, J. (1996) Desertification narratives: winners and losers. In *The Lie of the Land: challenging received wisdom on the African environment*, eds M. Leach & R. Mearns, pp. 73–90. James Currey/Heinemann, London.

Western, D., Wright, R.M. and Strum, S.C. (eds) (1994) *Natural Connections: perspectives on community-based conservation*. Island Press, Washington, DC.

Woodward, R.T. & Bishop, R.C. (1997) How to decide when experts disagree: uncertainty-based choice rules in environmental policy. *Land Economics*, 73, 492–507.

Wunder, S. (2006) The efficiency of payments for environmental services in tropical conservation. *Conservation Biology*, 21, 48–58.

17

Conservation Trade-offs and the Politics of Knowledge

J. Peter Brosius

Center for Integrative Conservation Research, Department of
Anthropology, University of Georgia, Athens, Georgia, USA

Introduction

Complex trade-offs occur between human well-being and biodiversity conservation goals in specific places, and between conservation and other economic, political and social agendas at local, national and international scales. In researching these trade-offs,[1] a common response was that there is really nothing new left to say about trade-offs – that anything interesting that might be said about trade-offs had already been said 30 years ago, in the field of economics or political science, or in some other realm of inquiry. I would argue this is patently wrong. While the idea of trade-offs is well established in certain disciplines, or manifested in now-conventional analytical perspectives, there is still much new to be said about the concept of trade-offs. Furthermore, new insights about the concept of trade-offs are most likely to result from the series of theoretical perspectives that emerged mostly *after* all those old trade-offs questions were purportedly solved decades ago.

A key challenge now is identifying what some of those emergent theoretical perspectives might be. What are the new questions that need to be asked about trade-offs, and what kinds of theoretical or conceptual perspectives will

---

[1] 'Advancing Conservation in a Social Context' (ACSC) research initiative, see: www.tradeoffs.org.

Trade-offs in Conservation: Deciding What to Save, 1st edition. Edited by N. Leader-Williams,
W.M. Adams and R.J. Smith. © 2010 Blackwell Publishing Ltd.

provide the most innovative and productive new lines of inquiry? This chapter offers an argument for why paying attention to the politics of knowledge might teach us something new about trade-offs in general, and conservation and development trade-offs in particular.

Background issues: conservation and trade-offs

Before defining the meaning for 'politics of knowledge' I first make a number of very basic points about conservation and about the concept of trade-offs.

Conservation

Three key points need to be understood about conservation. First, conservation is constantly changing, a shifting configuration of institutions, initiatives, funding streams, alliances, practices, buzzwords and critiques. Second, conservation is a nexus of relationships – between large organizations and donors, between organizations and national governments, between national governments and local people, and more. All of these relationships are negotiated in various ways, and some actors are privileged and some actors are marginalized to varying degrees in the process. Third, conservation is inherently political in all the ways that it involves institutions, local communities, livelihoods, legal codes, knowledge-making practices, and more: establishing and enforcing boundaries; curtailing subsistence activities; negotiating benefits; making ecoregional maps; and applying reserve selection algorithms. But different actors – and different analytical perspectives – locate the politics in different domains of conservation practice.

Trade-offs

Three key points also need to be understood about trade-offs. First, people tend to view the concept of trade-offs through the lens particular to their field of inquiry, and fail to appreciate that the idea has a very diverse range of applications and meanings. The concept is part of the working vocabularies of several different fields of inquiry – economics, political science, ethics, biology – each of which have used the concept in different ways, at different times, in different debates, with different kinds of data. One can perceive a kind of spectrum. Some employ the concept in conjunction with a rigorous and highly formalized set of methods, yielding analyses that are presumably objective and precise: the life history approach in evolutionary theory; optimization models in behavioral ecology; cost–benefit analysis in economics; and decision support systems, to name just a few. Other disciplines or fields of practice employ the idea of trade-offs as an abstract concept or guiding assumption, without any necessary reference to method: the idea of rational choice in economics; the idea of intergenerational equity in the field of ethics; or the idea of compromise in the field of conflict resolution. Our understanding of trade-offs is complicated further when we recognize that the idea is embedded in implicit and eclectic ways in a huge variety of other domains – in polemics and debates (e.g. 'people versus parks'); in the outcomes of international fora (e.g. the Millennium Development Goals); in programmatic rhetorics (win–win) that appear in funding proposals and project documents; and in the practices of conservation actors and institutions (priority-setting approaches to conservation). We need to acknowledge that a lot of discussion about trade-offs is carried on through the medium of predicates. In virtually any context in which the idea of choice, decision, triage, cost, benefit, compromise or priority is invoked, explicitly or implicitly, the concept of trade-offs is present.

Second, the issue of *scale* is embedded in a multiplicity of ways in the analysis and implementation of conservation trade-offs, and any attempt to further any understanding of trade-offs must engage seriously with the matter of scale. The focus might for instance be on understanding the kinds of trade-offs that result from conservation practices operating at different *spatial* scales: site-based versus global priority-setting approaches, for example. Conservation trade-offs may also need to be understood over longer *temporal* scales, as conservation paradigms shift (Adams, this volume, Chapter 16). Furthermore, the focus might be on how different *institutional* scales or scales of governance impinge on the way trade-offs are made. Finally, recalling the point that the idea of trade-offs is embedded in our concepts, debates and forms of practice in a number of *implicit* ways, it is crucial to acknowledge the ways in which scalar trade-offs may be embedded in such implicit ways as well. Consider, for instance, how the idea of trade-offs might be embedded within ideas about 'scaling-up' or 'transaction costs'.

Third, the 'social context' in which conservation is carried out encompasses much more than the social realities that exist at specific sites of intervention.

Conventional thinking about conservation holds that assessing the success or failure of specific conservation projects requires attention to the local contexts in which those projects are focused. While important dynamics of conservation can be perceived through site-based research, there is much that cannot. Developing new understandings of how conservation trade-offs are identified, calculated, analyzed and negotiated requires that research focuses not only on particular sites of implementation, whether individual protected areas or specific ecoregions, but also on sites of organization, concept development and planning: conservation organizations, donors and others.

Politics of knowledge and analysis of conservation trade-offs

In the last few decades, scholars in several disciplines have devoted much effort to rethinking the issue of power. They have come to recognize that the contours of power are more convoluted, and more implicit, than once thought. Power, in this emergent view, is no longer just about coercion. It is also about who establishes the categories that we take to be natural and immutable, how those categories get produced, how they are reproduced through the production of knowledge, and how they operate through institutions (Foucault, 1980; Dirks *et al.*, 1993; Escobar, 1995; Mathews, 2005). Thus the theoretical landscape across a range of disciplines is defined by a concern with the links between knowledge and power. The boundaries between epistemology and politics are therefore much more problematic than once assumed. We recognize that *who* we are has a great deal to do with what we can claim to know and with how valid others take our knowledge to be. As a result, the argument goes, we can no longer take our categories, our knowledge-making practices or our representational conventions for granted.

This body of work on the politics of knowledge is premised on recognition that all forms of knowledge are inherently political and it challenges the assumption that there is any such thing as 'objective' knowledge (Haraway, 1988; Rouse, 1992). Scholars working in this area have demonstrated this in the context of medicine, education, development, environment, technology and in numerous other domains (Banuri, 1990; Pigg, 1992; Ferguson, 1994; Escobar, 1995; Epstein, 1996). They examine how knowledge is produced and who is empowered to produce it, how it circulates, how some forms of knowledge are taken to be authoritative while others are marginalized, and how some forms of knowledge are taken to be credible by certain categories of actors or contested by others (Franklin, 1995; Jasanoff, 2004; Nadasdy, 2005; Lowe, 2006). They have also explored how various forms of knowledge are represented – narratively, statistically, cartographically – and how various forms of visualization have unexpected implications (Scott, 1998; Brosius, 2006a).

Consequently, any effort to analyze the contemporary conservation domain must engage with the issue of knowledge (Figure 17.1). The effort by actors and institutions to address the erosion of global biodiversity intersects with the production, distribution and application of knowledge in myriad ways. There are enumerable manifestations of this: virtually all major conservation organizations stress that their efforts are firmly grounded in, and guided by, the production of credible scientific knowledge; IUCN, the International Union for the Conservation of Nature and other organizations have followed the lead of development institutions in promoting more rigorous 'knowledge management'; conservation organizations increasingly devote resources to the monitoring and evaluation of conservation initiatives; numerous individuals and institutions are linked through learning networks dedicated to sharing knowledge of best practices; and we have recently witnessed an increasing valorization of the role that indigenous knowledge/traditional ecological knowledge (IK/TEK) should play in conservation (Berkes, 1999; Stripen & DeWeerdt, 2002). In the global conservation realm, emergent systems of environmental governance are mostly produced through categories and knowledge-making practices promulgated by major organizations located in the global North (Brosius *et al.*, 1998, 2005; Lowe, 2006; West, 2006).

The way the politics of knowledge is understood in conservation is the key to understanding the processes by which trade-offs are identified, calculated, analyzed and negotiated through their embeddedness in broader configurations of power. The rhetoric of trade-offs in the field of conservation and development is premised on assumptions about the existence of certain categories of 'key actors' who effect, or are affected by, decisions made regarding trade-offs. These include 'policy makers', 'decision makers', 'stakeholders', the 'scientific community' and 'local communities', among others.

Embedded within these categories is a set of implicit assumptions about how agency is, or should be, exercised and who is empowered to make decisions. The concept of agency is one of the foundations of contemporary social theory across a range of academic fields. This concept draws attention '*to the role of*

Figure 17.1 **An American graduate student David Himmelfarb conducting an interview on conservation conflicts with the farmer Jackson Chesang on the edge of Mt Elgon National Park, Uganda. The power of knowledge is recognized to be increasingly important in conservation discourse. (Photograph by kind permission of William Cheptegei.)**

the human actor as individual or group in directing or effectively intervening in the course of history' (Brooker, 1999: 3). According to Ashcroft *et al.* (1998), *'Agency refers to the ability to act or perform an action. In contemporary theory, it hinges on the question of whether individuals can freely and autonomously initiate action, or whether the things they do are in some sense determined by the ways in which their identity has been constructed'*. Agency, in short, refers to the ability of different kinds of actors to act given the cards they are dealt (or the cards they are able to manufacture).

Regardless of whether trade-offs are being referred to explicitly, or referred to as one of its predicates, whether choice, decision making or priority setting, a series of categories is being reproduced that obscure the ways in which agency is exercised in the conservation realm. The question for conservation practitioners, then, is how conventional forms of conservation science and practice reproduce certain configurations of power through the production of knowledge, and how does knowing this inform our understanding of trade-offs? The answer lies, in part, in framing a series of more proximate questions pertaining to conservation trade-offs and the politics of knowledge, such as are outlined below:

- What are the most significant forms of knowledge production in the field of conservation and development that can be linked to understanding trade-offs?
- How does knowledge circulate within the conservation domain, and how does this affect how trade-offs are made?
- What forms of agency are embedded or elided in different renderings and predicates of trade-offs?
- What are some of the key trends in global conservation practice, what forms of knowledge production are these linked to, and how do these shape how trade-offs are made?
- How have various forms of conservation knowledge production been shaped or manipulated to fit larger political/institutional interests, and how do these shape how trade-offs are understood and negotiated?
- What are some of the key nodes or sites salient to understanding global conservation trends? What kinds of analytical tools exist for understanding trade-offs that occur at these nodes/sites as well as the links (scalar, networks) between them?
- What are some of the key predicates of the idea of trade-offs (choice, compromise, priority), and how can we better understand them in the context of conservation?
- How does the imperative for model-driven conservation affect the possibilities for understanding and negotiating trade-offs?
- What are some of the existing or emerging structures of accountability in conservation, and how do these structures condition the processes by which trade-offs are made?

These are only some of the kinds of questions that might be asked in seeking to understand conservation trade-offs and the politics of knowledge.[2]

---

[2] In the last decade, there has been a remarkable proliferation of academic studies of conservation that address these kinds of questions, particularly by anthropologists and geographers. Exemplars of this body of work include Anderson and Berglund (2003), Brosius (2006b), Haenn (2005), Lowe (2006), Nadasdy (2005) and West (2006).

Other avenues to analyze conservation trade-offs and politics of knowledge

A number of other avenues may be followed through which conservation and development trade-offs might be analyzed with reference to the politics of knowledge. In the following, I suggest a few. These examples are intended only to indicate some of the breadth of approaches possible; clearly there are numerous other lines of inquiry that might shed light on conservation and development trade-offs.

Technologies of visualization

Conservation initiatives take shape because certain places, or the species that inhabit them, are perceived to be at risk. What lies behind this deceptively simple observation is a series of questions concerning how and by whom such perceptions of risk are formed. What technologies of visualization – such as geographic information systems (GIS), rapid ecological assessment, gap analysis and rapid rural appraisal (RRA) – are used in conservation to make natural and cultural communities *legible*? In invoking the idea of legibility, I draw on the work of James Scott in *Seeing Like a State* (1998). Legibility for Scott was achieved through a series of 'state simplifications' designed to reduce the opacity of the local, and is the 'central problem of statecraft'. It is also the central problem of conservation.

Recent scholarship by geographers and others has shown us how maps and other forms of visualization reinforce certain configurations of power (Harley, 1988; Orlove, 1991; Brosius & Russell, 2003; Brosius, 2006b). How, they ask, are the many strategic mediations involved in the making of maps, each subject to extensive commentary by their designers – decisions about scale, biogeographic boundaries and 'potential natural vegetation' – made to disappear in the final product, which is then taken to represent the actual or potential natural state of the world? How do maps overwrite human presence in the landscape and efface histories? Such effacements occur in many ways. One is through the cartographic privileging of natural boundaries over political boundaries. Another is the overwriting of human use of the environment in the production of both maps and the strategic plans that they generate, whether in the guise of land classification categories, assumptions about 'potential natural vegetation', 're-wilding' or the establishment of corridors.

Another form of effacement is the coding of people as threats. What happens when all the cartographic and algorithmic visualizing power is turned on the assessment of threat, and when 'disturbed' establishes itself as a stand-in for 'inhabited'? Each of these raises the question of how assumptions about human communities become coded cartographically in maps, and how this in turn produces capillary processes of power by which visualizations are transferred from map to ground.

Maps are not the only technologies of visualization employed in contemporary conservation and development interventions. We might ask what are some of the other key technologies of visualization being used today – economic models such as those that underlie Conservation International's (CI) Enforcement Economics initiative, reserve selection algorithms or decision support models – and how do the visualizations that result from using them influence the ways in which trade-offs are identified, calculated, analyzed and negotiated?

The positioning of social science knowledge in conservation

Technologies of visualization also include the kinds of social research methods employed in conservation and development initiatives. In recent years, the social sciences have played an increasingly important role in the planning and implementation of conservation projects (Mascia *et al.*, 2003; Russell & Harshbarger, 2003; Campbell, 2005; Agrawal & Ostrom, 2006; Brosius, 2006b; Fox *et al.*, 2006; Adams, 2007; Buscher & Wolmer, 2007). Most conservation initiatives mandate some sort of 'social science' or 'socioeconomic' component, although this can mean any number of things and encompass a range of disciplines.

Two observations can be made about the positioning of the social sciences within conservation. First, a shift has been witnessed toward rapid and/or formalistic protocols for social research that are primarily aimed at providing information on populations around sites, and at identifying the threats posed by local communities. The literature on social research in conservation is replete with discussions of rapid methods, and they are given primacy within most conservation and development organizations. To the extent that such analyses do consider the concerns of local communities, it is through the lens of a stakeholder-based approach that reduces most needs, concerns and sentiments to 'interests'. Those who rely on quantitative rapid/survey methods are much better positioned to provide the sorts of deliverables that conservation organizations find useful. However, one might argue that rapid methods are problematic in describing local communities because they provide a false sense of reliability. They in fact 'mis-read' local realities in numerous ways. The fundamental problem with these types of research methods is that they are designed for the convenience of researchers and implementers.

Second, social science in conservation gets positioned as the science of 'the local'. That is to say, the social sciences are seen as operating primarily at the local level, in communities where conservation interventions occur. Social scientists are viewed as offering expertise in the production of local-level data. Sometimes they are included in the context of monitoring and evaluation exercises; at other times they are brought in to help conservation practitioners understand and ameliorate conflict. The important question for us is how the use of various social science methods or approaches condition perceptions about local 'stake-holders', and how this in turn influences the ways in which trade-offs are identified, calculated, analyzed and negotiated.

Responsibility for articulating the potential contribution of the social sciences to understanding conservation and development trade-offs lies in part with social scientists themselves, who must be more forthright in demonstrating how they can address trade-offs and decision making by conservation actors operating at or across different scales. This is particularly crucial given the recent proliferation of market-based conservation strategies that engage more directly with macroeconomic structures and processes. Doing so would also provide a necessary corrective to recent critical social science treatments of conservation, which often treat major conservation organizations as powerful actors in conservation decision making while taking inadequate account of the role of other power actors such as extractive industries.

The politics of translation

At a time when English has emerged as a globally dominant language, it is worth considering what role English language dominance plays in transnational conservation and development projects, particularly as these pertain to the calculation of trade-offs. Global conservation practices are increasingly reproduced through categories and managerial rhetorics that emanate from the global North (Sachs, 1993; Adams & Mulligan, 2003; Brosius, 2004; MacDonald, 2004; Lowe, 2006; West, 2006; Zimmerer, 2006; Campbell, 2007). Or as Peruvian lawyer Bruno Monteferri, a contributor to this volume (Chapter 13), said in a recent email, they emerge in the North as debate, and land in the South as mandate. At what point do we encounter the limits of commensurability in the process of translating an idea such as trade-offs across a broad spectrum of languages? At a recent meeting, [3] two Peruvians noted the difficulty they were having in translating the term 'trade-offs' into Spanish (Figure 17.2). Their Vietnamese colleagues

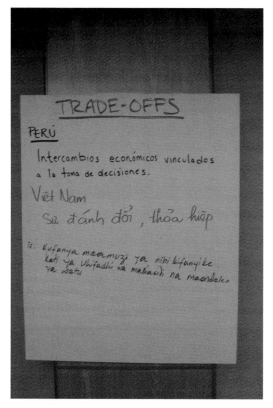

Figure 17.2 **The difficulty faced by delegates from Peru, Vietnam and Tanzania in translating the term 'trade-offs' into their own languages, at a workshop organized by Advancing Conservation in a Social Context and held in October 2007. The hegemony of English is often underestimated in debates on conservation policy. (Photograph by Peter Brosius.)**

[3] ACSC meeting in February 2007: Manuel Pulgar-Vidal and Bruno Monteferri, both from the *Sociedad Peruana de Derecho Ambiental*.

immediately responded that they were having the very same problem, and that the only Vietnamese term they could come up with was one that meant having to 'compromise from a position of weakness'. Conservation practitioners have not adequately acknowledged the linguistic hegemony of English that permeates global circuits of conservation knowledge production and practice, nor the politics of translation that emerge as conservation actors in the global North and South engage across the differences of language in the conceptualization and enactment of conservation trade-offs.

Indigenous knowledge

In the past decade, we have seen an increasing valorization of indigenous knowledge as being relevant to conservation and resource management decision making (Sillitoe, 1998; Agrawal, 2002; Brosius, 2006a). Reference to indigenous or local knowledge, often with the acronyms IK/TEK, is generally applied to knowledge of the natural world: what such groups know

Figure 17.3   The nomadic Penan headman Tebaran Agut illustrating his knowledge of the rainforest by describing the location of sago palm groves in the surrounding watershed. Many in the policy and scientific arenas do not regard indigenous knowledge as highly as formal scientific knowledge. (Photograph by Peter Brosius.)

about the resources they exploit and how these societies cognize or interpret natural processes (Figure 17.3). That the value of local/indigenous knowledge is at last being recognized, rather than dismissed as anecdotal or irrelevant, is clearly a positive development. But that we limit our valorization of knowledge largely to that which pertains to the natural world, consigns that knowledge to a form of knowledge that is subordinate to the forms of knowledge possessed by decision makers (Brosius, 2006a; Knight & Cowling, this volume, Chapter 15). Further, one can draw a distinction between indigenous/local knowledge mediated by the research activities of social scientists and that articulated by local/indigenous activists/advocates themselves. One speaks in the passive voice of science – translating indigenous ways of knowing into forms intelligible to practitioners and decision makers; the other speaks in the active voice of advocacy. Making this distinction draws our attention to the question of how local/indigenous perspectives and ways of knowing are elicited and translated between scales, and how the link is made between this knowledge and the policy domain in negotiating trade-offs.

Conclusions: recognizing regimes of credibility

The effort by key actors and institutions to address the erosion of global biodiversity intersects with the production, distribution and application of knowledge in myriad ways. How the politics of knowledge is understood in conservation is key to understanding the processes by which trade-offs are identified, calculated, analyzed and negotiated through their embeddedness in broader configurations of power.

A fundamental starting point for the analysis of conservation trade-offs is to understand how particular problems or issues are formulated and bounded in the first place. Problem definition and bounding lay the foundation for any subsequent analysis of trade-offs. That is to say, the analysis of trade-offs should begin not at the end of the policy sequence, when those trade-offs are actually being made, but at the beginning, when the terms of trade-offs are being determined as the scope of the problem is being defined (Adams, this volume, Chapter 16).

It is also crucial to acknowledge the problem of unequal knowledge. Too often frameworks for addressing conservation trade-offs assume the parity and equity of bargainers, but this assumption is highly problematic. In reality, access to knowledge and the ability to utilize knowledge is inevitably unequal.

Thus the ability of different groups to imagine the implications of particular choices for the future is also quite variable. It is not difficult to imagine contexts in which people 'on the ground' may not even know they are actors in trade-offs being made at other scales, making it difficult to determine whether a trade-offs is being negotiated or imposed. A framework of trade-offs may thus be inherently unjust unless these knowledge inequalities are acknowledged and addressed at a fundamental level within the framework.

Finally, it is imperative to recognize that the making of trade-offs is inevitably highly political, but that the politics may be submerged in a technical process of planning and implementation – what Ferguson (1994) referred to as 'anti-politics', those paradigms and analytical frameworks that conceal politics by defining the problem as technical. Too often the framing of conservation trade-offs misses or conceals the politics of argument that determines what the outcome is, and assumes the need to get rid of disagreements and flatten incommensurabilities of value in order to carry out trade-offs.

Acknowledging these various manifestations of the politics of knowledge, this chapter offers an alternate strategy for thinking about knowledge politics in the making of conservation trade-offs. As noted, virtually all major conservation organizations stress that their efforts are firmly grounded in, and guided by, the production of credible scientific knowledge. For most conservation and development practitioners, it is taken as an article of faith that if their work is to be taken seriously by the scientific community and by policy makers, it must be credible. By this they mean that we must build a firewall between science as the domain of facts and advocacy as the domain of politics. It is as if credibility is a pure, singular, free-floating, abstract entity that exists outside the human sphere.

An alternative approach is to recognize that credibility is plural and contextual, rather than singular and abstract. Thus, the analysis of conservation and development trade-offs must be premised on recognition of the significance of the *multiple regimes of credibility* that exist in the contexts in which academics, practitioners, state authorities, community members and other actors work. Many kinds of actors are weighing in on conservation decision making involving trade-offs, and it is no longer possible to write for a single audience even if that was appropriate. Thus, research findings that are salient and credible to the scientific community, to policy makers or to donors may not be salient or credible to those in the communities where conservationists work. Insistence that credibility pertains exclusively to the production of certain kinds of scientific information ignores or disregards forms of credibility that are important to multiple other kinds of actors, including local communities. Credibility cannot only 'look up' to members of the scientific community, but it must also look down and around to other actors or communities. Accepting this proposition entails recognizing that establishing credibility requires much more than just doing good science. Credibility is first and foremost a form of relationship, premised on the trust that one set of actors has in the integrity, reliability and legitimacy of information provided by another. The goal, then, should be to establish credibility not only with the scientific community, but with as broad a range of actors as possible (Knight & Cowling, this volume, Chapter 15). Acknowledging the plurality of regimes of credibility entails an engagement with the complexities of the multiple contexts in which the production and dissemination of knowledge, and the calculation of trade-offs, takes place today.

This may not be a solution to all the issues raised by a politics of knowledge perspective. However, even recognizing the multiple forms of credibility that characterize the contexts in which conservation practitioners work at least provides a starting point for a more adequate understanding of the ways in which conservation trade-offs are identified, calculated, analyzed and negotiated.

Acknowledgments

This chapter results mostly from my engagement in a collaborative effort to define a research programme focused on conservation and development trade-offs in conjunction with the 'Advancing Conservation in a Social Context' (ACSC) research initiative. ACSC is a 3-year, international, interdisciplinary research programme initiated by the MacArthur Foundation to investigate the complex trade-offs between human well-being and biodiversity conservation goals in specific places, and between conservation and other economic, political and social agendas at local, national and international scales (see www.tradeoffs.org). This chapter draws extensively on an ACSC workshop held in late 2007 on the politics of knowledge.

References

Adams, W.M. (2007) Thinking like a human: social science and the two cultures problem. *Oryx*, 41, 275–276.
Adams, W.M. & Mulligan, M. (eds) (2003) *Decolonizing Conservation: strategies for conservation in a post-colonial era*. Earthscan, London.

Agrawal, A. (2002) Indigenous knowledge and the politics of classification. *International Social Science Journal*, 173, 287–297.

Agrawal, A. & Ostrom, E. (2006) Political science and conservation biology: a dialogue of the deaf? *Conservation Biology*, 20, 683–685.

Anderson, D.G. & Berglund, E. (2003) *Ethnographies of Conservation: environmentalism and the distribution of privilege*. Berghahn Books, New York.

Ashcroft, W., Griffiths, G. & Tiffin, H. (1998) *Key Concepts in Post-Colonial Studies*. Routledge, London.

Banuri, T. (1990) Development and the politics of knowledge: a critical interpretation of the social role of modernization theories in the development of the Third World. In *Dominating Knowledge: development, culture, and resistance*, eds F. Apffel-Marglin & S.A. Marglin, pp. 73–101. Oxford University Press, Oxford.

Berkes, F. (1999) *Sacred Ecology: traditional ecological knowledge and resource management*. Taylor & Francis, Philadelphia.

Brooker, P. (1999) *A Concise Glossary of Cultural Theory*. Edward Arnold, London.

Brosius, J.P. (2004) Indigenous peoples and protected areas at the World Parks Congress. *Conservation Biology*, 18, 609–612.

Brosius, J.P. (2006a) What counts as local knowledge in global environmental assessments and conventions? In *Bridging Scales and Knowledge Systems: concepts and applications in ecosystem assessment*, eds W.V. Reid, F. Berkes, T.J. Wilbanks, & D. Capistrano, pp. 129–144. Island Press, Washington, DC.

Brosius, J.P. (2006b) Seeing communities: technologies of visualization in conservation. In *The Seductions of Community: emancipations, oppressions, quandries*, ed. G.W. Creed, pp. 227–254. School of American Research Press, Santa Fe, NM.

Brosius, J.P. & Russell, D. (2003) Conservation from above: an anthropological perspective on transboundary protected areas and ecoregional planning. *Journal of Sustainable Forestry*, 17, 39–65.

Brosius, J.P., Tsing, A. & Zerner, C. (1998) Representing communities: histories and politics of community-based natural resource management. *Society and Natural Resources*, 11, 157–168.

Brosius, J.P., Tsing, A. & Zerner, C. (eds) (2005) *Communities and Conservation: histories and politics of community-based natural resource management*. Altamira Press, Lanham, MD.

Buscher, B. & Wolmer, W. (2007) Introduction: the politics of engagement between biodiversity conservation and the social sciences. *Conservation and Society*, 5, 1–21.

Campbell, L. (2005) Overcoming obstacles to interdisciplinary research. *Conservation Biology*, 19, 574–577.

Campbell, L. (2007) Local conservation practice and global discourse: a political ecology of sea turtle conservation. *Annals of the Association of American Geographers*, 97, 313–334.

Dirks, N., Eley, G. & Ortner, S. (eds) (1993) *Culture/Power/History: a reader in contemporary social theory*. Princeton University Press, Princeton, NJ.

Epstein, S. (1996) *Impure Science: AIDS, activism, and the politics of knowledge. Medicine and Society Volume 7*. University of California Press, Berkeley, CA.

Escobar, A. (1995) *Encountering Development: the making and unmaking of the Third World*. Princeton University Press, Princeton, NJ.

Ferguson, J. (1994) *The Anti-politics Machine: 'development', depoliticization and bureaucratic power in Lesotho*. University Minnesota Press, Minneapolis, MN.

Foucault, M. (1980) *Power/Knowledge: selected interviews and other writings, 1972–77*, translated by C. Gordon. Pantheon, New York.

Fox, H.E., Christian, C., Cully Nordby, J. *et al.* (2006) Perceived barriers to integrating social science and conservation. *Conservation Biology*, 20(6), 1817–1820.

Franklin, S. (1995) Science as culture, cultures of science. *Annual Review of Anthropology*, 24, 163–184.

Haenn, N. (2005) *Fields of Power, Forests of Discontent: culture, conservation and the state in Mexico*. University of Arizona Press, Tuscon, AZ.

Haraway, D. (1988) Situated knowledges: the science question in feminism and the privilege of partial perspective. *Feminist Studies*, 14, 575–599.

Harley, J.B. (1988) Maps, knowledge, and power. In *The Iconography of Landscape*, eds D. Cosgrove & S. Daniels, pp. 277–312. Cambridge University Press, Cambridge.

Jasanoff, S. (2004) *States of Knowledge: the co-production of science and social order*. Routledge, New York.

Lowe, C. (2006) *Wild Profusion: biodiversity conservation in an Indonesian Archipelago*. Princeton University Press, Princeton, NJ.

MacDonald, K. (2004) Developing 'nature': global ecology and the politics of conservation in northern Pakistan. In *Confronting Environments: local understanding in a globalizing world*, ed. J. Carrier, pp. 71–96. Altamira Press, Walnut Creek, CA.

Mascia, M., Brosius, J.P., Dobson, T. *et al.* (2003) Conservation and the social sciences. *Conservation Biology* 17, 649–650.

Mathews, A.S. (2005) Power/knowledge, power/ignorance: forest fires and the state in Mexico. *Human Ecology*, 33, 795–820.

Nadasdy, P. (2005) *Hunters and Bureaucrats: power, knowledge and aboriginal–state relations in the southwest Yukon*. University of Washington Press, Seattle, WA.

Orlove, B. (1991) Mapping reeds and reading maps: the politics of representation in Lake Titicaca. *American Ethnologist*, 18, 3–38.

Pigg, S. (1992) Constructing social categories through place: social representations and development in Nepal. *Comparative Studies in Society and History*, 34, 491–513.

Rouse, J. (1992) What are cultural studies of scientific knowledge? *Configurations*, 1, 57–94.

Russell, D. & Harshbarger, C. (2003) *Groundwork for Community-Based Conservation*. Altamira Press, Walnut Creek, CA.

Sachs, W. (ed.) (1993) *Global Ecology: a new arena of political conflict*. Zed Books, London.

Scott, J. (1998) *Seeing Like a State*. Yale University Press, New Haven, CT.

Sillitoe, P. (1998) The development of indigenous knowledge. *Current Anthropology*, 39, 223–252.

Stripen, C. & DeWeerdt, S. (2002) Old science, new science: incorporating traditional ecological knowledge into contemporary management. *Conservation in Practice*, 3, 20–27.

West, P. (2006) *Conservation is Our Government Now: the politics of ecology in Papua New Guinea*. Duke University Press, Durham, NC.

Zimmerer, K. (2006) *Globalization and New Geographies of Conservation*. University of Chicago Press, Chicago.

Part V
Future Challenges

18

Climate Change and Conservation

Stephen G. Willis, David G. Hole and Brian Huntley

Centre for Ecosystem Sciences, School of Biological and Biomedical
Sciences, Durham University, Durham, UK

Introduction

In a world of increasingly isolated and fragmented populations of individual
species, protected areas are now vital to maintain biodiversity. However,
recently observed changes in global climate, and future forecasts, are likely
to result in large range shifts for many taxa. In turn, this may lead to high
turnover of species in protected areas (Thuiller *et al.*, 2006; Hannah *et al.*, 2007;
Hole *et al.*, 2009) or local extinctions of less mobile and isolated populations.
Protected area networks as a whole tend to cover available climate space well
at a continental scale, suggesting that they could be robust to climatic change.
However, the abilities of different species to move between protected areas will
vary. Furthermore, such moves will also depend on the presence of other key
species and habitats in newly suitable climatic areas. Combined, these factors
could make it uncertain whether shifts will be realized.

   In this chapter, we explore these issues and discuss the inevitable trade-offs
that result from the impacts of climate change on biodiversity. We start by
reviewing the literature on these impacts and discuss how they will affect
the main strategies currently used by conservationists. We then suggest an
integrated approach for mitigating the impacts of climate change and describe
the trade-offs that this will involve. Given the potential for current protected
area networks as a whole to conserve species, we suggest a paradigm shift is
needed in the designation of protected areas. This shift will need to emphasize

Trade-offs in Conservation: Deciding What to Save, 1st edition.   Edited by N. Leader-Williams,
W.M. Adams and R.J. Smith.   © 2010 Blackwell Publishing Ltd.

the importance of protected areas for harbouring a range of species both now and in the future, rather than the designation of protected areas being based solely on their current complement of species, as is currently the case.

Global climate change

It is widely acknowledged that the global climate is changing (IPCC, 2007) and at a rate more rapid than in the last several millennia (Jansen et al., 2007). Climatic changes are not occurring evenly across the globe. Temperature increases have been more extreme over land masses, while regions at high latitudes have warmed to a greater extent than regions at lower latitudes. This has led to speculation that climate change will be a greater threat to biodiversity at high latitudes, whereas habitat loss will remain the primary threat in tropical and subtropical regions (Jetz et al., 2007). The mean of projected global warming for the end of the 21st century lies in the range 1.8–4.0°C, with an overall likely range of 1.1–6.4°C, based upon multi-model averaged projections (IPCC, 2007). Polar regions, and the northern polar region in particular, are expected to experience temperature increases of ~4–7°C by the end of the current century, while regions at low latitudes are projected to experience temperature increases of ~1.5–4°C. Precipitation is expected generally to increase at high latitudes and to decrease in subtropical land regions.

Most evidence on responses to past climate changes points to species shifting their ranges to remain within the bounds of their climatic tolerances (Huntley & Webb, 1989). Therefore, it should be expected that most species will respond in the same way to future changes. However, two major factors may constrain the abilities of species to respond. First, the change in climate is much more rapid than any experienced in the last 10 000 years and is very likely to be more than 10 times faster than the global warming that occurred between the last glacial maximum and the Holocene (Jansen et al., 2007). Second, natural landscapes are now more fragmented as a result of human land use modification than in the past. Consequently, many species may find it more difficult to achieve the required range shifts.

Impacts of recent climate change on species

Climate change can have direct effects upon species, for example by causing pollination or seed production failure in plants (Pigott & Huntley, 1981) or through increased mortality of overwintering bird species during prolonged periods of harsh weather (Greenwood & Baillie, 1991; Peach *et al.*, 1995). A recently documented extinction caused directly by climate change is that of the golden toad *Bufo periglenes*. This Costa Rican cloud forest species became extinct as climatic warming forced the moist environments upon which the species depended above the level of its mountainous forest habitat (Pounds *et al.*, 1999). However, not all direct impacts are deleterious. The Dartford warbler *Sylvia undata* has expanded its range, and its survival has increased, as a result of recent milder winters in the UK (Westerhoff & Tubbs, 1999). Furthermore, extended growing seasons permit butterflies to emerge earlier in the year (Roy & Sparks, 2000).

It is more common for species to be impacted indirectly by climate change. There are many examples of species that have shifted their ranges, or whose population size is altering, as an indirect result of climate change. However, unravelling the effects of climate change from habitat loss and fragmentation can be difficult. Well-documented examples of indirect climatic change impacts include the mistiming of breeding by European woodland birds in relation to the peak abundance of food for their chicks. In these cases, warming has triggered an earlier onset of the annual cycle of insect prey, whereas the onset of breeding by the birds is unresponsive to warmer conditions (Both *et al.*, 2006). There are many documented examples of species altering their ranges in response to recent climate changes (Walther *et al.*, 2002; Parmesan & Yohe, 2003; Hickling *et al.*, 2006). There is also increasing evidence of recent climate change altering the abundances of species (Gregory *et al.*, 2009). However, elucidating which of these changes have occurred as a result of direct, rather than indirect, effects is difficult.

Potential problems arising from future climate change

In future, it is anticipated that many species will alter their ranges substantially in response to changing climate. On average, European birds are projected to shift their ranges to the north or northeast by 258–882 km by the end of the 21st century (Huntley *et al.*, 2008), resulting in reduced range sizes for many species. Even if European birds track suitable climate perfectly, their ranges are projected to decline on average to 72–89% of their current size. Perhaps more importantly, the mean overlap between areas projected to be climatically suitable for European bird species currently and by the end of the 21st century is only 31–53% of their current range extent. From a conservation perspective, species of greatest conservation concern often fare worst, especially those with small range extents. For species with limited dispersal capability, range sizes could shrink considerably and some species may be threatened with extinction. Studies on other taxa in Europe and elsewhere have produced similar projections (Huntley *et al.*, 1995; Sykes, 1997; Hill *et al.*, 2003; Araújo *et al.*, 2004, 2006; Thuiller *et al.*, 2006). Projections of climate change in the tropics and sub-tropics suggest that similar impacts might be expected in these regions (Plate 18.1).

Additional problems for migratory species

Migratory species, particularly long-distance migrants, face the additional complication that climate change is likely to occur in both their breeding and non-breeding ranges, although the changes will not necessarily be of the same nature or in the same direction in both ranges (Doswald *et al.*, 2009). Many trans-Saharan migrant bird populations (Figure 18.1) are already known to be affected by climate in both their breeding and non-breeding ranges (Peach *et al.*, 1991; Marchant, 1992). For trans-equatorial migrants, breeding and non-breeding ranges may move further apart as a result of a general poleward shift of ecosystems in both hemispheres. This will entail migrating longer distances and may necessitate an increase in stopover sites used between the two end points of migratory journeys. Species whose ranges shift longitudinally may also in future require new stopover points to support an altered migratory route, as well as to increase the overall distances across which they migrate. Alternatively, such species may maintain traditional routes, and subsequently continue their migration to newly suitable grounds (Ruegg *et al.*, 2006). Conversely, some species may reduce the distances across which they migrate in future. For example, some current trans-Saharan migrants may be able to spend the non-breeding season in the Mediterranean region (Plate 18.2). In future, some currently migratory species may become residents in their breeding area. There is evidence that some species are already becoming increasingly sedentary in areas where they have previously been predominantly migratory (Rivalan *et al.*, 2007), which may impact upon resident species.

Figure 18.1   Breeding populations of the sedge warbler *Acrocephalus schoenobaenus* in Britain have been strongly influenced by the wet season rainfall on their non-breeding grounds in sub-Saharan West Africa (Peach *et al.*, 1991). (Photograph by Stephen Willis.)

Strategies to conserve biodiversity

In light of potential changes in Earth's climate, current conservation strategies must be adapted to maximize the long-term persistence of biodiversity. Given the forecast potential for extensive range shifts by most taxa across many regions, it is imperative that conservationists forward plan to ensure that their strategies for triage are as effective as possible (Samways, this volume, Chapter 6). This will inevitably involve trade-offs in deciding which species or ecosystems to prioritize at any time and in any region. Conservation practitioners currently tend to follow one of four main strategies to conserve biodiversity, and we now discuss the relevance of each strategy to averting extinctions through climatic change.

Protected area networks

Many conservationists currently consider protected areas as an effective way to protect biodiversity (Bruner *et al.*, 2001). In future, however, protected areas may no longer hold the species for which they were originally designated. If such species disappear from a protected area, it may no longer satisfy the criteria required for its designation, leading to the potential for it to be declassified. However, protected areas should not be considered on an individual basis, but rather as networks that together protect a wide range of species. With a changing climate, these networks will probably still serve to protect the vast majority of species for which the individual protected areas in the network were together originally designated (Araújo *et al.*, 2004; Huntley, 2007; Hole *et al.*, 2009), although not necessarily in the protected areas in which they currently occur.

Conservationists and legislators therefore require a paradigm shift in their definition of what constitutes a protected area. It is necessary to move away from designating areas for particular species or assemblages and an implicitly static view of biodiversity preservation. Instead, a more dynamic approach is required in which protected areas are seen as forming networks of sites that together protect a sufficient extent of the full range of habitat types required to conserve the biodiversity of a region. Any individual protected area in such a network will serve at different times to protect a changing suite of species of conservation concern, thus ensuring their survival into the future.

Managing the wider landscape

The degree of connectivity between individual protected areas is a major problem for protected area networks facing climatic changes. Currently, many protected areas are too isolated or remote from one another to permit the majority of taxa to move readily between them. Although much theoretical discussion in the field of conservation biology has focused on connectivity and habitat corridors, relatively little has been delivered in terms of improved connectivity in the real world, particularly in developed regions such as western Europe. In regions with lower human population density, and hence less land use pressure, corridor schemes can be implemented more easily. However, in landscapes that are heavily altered and/or occupied by people, such strategies are not generally feasible. Furthermore, given uncertainty about the precise nature and complexity of future regional climatic changes, and the individualistic spatial response of species to such changes (Huntley *et al.*, 2007, 2008), it is probably inappropriate to use corridors as the principal means of enabling species to shift their ranges in response to climate change.

A better approach for achieving connectivity is to increase landscape permeability. This would involve providing a functional network of sites with a range of suitable habitats capable of maintaining populations of different species. These areas should also be sufficiently close to one another to support the spatial dynamics of populations as they respond to climatic change. Such networks will be vital to enable key species of conservation concern to shift between what frequently are relatively remote protected areas, as the suitability of protected areas occupied by the species declines, and as currently unoccupied protected areas become increasingly suitable. The habitats required to form such networks may already exist but currently lack protection, or they may have to be newly created. Some newly protected areas arguably might require only temporary designation during a 'protected areas bottleneck' period for a key species, and be returned to other uses once that species has achieved the range shift necessary to occupy newly suitable protected areas. However, this approach is predicated upon the 'static' view of area protection that has long prevailed in conservation biology. In reality, the effects of anthropogenic climatic change must be expected to continue to impact upon biodiversity beyond the end of the present century.

Translocation and assisted colonization

In many regions, the development of a functional network of subsidiary sites that permit permeability will not be achieved sufficiently quickly. In these situations, it may be necessary to translocate species between protected areas (McLachlan *et al.*, 2007; Willis *et al.*, 2009). This option will apply principally to species of conservation concern, whose populations are largely restricted to protected areas. It will be especially important where the species of concern is relatively less mobile or dispersive. Translocation sites may lie within former ranges from which the species has been extirpated as a consequence of a historical factor that no longer operates. Alternatively, translocation sites may be a novel area in which the species has never occurred in historical times. This approach is already used widely to facilitate colonization of areas unlikely to be reached naturally. The widespread translocation of large mammals between game reserves within East and southern Africa is one such example. Another is the reintroduction of species to areas from which they were previously exterminated; examples include the successful introduction of the red kite *Milvus milvus* across the United Kingdom and of the wolf *Canis lupus* to Yellowstone National Park. Lessons learnt from recent reintroductions and translocations could be applied to enhance the success of assisted colonizations required to mitigate the effects of climatic change. However, such schemes are often expensive and pose ethical dilemmas (Hunter, 2007; Harrop, this volume, Chapter 7), some of which take much time to resolve. For this and many other reasons, it is preferable to encourage the natural movement of species wherever feasible. Translocation should be considered an option of last resort, albeit a potentially vital one for some species.

Ex situ *conservation*

The establishment of populations of species in captivity is one of the most expensive conservation strategies. Therefore, when weighing up where trade-offs between conservation and financial resources should occur, *ex situ* conservation should be considered only as a final resort when all other options, including translocation, have been excluded as viable possibilities. Nonetheless, *ex situ* conservation could prove valuable as a security measure for declining species. It may also be necessary to prevent the extinction of rare and geographically restricted species faced with climatic bottlenecks, during which regions of suitable climate and appropriate habitat do not overlap for some period of time.

Climate change impacts on these various approaches

Flexibility will be fundamental to conserving species using all of the above strategies. The management plans for protected areas, wider countryside stewardship schemes, translocations and *ex situ* conservation should all be designed to be adaptable and proactive in the future. The speed at which climate may affect some species will mean that a degree of forward planning, and some idea of which species are likely to become threatened, and where, will be vital in successfully conserving species under climatic change. Dispersal ability and habitat availability in climatically suitable new areas will be key to the responses of different species. Therefore, it will become of paramount importance to mainstream conservation in protected areas with stewardship and management of the wider countryside outside formally protected areas (Rosser & Leader-Williams, this volume, Chapter 8).

It must also be borne in mind that genetic (and potentially behavioural) diversity exists among populations of the same species. Consequently, some populations may be better adapted to particular conditions but may not fare well in novel climates or environments. Climate change is likely to lead to the loss of intraspecific genetic diversity, as populations adapted to particular climatic conditions, especially those near the limits of a species' range of tolerance, die out. Such die-offs may occur either because populations at the margins are unable to shift their range sufficiently rapidly to adjust to the climate change, or are swamped by immigration of individuals from populations with tolerances better suited to the new climatic conditions, or because the climatic conditions to which they are adapted are no longer available (Williams *et al.*, 2007). In turn, such loss of genetic diversity is likely to limit species' abilities to cope with further changes. Therefore, conservation strategies should, wherever possible, be designed to maintain populations of species throughout the full range of climatic conditions that they are able to occupy.

Projecting changes due to climate change

'Climate envelope' models can be used to simulate potential range shifts for individual species, given different scenarios of climate change. Climate envelopes are an important tool with which to assess probable threats to species resulting from future climatic change, enabling some degree of forward planning by conservationists. Such models provide useful estimates of the extent to which the distribution and abundance of individual species are likely to be affected by a given climatic change. In addition, climate envelopes provide an indication of the likely direction and magnitude of range shifts. Climate envelope models can also be used to highlight regions that are projected to remain suitable for a species in the future, and hence to identify regions or protected areas that will be of most benefit in the long term (Willis *et al.*, 2008). It has been suggested that robust bioclimatic representation within protected areas can be achieved at negligible marginal costs (Pyke & Fischer, 2005).

It has repeatedly been shown that observed range shifts for several species have lagged behind recent climatic changes (Warren *et al.*, 2001; Archaux, 2004; Menendez *et al.*, 2006). Such lags may reflect inherent limitations of a species resulting from its life history and/or dispersal traits, or may result from limited habitat availability and/or relative isolation of suitable habitat patches. Climate envelope models, being static, are insufficient alone to forecast future range dynamics, including potential lags. Therefore, an urgent need is to develop hybrid models that combine population processes, dispersal processes, patterns of habitat availability and changing climatic suitability (Ibanez et al., 2006; Midgley et al., 2006; Thuiller, 2007; Zimmer, 2007). In addition, it is vital that the inherent uncertainties, both of current climate envelope simulations of potential future ranges, and of forecasts made using newly developed models and methods, are quantified and fully understood.

Combined threats of climate change and habitat loss

Climate change will not act alone to threaten biodiversity. Until recently, the primary threats to biodiversity conservation have been habitat loss, fragmentation and degradation, along with problems associated with a growing and increasingly mobile human population. In addition to these threats, introductions of invasive species and exploitation are considered to be the most important threats to globally threatened birds (BirdLife International, 2000). Although this chapter deals primarily with the threats and necessary trade-offs resulting from climatic change, it is important to acknowledge that unless suitable habitats remain to protect species in the short to medium term, there is little point in long-term planning for climatic change impacts. Therefore, any proposed large-scale trade-offs of resources currently allocated to protect existing biodiversity into new strategies to cope with climatic change impacts must be evaluated very carefully. Any strategy for the protection of current biodiversity should be carefully examined to ensure that it is 'future proof' and robust to projected changes in climate.

Lags in climate change impacts

Although climate envelope models have been used to assess the potential extinction risk for various groups of species (Thomas et al., 2004), such projections are of the potential long-term outcomes of climate change. Thus, a given scenario of climatic change may commit many species to eventual regional or even global extinction. However, for most of those species there will be a substantial lag between climatic suitability deteriorating to a level where the species population enters a potentially terminal decline and the final extinction of the species. In turn, this generates a climatic 'extinction debt' (Tilman *et al.*, 1994). This is especially likely to occur where climate acts indirectly to determine the distribution of a species that is restricted to a particular habitat type or biome, the extent and location of which is climatically determined. Range shifts in that habitat type or biome may lag considerably behind climatic change as a result of the longevity of the component plants and/or the requirement for disturbance to trigger transformation of the vegetation (Archaux, 2004).

Of the 1186 globally threatened bird species worldwide, ~900 are dependent upon forest habitats. Of these, 94% occur in tropical forests (BirdLife International, 2000). However, many tropical forests are simulated to decline substantially in extent during the present century as a result of climate change (Malhi *et al.*, 2008). Therefore, the situation could be bleak for many tropical forest species. However, the primary threat to tropical forest currently is clearance. If the clearance of important tropical forests can be stemmed, one glimmer of hope for conservationists is that many tropical forest species may be limited only indirectly by climate. In other words, the presence of the forest habitat may be more critical for them than any climatic change. Even if climate change leads to the eventual loss of forest, it may not do so by killing the trees forming the present forest canopy, but instead by limiting their regeneration. Consequently, there may be a substantial lag before these forests disappear as a result of climate change. This window of opportunity may allow more time for forest-dependent species to shift their ranges into newly suitable areas than if the climatic change had an immediate impact upon the extent of forest. However, the ecology and interspecific interactions of most species are so poorly understood that it cannot be predicted with confidence those species that are likely to benefit from this extended window of opportunity. Therefore, in the vast majority of cases, it will not be possible to potentially trade-off, for example, concentrating conservation resources on species more urgently threatened with local extinction.

An integrated and holistic approach to developing conservation strategy

One thing is clear about seeking to incorporate the potential impacts of projected future climate change into the development of conservation strategies:

there is no single 'silver bullet' solution to the problem. Instead, any strategy needs to combine a series of complementary components that together offer the best hope of successful conservation of a substantial fraction of present biodiversity. Principal amongst the more essential components are the following:

- To maintain present protected area networks, and to ensure that as many species as possible continue to be protected, while also maintaining as much genetic diversity as possible within and among populations of species.
- In the short to medium term, to strategically enlarge present protected areas, add new protected areas to the network, and enhance landscape permeability through the development of functional networks of sites offering habitat and/or favourable management for the widest possible range of wild species, to enable species to achieve the range shifts necessary in response to climatic change.
- To identify species at greatest risk from climatic change, especially those at risk of regional or global extinction. Some of the latter species may require assisted migration in order to ensure their persistence (Willis *et al.*, 2009).
- To forward plan, and adaptively manage, integrated biodiversity conservation strategies. In order to ensure the greatest effectiveness and improve returns on resources, such strategies will need to consider trade-offs between competing priorities to ensure that effort is focused where it will achieve most (Wilson *et al.*, this volume, Chapter 2; Murdoch *et al.*, this volume, Chapter 3). Such trade-offs can only be evaluated adequately by adopting a holistic approach to the planning and development of conservation strategies rather than, as is often the case at present, considering and addressing the protection of individual species, ecosystems or protected areas on a piecemeal and case-by-case basis.

Trade-offs and prioritization

Trade-offs between competing priorities will be an inevitable part of future conservation strategies that take account of projected climatic changes. However, it will be difficult to determine priorities at the species level. Where priorities can be identified, they will be for those species likely to be most immediately and severely affected by climatic change. These will include: species whose habitat is particularly susceptible to rapid alteration as climate changes, for example species restricted to tropical montane cloud forests,

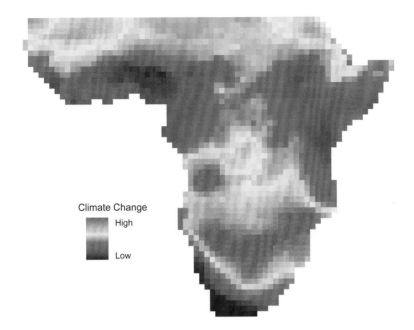

Plate 18.1 **Euclidean distance between recent and projected future climates of sub-Saharan Africa. Euclidean distance is calculated in the space of four climate variables: coldest month mean temperature; warmest month mean temperature; annual ratio of actual to potential evapotranspiration; and dry season intensity (Huntley *et al.*, 2006). Dark blue represents least change, green an intermediate level of change and dark red the largest change. Those regions whose climates are projected to become most dissimilar by the end of the current century are principally in the semi-arid southern tropical zone. The projected future climate used is a 30-year mean for 2070–2099 derived from a transient simulation made using HadCM3 GCM (Gordon *et al.*, 2000) for the SRES B2a emissions scenario (Nakicenovic & Swart, 2000).**

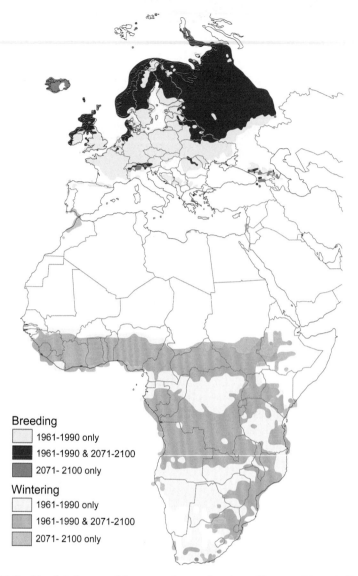

Breeding
☐ 1961-1990 only
■ 1961-1990 & 2071-2100
▨ 2071- 2100 only

Wintering
☐ 1961-1990 only
▨ 1961-1990 & 2071-2100
▨ 2071- 2100 only

Plate 18.2   Simulated potential range change for the willow warbler *Phylloscopus trochilus*. The willow warbler is a widespread trans-Saharan migrant, breeding principally from Europe eastwards across Siberia, and having an extensive non-breeding range in wooded savanna and forest edge habitat in sub-Saharan Africa. Potential changes are illustrated between simulations of both its breeding and non-breeding ranges for the recent past (1961–1990) and potential future (2071–2100). The potential future climate for 2071–2100 was derived from a simulation made using HadGEM1 GCM (Johns *et al.*, 2006) for the A1b SRES emissions scenario (Nakicenovic & Swart, 2000). Simulations were made following Huntley *et al.* (2006) and using the four climate variables from Plate 18.1. Red, yellow and dark blue in Europe indicate the simulated breeding range; red, yellow and light blue in Africa (the latter extending into southern Europe) indicate the simulated non-breeding range. Red indicates areas simulated as suitable for both periods, yellow areas are simulated as suitable for the first period but unsuitable for the second (i.e. potential range contraction), and blue areas are simulated as unsuitable for the first period but suitable for the second (i.e. potential range expansion). Note the potential in future for the non-breeding range to include areas around the Mediterranean basin.

Arctic tundra species and species of the polar ice caps such as the polar bear *Ursus maritimus*; species that are already spatially 'on the edge', for example mountain-top species and species restricted to the poleward extremities of land masses; and any species with small overlap between their current and potential future ranges. Furthermore, many such species will already be of conservation concern because of their spatially restricted ranges and/or small populations. In contrast, widespread and generalist species are widely expected to be more resilient to the impacts of climate change (Juillard *et al.*, 2003; Jiguet *et al.*, 2006).

Given the difficulty in identifying priorities other than for those few species at most obvious risk from climate change, integrated conservation strategies should be designed so as to provide protection for the maximum number of species (Moilanen *et al.*, 2005). Economic costs must also be integrated into this prioritization in order to ensure the greatest benefit from limited resources (Naidoo *et al.*, 2006). Although conservation biologists have identified the general attributes of strategies that will maximize the number of species protected, and some simple rules of thumb can be applied to develop strategies that will maximize the number of species protected in a world of changing climates, decisions about trade-offs and prioritization inevitably will be contentious and thus not easily made.

The most obvious trade-off that society can make is to trade-off dependence upon fossil fuels, which accelerates global warming, against the rate and magnitude of future climate change. Because climatic change impacts upon biodiversity, and hence upon the capacity to deliver ecosystem services upon which society depends, reducing the rate and magnitude of future climatic change will reduce the costs to society arising from losses of ecosystem services. Every degree or half degree by which society is able to reduce future global temperatures will serve to make the task of conserving species in the future so much easier, and will render adaptations and range shifts of species and communities more achievable. In turn, this will reduce the impact upon vital ecosystem services. This is the ultimate trade-off, but it is in many ways the easiest means by which to preserve global biodiversity and its associated ecosystem services.

Acknowledgments

We are grateful to Nathalie Doswald for allowing us to use Plate 18.2 and to Dr Yvonne Collingham for general computing support. The work presented was partly funded by the Royal Society for the Protection of Birds. Lincoln Fishpool and Stuart Butchart at BirdLife International provided help and advice.

References

Araújo, M.B., Cabeza, M., Thuiller, W., Hannah, L. & Williams, P.H. (2004) Would climate change drive species out of reserves? An assessment of existing reserve-selection methods. *Global Change Biology*, 10, 1618–1626.

Araújo, M.B., Thuiller, W. & Pearson, R.G. (2006) Climate warming and the decline of amphibians and reptiles in Europe. *Journal of Biogeography*, 33, 1712–1728.

Archaux, F. (2004) Breeding upwards when climate is becoming warmer: no bird response in the French Alps. *Ibis*, 146, 138–144.

BirdLife International (2000) *Threatened Birds of the World*. Lynx Edicions, Barcelona, and BirdLife International, Cambridge.

Both, C., Bouwhuis, S., Lessells, C.M. & Visser, M.E. (2006) Climate change and population declines in a long-distance migratory bird. *Nature*, 441, 81–83.

Bruner, A.G., Gullison, R.E., Rice, R.E. & da Fonseca, G.A.B. (2001) Effectiveness of parks in protecting tropical biodiversity. *Science*, 291, 125–128.

Doswald, N., Huntley, B., Collingham, Y.C., Pain, D.J., Green, R.E. & Willis, S.G. (2009) Potential consequences of climate change on the migration of European Sylvia warblers. *Journal of Biogeography*, 36, 1194–1208.

Gordon, C., Cooper, C., Senior, C.A. *et al.* (2000) The simulation of SST, sea ice extents and ocean heat transports in a version of the Hadley Centre coupled model without flux adjustments. *Climate Dynamics*, 16, 147–168.

Greenwood, J.J.D. & Baillie, S.R. (1991) Effects of density-dependence and weather on population changes of English passerines using a non-experimental paradigm. *Ibis*, 133, S121–S133.

Gregory, R.D., Willis, S.G., Jiguet, F. *et al.* (2009) An indicator of the impacts of climate change on European bird populations. *PLoS One*, 4, e4678.

Hannah, L., Midgley, G., Andelman, S. *et al.* (2007) Protected area needs in a changing climate. *Frontiers in Ecology and the Environment*, 5, 131–138.

Hickling, R., Roy, D.B., Hill, J.K., Fox, R. & Thomas, C.D. (2006) The distributions of a wide range of taxonomic groups are expanding polewards. *Global Change Biology*, 12, 450–455.

Hill, J.K., Thomas, C.D. & Huntley, B. (2003) Modelling present and potential future ranges of European butterflies using climate response surfaces. In *Butterflies: ecology and evolution taking flight*, eds C.L. Boggs, W.B. Watt & P.R. Ehrlich, pp. 149–167. University of Chicago Press, Chicago.

Hole, D.G., Willis, S.G., Pain, D.J. *et al.* (2009) Projected impacts of climate change in a continent-wide protected area network. *Ecology Letters*, 12, 420–431.

Hunter, M.L. (2007) Climate change and moving species: furthering the debate on assisted colonization. *Conservation Biology*, 21, 1356–1358.

Huntley, B. (2007) *Climatic Change and the Conservation of European Biodiversity: towards the development of adaptation strategies*. T-PVS/Infy (2007)3. Council of Europe, Convention of the Conservation of European Wildlife and Natural Habitats, Strasbourg. Available at: http://www.coe.int/t/dg4/cultureheritage/conventions/bern/T-PVS/sc27_inf03_en.pdf.

Huntley, B. & Webb, T., III (1989) Migration: species' response to climatic variations caused by changes in the earth's orbit. *Journal of Biogeography*, 16, 5–19.

Huntley, B., Berry, P.M., Cramer, W.P. & McDonald, A.P. (1995) Modelling present and potential future ranges of some European higher plants using climate response surfaces. *Journal of Biogeography*, 22, 967–1001.

Huntley, B., Collingham, Y.C., Green, R.E., Hilton, G.M., Rahbek, C. & Willis, S.G. (2006) Potential impacts of climatic change upon geographical distributions of birds. *Ibis*, 148, 8–28.

Huntley, B., Collingham, Y.C., Willis, S.G. & Green, R.E. (2008) Potential impacts of climate change on European breeding birds. *PLoS One*, 3, e1439.

Huntley, B., Green, R.E., Collingham, Y.C. & Willis, S.G. (2007) *A Climatic Atlas of European Breeding Birds*. Lynx Edicions, Barcelona.

Ibanez, I., Clark, J.S., Dietze, M.C. *et al.* (2006) Predicting biodiversity change: outside the climate envelope, beyond the species–area curve. *Ecology*, 87, 1896–1906.

IPCC (Intergovernmental Panel on Climate Change) (2007) *Climate Change 2007: the physical science basis. Contribution of Working Group I to the Fourth Assessment Report of the Intergovernmental Panel on Climate Change*, eds S. Solomon, D. Qin, M. Manning *et al.* Cambridge University Press, Cambridge and New York.

Jansen, E., Overpeck, J., Keith R.B. *et al.* (2007) Paleoclimate. In *Climate Change 2007: the physical science basis. Contribution of Working Group I to the Fourth Assessment Report of the Intergovernmental Panel on Climate Change*, eds S. Solomon, D. Qin, M. Manning *et al.*, pp. 433–497. Cambridge University Press, Cambridge and New York.

Jetz, W., Wilcove, D.S. & Dobson, A.P. (2007) Projected impacts of climate and land-use change on the global diversity of birds. *PLoS Biology*, 5, 1211–1219.

Jiguet, F., Julliard, R., Thomas, C.D., Dehorter, O., Newson, S.E. & Couvet, D. (2006) Thermal range predicts bird population resilience to extreme high temperatures. *Ecology Letters*, 9, 1321–1330.

Johns, T.C., Durman, C.F., Banks, H.T. *et al.* (2006) The new Hadley Centre Climate Model (HadGEM1): evaluation of coupled simulations. *Journal of Climate*, 19, 1327–1353.

Julliard, R., Jiguet, F. & Couvet, D. (2003) Common birds facing global changes: what makes a species at risk? *Global Change Biology*, 10, 148–154.

Malhi, Y., Roberts, J.T., Betts, R.A., Killeen, T.J., Li, W. & Nobre, C.A. (2008) Climate change, deforestation and the fate of the Amazon. *Science*, 319, 169–172.

Marchant, J.H. (1992) Recent trends in breeding populations of some common trans-Saharan migrant birds in northern Europe. *Ibis*, 134, S113–S119.

McLachlan, J.S., Hellmann, J.J. & Schwartz, M.W. (2007) A framework for debate of assisted migration in an era of climate change. *Conservation Biology*, 21, 297–302.

Menendez, R., Megias, A.G., Hill, J.K. *et al.* (2006) Species richness changes lag behind climate change. *Proceedings of the Royal Society of London, Series B*, 273, 1465–1470.

Midgley, G.F., Hughes, G.O., Thuiller, W. & Rebelo, A.G. (2006) Migration rate limitations on climate change-induced range shifts in Cape Proteaceae. *Diversity and Distributions*, 12, 555–562.

Moilanen, A., Franco, A.M.A., Eary, R.I., Fox, R., Wintle, B. & Thomas, C.D. (2005) Prioritising multiple-use landscapes for conservation: methods for large multi-species planning problems. *Proceedings of the Royal Society of London, Series B*, 272, 1885–1891.

Naidoo, R., Balmford, A., Ferraro, P.J., Polasky, S., Ricketts, T.H. & Rouget, M. (2006) Integrating economic costs into conservation planning. *Trends in Ecology and Evolution*, 21, 681–687.

Nakicenovic, N. & Swart, R. (2000) *Emissions Scenarios. Special report of the Intergovernmental Panel on Climate Change*. Cambridge University Press, Cambridge.

Parmesan, C. & Yohe, G. (2003) A globally coherent fingerprint of climate change impacts across natural systems. *Nature*, 421, 37–42.

Peach, W., Baillie, S. & Underhill, L. (1991) Survival of British sedge warblers *Acrocephalus schoenobaenus* in relation to West African rainfall. *Ibis*, 133, 300–305.

Peach, W.J., du Feu, C. & McMeeking, J. (1995) Site tenacity and survival rates of wrens *Troglodytes troglodytes* and treecreepers *Certhia familiaris* in a Nottinghamshire wood. *Ibis*, 137, 497–507.

Pigott, C.D. & Huntley, J.P. (1981) Factors controlling the distribution of *Tilia cordata* at the northern limits of its geographical range. III. Nature and causes of seed sterility. *New Phytologist*, 87, 817–839.

Pounds, J.A., Fogden, M.P.L. & Campbell, J.H. (1999) Biological response to climate change on a tropical mountain. *Nature*, 398, 611–615.

Pyke, C.R. & Fischer, D.T. (2005) Selection of bioclimatically representative biological reserve systems under climate change. *Biological Conservation*, 121, 429–441.

Rivalan, P., Frederiksen, M., Lois, G. & Julliard, R. (2007) Contrasting responses of migration strategies in two European thrushes to climate change. *Global Change Biology*, 13, 275–287.

Roy, D.B. & Sparks, T.H. (2000) Phenology of British butterflies and climate change. *Global Change Biology*, 6, 407–416.

Ruegg, K.C., Hijmans, R.J. & Moritz, C. (2006) Climate change and the origin of migratory pathways in the Swainson's thrush, *Catharus ustulatus*. *Journal of Biogeography*, 33, 1172–1182.

Sykes, M.T. (1997) The biogeographic consequences of forecast changes in the global environment: individual species' potential range changes. In *Past and Future Rapid Environmental Changes: the spatial and evolutionary responses of terrestrial biota*, eds B. Huntley, W. Cramer, A.V. Morgan, H.C. Prentice & J.R.M. Allen, pp. 427–440. NATO ASI Series I: Global Environmental Change No. 47. Springer-Verlag, Berlin.

Thomas, C.D., Cameron, A., Green, R.E. *et al.* (2004) Extinction risk from climate change. *Nature*, 427, 145–148.

Thuiller, W. (2007) Biodiversity – climate change and the ecologist. *Nature*, 448, 550–552.

Thuiller, W., Broennimann, O., Hughes, G., Alkemade, J.R.M., Midgley, G.F. & Corsi, F. (2006) Vulnerability of African mammals to anthropogenic climate change under conservative land transformation assumptions. *Global Change Biology*, 12, 424–440.

Tilman, D., May, R.M., Lehman, C.L. & Nowak, M.A. (1994) Habitat destruction and the extinction debt. *Nature*, 371, 65–66.

Walther, G.R., Post, E., Convey, P. *et al.* (2002) Ecological responses to recent climate change. *Nature*, 416, 389–395.

Warren, M.S., Hill, J.K., Thomas, J.A. *et al.* (2001) Rapid response of British butterflies to opposing forces of climate and habitat change. *Nature*, 414, 65–69.

Westerhoff, D. & Tubbs, C.R. (1999) Dartford warblers *Sylvia undata*, their habitat and conservation in the New Forest, Hampshire, England in 1988. *Biological Conservation*, 56, 89–100.

Williams, J.W., Jackson, S.T. & Kutzbach, J.E. (2007) Projected distributions of novel and disappearing climates by 2100 AD. *Proceedings of the National Academy of Sciences of the USA*, 104, 5738–5742.

Willis, S.G., Hill, J.K., Thomas, C.D. *et al.* (2009) Assisted colonization in a changing climate: a test-study using two UK butterflies. *Conservation Letters*, 2, 46–52.

Willis, S.G., Hole, D.G., Collingham, Y.C., Hilton, G., Rahbek, C. & Huntley, B. (2008) Assessing the impacts of future climate change on protected area networks: a method to simulate individual species' responses. *Environmental Management*, 5, 836–845.

Zimmer, C. (2007) Predicting oblivion: are existing models up to the task? *Science*, 317, 892–893.

Drivers of Biodiversity Change

Georgina M. Mace

Centre for Population Biology, Imperial College London, Ascot,
Berkshire, UK

Introduction

An important objective of conservation management is to ensure the persistence of species and ecosystems in the face of global change. This will require both an analysis of the *status quo*, which may or may not indicate a current problem, and a forward look at what may happen next. During the 21st century, this forward look will become increasingly important as climate change and other emerging aspects of global change take hold. This chapter discusses how analysis of present and future drivers of biodiversity loss will become an essential prerequisite to conservation planning in a world of trade-offs. First, it presents an overview of the main drivers of biodiversity loss at the global level. Second, it outlines information needs to allow better predictions and management. Third, it discusses trends in drivers and how our lack of understanding of the dynamic nature of these systems is often limiting. Fourth, it examines emerging threats from climate change. Fifth, it discusses the knowledge needed to proactively address emerging aspects of global change. Sixth, it outlines some really hard problems. Finally, the chapter concludes by suggesting some immediate steps that need to be taken to mitigate future losses of biodiversity.

Trade-offs in Conservation: Deciding What to Save, 1st edition.   Edited by N. Leader-Williams,
W.M. Adams and R.J. Smith.   © 2010 Blackwell Publishing Ltd.

The main drivers of biodiversity loss

Despite the fact that our understanding of the causes of biodiversity loss and ecosystem change has increased greatly in recent years, some significant knowledge gaps remain. Even in situations where something is known about future trends in drivers of change, it still may not be possible to reliably predict the impacts of those changes on species and natural ecosystems, nor to mitigate their damaging effects. However, conservation planners should pay attention to these topics since proactive responses will be more resource efficient and effective, and will achieve a necessary maturing of conservation science, moving from a focus on cures to a focus on prevention (Balmford *et al.*, 1998; Mace *et al.*, 1998; Pressey *et al.*, 2007).

Direct and indirect drivers

A complex chain of interacting effects ultimately leads to the loss of biodiversity. Direct drivers, defined as those causal processes that act directly on species and ecosystems, lie at the proximate end of this chain. Direct drivers are relatively well characterized and are now generally grouped into the five major categories of: land use change leading to habitat degradation; overexploitation; impacts of invasive species and pathogens; pollution; and climate change (Purvis *et al.*, 2000; Mace *et al.*, 2005; Brook *et al.*, 2008). These five direct drivers are often inter-related and their impacts may well be synergistic, leading to worse outcomes than suggested by considering any of them individually (Didham *et al.*, 2007; Brook *et al.*, 2008).

Some direct drivers are easier to address than others. For example, overexploitation, especially that arising from recreational or local hunting pressures, can be relatively easily reversed given strong political will and effective implementation, as several iconic conservation successes bear testament, including the Arabian oryx *Oryx leucoryx* and southern white rhino *Ceratotherium simum simum*. The dramatic recovery of many raptors following analysis of the effects of DDT and elimination of the pollutants (Cade & Burnham, 2003) is another example of success following the identification of a clearly defined and reversible threat. There are other cases where well-directed local actions to eliminate direct threats have led to measurable recoveries (Butchart *et al.*, 2006). Unfortunately, most direct threats are not so simple or easily isolated from other threatening or societal processes, making such focused actions difficult in many cases.

The direct drivers are themselves a consequence of broader environmental and societal changes, referred to *inter alia* as indirect drivers (MA, 2005) or underlying (Xu & Wilkes, 2004) or contributory factors (Salafsky *et al.*, 2008). The indirect drivers include processes such as population and economic growth, technological development, land use change, and social and political change. These do not directly impact biodiversity and ecosystems but act through the direct drivers. Most conservation action addresses direct, rather than indirect, drivers. However, sustainable and long-term reversals in biodiversity loss are most likely to be achieved through reducing the impact of indirect drivers.

Rather different approaches underlie efforts to reduce the deleterious impacts of direct and indirect drivers. Direct drivers can be effectively addressed *in situ* through actions to limit their impact, for example through species- and habitat-specific actions aimed at reducing the level of exploitation or through the creation of protected areas that conserve key habitats and communities. These kinds of species- or habitat-specific responses tend to be somewhat piecemeal, but can be very effective, especially over the short term (Butchart *et al.*, 2006). However, such activities generally require ongoing attention to avoid becoming vulnerable to lapses or waning commitment. In contrast, interventions higher up the causal chain may take longer to develop and implement, but may have more lasting effects. For example, major policy mechanisms to limit the trade in endangered wildlife, such as the Convention on International Trade in Endangered Species of Wild Fauna and Flora (CITES), or agreements to protect wetlands such as the Convention on Wetlands of International Importance (Ramsar Convention), can provide long-term improvements to a much broader set of locations and species. Nevertheless, the implementation and effectiveness of these mechanisms also requires continued vigilance.

Ultimately, the most effective responses will come from widespread societal changes in attitudes or practices relating to the indirect drivers of biodiversity loss. Some changes in policy, and even in societal attitude, can occur surprisingly quickly. For example, efforts to clean up urban air quality following devastating smogs in the major cities of Europe in the 1950s were accomplished effectively within 20 years through a combination of public education, legislation and new technology. At an international scale, the implementation of the Montreal Protocol to limit emissions of ozone-depleting substances was drafted, agreed and signed by over 150 countries within 10 years. Furthermore, the Protocol rapidly achieved substantial reductions of ozone-depleting sub-stances, and current projections suggest that zero emissions can be achieved by 2020 (UNEP, 2007). Unfortunately, current initiatives to achieve consensus even on the basic goals of international agreements to limit global climate change and biodiversity loss are at best weak and mostly stalled, despite clear evidence that proactive adoption of environmental policies could benefit many people through improved ecosystem services (MA, 2005). Therefore, at present most actions to address the drivers of biodiversity loss continue to be designed and implemented locally or nationally, and to focus on direct drivers.

Approaches to conservation action planning

Although addressing indirect drivers of biodiversity loss may be more effective in the long term, short-term conservation planning is essential to stem the currently increasing rate of loss of global biodiversity. In practice, the analysis of the causal processes behind observed declines and collapses is essential to designing effective restorative actions (Caughley, 1994). This process relies on sound analysis and integration of different lines of evidence, as has been well demonstrated in recent analyses of declining bird populations at national (e.g. Newton, 2004; Xu & Wilkes, 2004) and international (e.g. Green *et al.*, 2004) levels. These causal pathways may often be context-specific and even idiosyncratic. In turn this raises the question of how useful generalized frameworks will be for analysis at this level. Rather, the main justification for detailed analytical frameworks lies in the analysis of conservation actions – to discover general principles for effective interventions. A variety of current programmes that will synthesize and hopefully lead to generalizations for conservation planning are now underway (Kapos *et al.*, 2008; Salafsky *et al.*, 2008; Kapos *et al.*, this volume, Chapter 5). This process will also feed usefully into new methods for assessing conservation actions in an explicit cost–benefit framework that should lead to improved efficiency and outcomes from conservation expenditure (McBride *et al.*, 2007; Murdoch *et al.*, 2007; Wilson *et al.*, 2007; Wilson *et al.*, this volume, Chapter 2; Murdoch *et al.*, this volume, Chapter 3). Hence, processes to design and implement actions to deal with direct drivers are considerably more advanced than is the case for indirect drivers.

Trends in drivers

The best available information on historical trends in species and ecosystems suggests that direct drivers of biodiversity loss have varied in importance over the past several hundred years, and will continue to do so into the near future. Thus, extinctions of birds, mammals and amphibians recorded since 1600 largely occurred on islands and were caused, in decreasing order of importance, by invasive species, overexploitation and habitat change (IUCN, 2004). These extinctions were a direct result of the major phase of human exploration and spread in the 18th and 19th centuries, when many remote areas were discovered and settled. People were often accompanied by domestic or pest species, and frequently hunted or gathered newly found species for food, fuel or fibre, leading to extinctions of many native species. This phase of exploration is probably now over, and species most vulnerable to immediate impacts from people are either extinct or persist only in small areas that are protected or remote.

Instead, the most common current cause of threat to species now listed on the Red Lists of IUCN, the International Union for the Conservation of Nature is habitat loss and degradation (Vié et al., 2009). Expanding human communities, infrastructure and urban spread, and especially the spread of agriculture has led to rapid and extensive conversion and fragmentation of natural habitats (Foley et al., 2005). Croplands and pastures now occupy around 40% of the terrestrial surface area in continental areas (MA, 2005). Most projections and scenario-based analyses suggest that the effects of land use change will continue to dominate biodiversity loss, at least until the mid-21st century, especially in tropical and mid-latitude continental areas. The effects of climate change, already important at high latitudes and in some coastal areas, will gradually spread in relative impact (Jetz et al., 2007; Lee & Jetz, 2008).

Marine systems differ slightly from terrestrial systems. Human impacts on marine systems have tended to lag behind those on terrestrial biomes, but are now extremely severe (Jackson et al., 2001). Overexploitation is still widely regarded as the strongest direct driver in marine ecosystems (Pauly et al., 1998, 2002), although invasive species, climate change, pollution and disease are all also important (Dulvy et al., 2003) and expected to worsen over coming decades (Verity et al., 2002). Overfishing has been most severe for the large predatory fishes (Christensen et al., 2003; Myers & Worm, 2005), and cascading effects on marine biological communities have led to major ecosystem changes clearly attributable to overfishing (Jackson *et al.*, 2001; Pauly *et al.*, 2003). The recent and widespread decline in many fisheries species has been associated with habitat loss and physical damage to coastal ecosystems from fishing gear. Furthermore, the complex life histories of many species contribute to slow or non-recovery of species following a cessation of fishing pressure (Hutchings & Reynolds, 2004). Increasingly, other anthropogenic processes are now having major impacts on marine systems. These include pollution from agricultural run-off and the exponential rise in urbanization of coastal zones, causing organic pollution as well as structural changes to coastal habitats from coastal amenities and infrastructure development.

Moving to the future, several assessments concur that the next major wave of species extinctions will be caused by climate change, while nitrogen and phosphorus loading and the effects of introduced species will also remain high-risk processes in certain areas (Sala *et al.*, 2000; MA, 2005). Therefore, a shift in dominant processes will continue from the past couple of centuries into the next couple of centuries (Table 19.1). In turn, this will make natural and managed adaptation all the more complex. Each time a new challenge arrives, natural communities will lose vulnerable components to that process, while those that are resilient or manage to adapt will persist. However, resilience and adaptation are often process-specific, leading to extinction filters that gradually deplete natural communities (Balmford, 1996). The rate and nature of changes in direct drivers presents contemporary natural communities with multiple jeopardies (Brook *et al.*, 2008). In particular, climate change presents some extreme challenges that will require very different kinds of responses from species and communities (Willis *et al.*, this volume, Chapter 18).

Table 19.1 **Trends in the direct drivers of biodiversity loss over time. The larger the number of + symbols in each box, the greater the relative importance of each process listed on the left in each time period, based on reviews in IUCN (2004), Vié *et al.* (2009), Sala *et al.* (2000) and MA (2005).**

Threat	Past 200 years	Present	Next 200 years
Habitat change	++	+++	+++
Invasives	+++	++	?
Overexploitation	+++	+	Very context-dependent
Pollution	++	+++	+++
Climate change	−	+	++++

Emerging threats from climate change

Climate change is a particularly problematic new threatening process for several reasons. First, its impacts are global and will affect different biomes and ecosystems to varying degrees, but prediction at scales appropriate for management are very uncertain. Second, the rate and extent of climate changes are unprecedented, at least within human history (Jackson & Overpeck, 2000). Third, substantial areas of the world expected to experience entirely new or disappearing local climates (Williams *et al.*, 2007). Fourth, it strikes at a time when natural communities are already reduced and fragmented. Indeed, there is already pervasive evidence for climate effects on species' distributions on land and in the sea (Parmesan & Yohe, 2003; Root *et al.*, 2003; Perry *et al.*, 2005). Likewise, there is evidence of climatic effects on the timing of movements or life history stages in species groups in temperate areas (Roy & Sparks, 2000), sometimes with demonstrable deleterious impacts on the synchronization of key life history events (Visser & Holleman, 2001; Bradshaw & Holzapfel, 2007). Recent work has linked increased rates of population decline among bird species to those that have not apparently shown phenological responses to recent climate change since 1990 (Moller *et al.*, 2008), suggesting that selection may be intense on species unable to show phenotypic responses to changing climates. Future projections suggest that ~15–40% of species will be committed to extinction by 2050 under moderate near-term emission scenarios (Thomas *et al.*, 2004; Fischlin & Midgley, 2007).

Successful responses to climate change will be very different from traits that have allowed species to persist in the face of the major drivers of the recent past. As local climates change, species will survive only by adapting or by dispersing to track their current climate. Adaptation may be achieved through phenotypic plasticity or through evolution, although current evidence is insufficient to demonstrate micro-evolutionary changes (Gienapp *et al.*, 2007). Rates of local climate change may mostly be too rapid for sustainable evolution. Therefore, species that survive will be dominated by: those that already occupy a broad range of habitats and that are not climate limited; those that are good dispersers; and those for which habitat corridors exist to allow tracking of climate. In a world where intact blocks of habitat are increasingly rare, it seems unlikely that dispersal will prove a realistic option except for the most vagile of species living in extensive areas of reasonably permeable habitats. Even then, species will also have to cope successfully with: new competitors, predators and pathogens; changes to the timing of events in relation to temperatures; altered precipitation schedules; and, probably, increasingly extreme climates.

Current failures to mitigate climate change make the science of predicting climate change impacts on biodiversity and devising effective interventions all the more urgent. Indeed, this has been described as the grand challenge for ecology (Thuiller, 2007). The dominant predictive methods based on niche modelling, have the advantage of providing spatially explicit and quantitative estimates of likely impacts (Araújo et al., 2005). However, these methods may not have high predictive accuracy (Pearson & Dawson, 2004) and do not take into account biological factors related to species interdependencies, life history and population structures, physiological tolerances or phenological processes (Keith et al., 2008). More specific models, including process-based models, and new approaches based around more specific susceptibility assessment should improve predictive accuracy.

There is still much debate on how to design and implement interventions that enhance the likelihood that biodiversity will persist in the face of rapid climate change. Protected areas may not be viable solutions. Instead, a radical shift may be needed to base conservation management around small-scale habitat mosaics, across which many species can move, and which will encourage adaptation to a range of new habitat types. Unfortunately, an obvious trade-off is that this sort of habitat will also be conducive to the rapid spread of alien invasive species and pathogens.

The knowledge needed to become proactive

New kinds of analytical approaches will be needed to inform preventative actions. First, the strength of the driver over time should be determined. Second, the relationship between the strength of the driver and the vulnerability of the species, community or ecosystem should be learned. These two pieces of information would enable the development of risk assessments for different drivers, according to the time frame and the biological community being considered. Then, in order to plan mitigation, estimates are needed of the effort required to achieve reductions in driver intensity. Coupled with the vulnerability analysis outlined above, this would allow models of the impact of alternative mitigation strategies on reducing biodiversity loss, and enable resources to be allocated efficiently according to time, place and biological community (Wilson et al., 2007; Wilson et al., this volume, Chapter 2).

While comprehensive assessments of this sort are a long way off, new data and approaches are starting to address such issues. Some particularly promising approaches include the modelling of latent risk where future high-intensity threats will affect vulnerable communities (Lee & Doughty, 2003; Cardillo *et al.*, 2006; Davies *et al.*, 2008; Lee & Jetz, 2008), new decision analysis approaches (Wilson *et al.*, 2006, 2007), and the assessment of vulnerable traits in relation to new drivers (Davies *et al.*, 2008). Along with better predictive approaches, probably based on scenario modelling (Peterson *et al.*, 2003; Carpenter *et al.*, 2006), such methods should provide robust tools to guide integrated planning such as that used in the Cape Floristic Region, which incorporates the preservation of both pattern and process (Rouget *et al.*, 2006). Over the coming years, these kinds of approaches should start to radically improve the extent to which conservation planning and management can be both explicit and predictive.

Some really hard problems

The previous section outlined a broad approach to better planning based around assessments of exposure, risk and reversibility. Some elements of this approach are more readily achievable than others, and some elements of this framework will prove particularly difficult.

Invasives and biotic interactions

Unlike climate change, invasive species are not a new problem and have been the focus of research for many years. Nevertheless, it has proven extremely hard to predict which species will invade and which species will become pests or pathogens. Nor has much progress been made in determining which species or ecosystems are most likely to be vulnerable to invasive aliens. Without such tools, predictive models will have extremely high uncertainty. As previously discussed, this problem will be exacerbated because management strategies that allow easy movements of species across the landscape will be preferable to mitigate the impacts of climate change but will also permit the broader spread of invasive species. Finally, globalization, trade and development are the indirect drivers of invasive alien species, so reversing this will require major changes to prevailing international trade and development.

Species and communities with high vulnerability

Under rapid climate change, as well as under other drivers such as habitat loss and fragmentation, it will prove very hard to find adequate management interventions to ensure the conservation of certain species and ecosystems. Currently, the conservation of species dependent on disappearing habitats appears an intractable problem, as the recent demise of the Yangtze river dolphin *Lipotes vexillifer* makes clear (Turvey *et al.*, 2008). Other well-known cases of large species facing severe habitat limitations include the Indian tiger *Panthera tigris*, the polar bear *Ursus maritimus* (Figure 19.1), and the Arctic fox *Alopex lagopus*. Unless these species radically alter their ecology, their conservation will either face severe conflicts or extremely high costs. Under rapid climate change, this same problem will be played out in many new situations, including: species in low-altitude islands and coastal zones; species whose climate envelope moves off the top of a mountain or the edge of a continent; or species that have lost their pollinator or symbionts. Likewise, there is currently no solution to the problems experienced by entire

Figure 19.1 **A polar bear mother and her cub on an ice flow near Edgeøya, Svalbard. Polar bears face loss of their key habitat as global warming increasingly melts Arctic ice. (Photograph by kind permission of Julian Dowdeswell.)**

359 of Biodiversity Change

coral communities facing increasingly acidic environments caused by the dissolution of elevated carbon dioxide from the atmosphere. In the recent past, conservationists have advocated an approach that does not give up even on the most difficult cases (Soulé, 1987). However, in future it seems likely that some difficult trade-offs will be faced and society will need to adopt more sophisticated approaches to deciding what they are going to prioritize (Bottrill *et al.*, 2008).

Predicting way beyond observations

One consequence of the current global change processes is that environments of the future will be qualitatively and quantitatively different from those of the present. In other words, many future predictive models parameter-ized from current or historical data will have to project way beyond the observations upon which they were based. While process-based models will help address this issue, non-linearities are commonly observed in ecological processes (Scheffer & Carpenter, 2003). Under continuously changing con-ditions, many species and ecosystems can show sudden or dramatic changes to new states, step changes and irreversible changes in state (Scheffer *et al.*, 2001). Furthermore, these changes are hard to foresee without detailed studies or a full understanding of the entire process. Under future environmental changes, predictions will increasingly need to be made outside observed limits. Approaches to this problem include the exploration of processes, or alter-native modelling approaches that may present different perspectives and at least allow improved representation of the uncertainties. In particular, the interactions among drivers may turn out to be especially complex.

Conclusions

Despite these many difficulties, progress has been and will continue to be made. It will be further enhanced through increased attention to other related environmental sciences and the adoption of improved tools for decision making and risk analysis. Some immediate steps that seem important include the following:

- Take an explicitly proactive approach to planning for the future: in particular, use, develop and test scenario-based approaches as a way of examining potential new threats and the policies and practices that are likely to reduce their impacts on important areas or species communities.

- Identify species and habitats that are especially vulnerable to future drivers and their impacts: in particular, consider vulnerability analyses as an alternative to threat assessments. The difference here is that the driver and the susceptibility of the species or ecosystems are explicitly distinguished. The overlay of the two highlights the most vulnerable areas and species (e.g. Turner *et al.*, 2003). The advantage of this approach is that it is more suitable for future scenarios and it can accommodate linked social and ecological systems.
- Start now on interventions appropriate to efficient mitigation of future drivers: for example, if future climate change will require new formulations of protected areas and new kinds and scales of habitat connectivity, then modifications need to be considered now to adjust plans that were developed under assumptions for a more stable future world.
- Determine the combination of policy versus local management interventions that will best meet identified needs: some kinds of interventions will best take place at local scales while changes to societal impacts need large-scale policy changes. For any particular conservation imperative it will be important to undertake work at all points on the continuum between the two. Short-term immediate actions will be essential for moderating impacts, but long-term sustainable solutions will also require societal and political changes.
- Develop environmental management plans that increase the potential for both evolutionary adaptation and phenotypic plasticity: in particular, the impacts of land use and climate change seem likely to be both profound and long lasting. Yet, there is currently no clear analysis tool, nor set of actions, to increase phenotypic plasticity and evolutionary potential among the species and communities most likely to be impacted. Neither have the risks of tinkering with these kinds of traits been assessed. Both these are urgent actions, given the impacts that are already being felt.
- Integrate biodiversity planning into spatial environmental planning: to date much conservation planning has been undertaken in isolation from broader environmental planning (Knight & Cowling, this volume, Chapter 15). New kinds of analyses based around ecosystem service trade-offs allow explicit decisions to be made from the costs and benefits of alternative interventions (Goldman *et al.*, this volume, Chapter 4). Such approaches will become increasingly important to evaluate trade-offs across temporal and spatial scales as well as among ecosystem services.

References

Araújo, M.B., Whittaker, R.J., Ladle, R.J. & Erhard, M. (2005) Reducing uncertainty in projections of extinction risk from climate change. *Global Ecology and Biogeography*, 14, 529–538.

Balmford, A. (1996) Extinction filters and current resilience: the significance of past selection pressures for conservation biology. *Trends in Ecology and Evolution*, 11, 193–196.

Balmford, A., Mace, G.M. & Ginsberg, J.R. (1998) The challenges to conservation in a changing world: putting processes on the map. In *Conservation in a Changing World*, eds G.M. Mace, A. Balmford & J.R. Ginsberg, pp. 1–28. Cambridge University Press, Cambridge.

Bottrill, M.C., Joseph, L.N., Carwardine, J. *et al.* (2008) Is conservation triage just smart decision making? *Trends in Ecology and Evolution*, 23, 649–654.

Bradshaw, W.E. & Holzapfel, C.M. (2007) Genetic response to rapid climate change: it's seasonal timing that matters. *Molecular Ecology*, 17, 157–166.

Brook, B.W., Sodhi, N.S. & Bradshaw, C.J.A. (2008) Synergies among extinction drivers under global change. *Trends in Ecology and Evolution*, 23, 453–460.

Butchart, S.H.M., Stattersfield, A.J. & Collar, N.J. (2006) How many bird extinctions have we prevented? *Oryx*, 40, 266–278.

Cade, T.J. & Burnham, W. (eds) (2003) *Return of the Peregrine: a North American saga of tenacity and teamwork*. Peregrine Fund, Boise, ID.

Cardillo, M., Mace, G.M., Gittleman, J.L. & Purvis, A. (2006) Latent extinction risk and the future battlegrounds of mammal conservation. *Proceedings of the National Academy of Sciences of the USA*, 103, 4157–4161.

Caughley, G. (1994) Directions in conservation biology. *Journal of Animal Ecology*, 63, 215–244.

Carpenter, S.R., Bennett, E.M. & Peterson, G.D. (2006) Scenarios for ecosystem services: an overview. *Ecology and Society*, 11, 29.

Christensen V., Guénette S., Heymans J.J. *et al.* (2003) Hundred-year decline of North Atlantic predatory fishes. *Fish and Fisheries*, 4, 1–24.

Davies, T.J., Fritz, S.A., Grenyer, R. *et al.* (2008) Phylogenetic trees and the future of mammalian biodiversity. *Proceedings of the National Academy of Sciences of the USA*, 105, 11556–11563.

Didham, R.K., Tylianakis, J.M., Gemmell, N.J., Rand, T.A. & Ewers, R.M. (2007) Interactive effects of habitat modification and species invasion on native species decline. *Trends in Ecology and Evolution*, 22, 489–496.

Dulvy, N.K., Sadovy, Y. & Reynolds, J.D. (2003) Extinction vulnerability in marine populations. *Fish and Fisheries*, 4, 25–64.

Fischlin, A. & Midgley, G.F. (2007) Ecosystems, their properties, goods and services. In *Climate Change 2007: impacts, adaptation and vulnerability*, eds M. Parry &

O. Canziani, pp. 215–272. Intergovernmental Panel on Climate Change, Washington, DC.

Foley, J.A., DeFries, R., Asner, G.P. *et al.* (2005) Global consequences of land use. *Science*, 309, 570–574.

Gienapp, P., Teplitsky, C., Alho, J.S., Mills, J.A. & Merila, J. (2007) Climate change and evolution: disentangling environmental and genetic responses. *Molecular Ecology*, 17, 167–178.

Green, R.E., Newton, I., Shultz, S. *et al.* (2004) Diclofenac poisoning as a cause of vulture population declines across the Indian subcontinent. *Journal of Applied Ecology*, 41, 793–800.

Hutchings, J.A. & Reynolds, J.D. (2004) Marine fish population collapses: consequences for recovery and extinction risk. *BioScience*, 54, 297–309.

IUCN (International Union for the Conservation of Nature) (2004) *A Global Species Assessment*. IUCN, Gland, Switzerland.

Jackson, J.B.C., Kirby, M.X., Berger, W.H. *et al.* (2001) Historical overfishing and the recent collapse of coastal ecosystems. *Science*, 293, 629–637.

Jackson, S.T. & Overpeck, J.T. (2000) Responses of plant populations and communities to environmental changes of the late Quaternary. *Paleobiology*, 26, 194–220.

Jetz, W., Wilcove, D.S. & Dobson, A.P. (2007) Projected impacts of climate and land-use change on the global diversity of birds. *PLoS Biology*, 5, 1211–1219.

Kapos, V., Balmford, A., Aveling, R. *et al.* (2008) Calibrating conservation: new tools for measuring success. *Conservation Letters*, 1, 155–164.

Keith, D.A., Akcakaya, H.R., Thuiller, W. *et al.* (2008) Predicting extinction risks under climate change: coupling stochastic population models with dynamic bioclimatic habitat models. *Biology Letters*, 4, 560–563.

Lee, M.S.Y. & Doughty, P. (2003) The geometric meaning of macroevolution. *Trends in Ecology and Evolution*, 18, 263–266.

Lee, T.M. & Jetz, W. (2008) Future battlegrounds for conservation under global change. *Proceedings of the Royal Society of London, Series B*, 275, 1261–1270.

Mace, G.M., Balmford, A. & Ginsberg, J.R. (eds) (1998) *Conservation in a Changing World*. Cambridge University Press, Cambridge.

Mace, G.M., Masundire, H., Baillie, J.E.M *et al.* (2005) Biodiversity. In *Millennium Ecosystem Assessment: current state and trends assessment*, Vol. 1, eds R. Hassan, R. Scholes & N. Ash, pp. 77–122. Island Press, Washington, DC.

McBride, M.F., Wilson, K.A., Bode, M. & Possingham, H.P. (2007) Incorporating the effects of socioeconomic uncertainty into priority setting for conservation investment. *Conservation Biology*, 21, 1463–1474.

MA (Millennium Ecosystem Assessment) (2005) *Ecosystems and Human Well-being: synthesis*. Island Press, Washington, DC.

Moller, A.P., Rubolini, D. & Lehikoinen, E. (2008) Populations of migratory bird species that did not show a phenological response to climate change are declining. *Proceedings of the National Academy of Sciences of the USA*, 105, 16195–16200.

Murdoch, W., Polasky, S., Wilson, K.A. *et al.* (2007) Maximizing return on investment in conservation. *Biological Conservation*, 139, 375–388.

Myers, R.A. & Worm, B. (2005) Extinction, survival or recovery of large predatory fishes. *Philosophical Transactions of the Royal Society of London, Series B*, 360, 13–20.

Newton, I. (2004) The recent declines of farmland bird populations in Britain: an appraisal of causal factors and conservation actions. *Ibis*, 146, 579–600.

Parmesan, C. & Yohe, G. (2003) A globally coherent fingerprint of climate change impacts across natural systems. *Nature*, 421, 37–42.

Pauly, D., Alder, J., Bennett, E., Christensen, V., Tyedmers, P. & Watson, R. (2003) The future for fisheries. *Science*, 302, 1359–1361.

Pauly, D., Christensen, V., Dalsgaard, J., Froese, R. & Torres, F., Jr (1998) Fishing down marine food webs. *Science*, 279, 860–863.

Pauly, D., Christensen, V., Guénette, S. *et al.* (2002) Towards sustainability in world fisheries. *Nature*, 418, 689–695.

Pearson, R.G. & Dawson, T.P. (2004) Bioclimate envelope models: what they detect and what they hide: response to Hampe. *Global Ecology and Biogeography*, 13, 471–473.

Perry, A.L., Low, P.J., Ellis, J.R. & Reynolds, J.D. (2005) Climate change and distribution shifts in marine fishes. *Science*, 308, 1912–1915.

Peterson, G.D., Cumming, G.S. & Carpenter, S.R. (2003) Scenario planning: a tool for conservation in an uncertain world. *Conservation Biology*, 17, 358–366.

Pressey, R.L., Cabeza, M., Watts, M.E., Cowling, R.M. & Wilson K.A. (2007) Conservation planning in a changing world. *Trends in Ecology and Evolution*, 22, 583–592.

Purvis, A., Mace, G.M. & Jones, K.E. (2000) Extinction. *BioEssays*, 22, 1123–1133.

Root, T.L., Price, J.T., Hall, K.R. *et al.* (2003) Fingerprints of global warming on wild animals and plants. *Nature*, 421, 57–60.

Rouget, M., Cowling, R.M., Lombard, A.T., Knight, A.T. & Graham, I.H.K. (2006) Designing large-scale conservation corridors for pattern and process. *Conservation Biology*, 20, 549–561.

Roy, D.B. & Sparks, T.H. (2000) Phenology of British butterflies and climate change. *Global Change Biology*, 6, 407–416.

Sala, O.E., Chapin, F.S., Armesto, J.J. *et al.* (2000) Biodiversity: global biodiversity scenarios for the year 2100. *Science*, 287, 1770–1774.

Salafsky, N., Salzer, D., Stattersfield, A.J. *et al.* (2008) A standard lexicon for biodiversity conservation: unified classifications of threats and actions. *Conservation Biology*, 22, 897–911.

Scheffer, M. & Carpenter, S.R. (2003) Catastrophic regime shifts in ecosystems: linking theory to observation. *Trends in Ecology and Evolution*, 18, 648–656.

Scheffer, M., Carpenter, S., Foley, J.A., Folke, C. & Walker, B. (2001) Catastrophic shifts in ecosystems. *Nature*, 413, 591–596.

Soulé, M.E. (ed.) (1987) *Viable Populations for Conservation*. Cambridge University Press, Cambridge.

Thomas, C.D., Cameron, A., Green, R.E. *et al.* (2004) Extinction risk from climate change. *Nature*, 427, 145–148.

Thuiller, W. (2007) Biodiversity: climate change and the ecologist. *Nature*, 448, 550–552.

Turner, B.L., Kasperson, R.E., Matson, P.A. *et al.* (2003) A framework for vulnerability analysis in sustainability science. *Proceedings of the National Academy of Sciences of the USA*, 100, 8074–8079.

Turvey, S.T., Pitman, R.L., Taylor, B.L., *et al.* (2008) First human-caused extinction of a cetacean species? *Biology Letters*, 3, 537–540.

UNEP (United Nations Environment Programme) (2007) *Global Environment Outlook GEO4*. UNEP, Nairobi.

Verity, P.G., Smetacek, V. & Smayda, T.J. (2002) Status, trends and the future of the marine pelagic ecosystem. *Environmental Conservation*, 29, 207–237.

Vié, J-C., Hilton-Taylor, C. & Stuart, S.N. (eds) (2009) *Wildlife in a Changing World – an analysis of the 2008 IUCN Red List of Threatened Species*. International Union for the Conservation of Nature, Gland, Switzerland.

Visser, M.E. & Holleman, L.J.M. (2001) Warmer springs disrupt the synchrony of oak and winter moth phenology. *Proceeding of the Royal Society of London, Series B*, 268, 289–294.

Williams, J.W., Jackson, S.T. & Kutzbacht, J.E. (2007) Projected distributions of novel and disappearing climates by 2100 AD. *Proceedings of the National Academy of Sciences of the USA*, 104, 5738–5742.

Wilson, K.A., McBride, M.F., Bode, M. & Possingham, H.P. (2006) Prioritizing global conservation efforts. *Nature*, 440, 337–340.

Wilson, K.A., Underwood, E.C., Morrison, S.A. *et al.* (2007) Conserving biodiversity efficiently: what to do, where, and when. *PLoS Biology*, 5, 1850–1861.

Xu, J.C. & Wilkes, A. (2004) Biodiversity impact analysis in northwest Yunnan, southwest China. *Biodiversity and Conservation*, 13, 959–983.

(20)

Another Entangled Bank: Making Conservation Trade-offs More Explicit

*Robert J. Smith[1], William M. Adams[2]
and Nigel Leader-Williams[1]*

[1]Durrell Institute of Conservation and Ecology, University of Kent,
Canterbury, UK
[2]Department of Geography, University of Cambridge,
Cambridge, UK

Introduction

As Charles Darwin (1859) noted, the natural world is a complex and highly interconnected place (Figure 20.1). Conservation researchers have catalogued this complexity and studied the cascading impacts of species and habitat loss on remaining biodiversity (Karieva & Levin, 2003), and this biological focus has produced important insights. However, this work alone is insufficient for addressing the current extinction crisis because conservation planning and decision making is also a highly entangled process. This book has explored the trade-offs that conservationists make and has emphasized that these interconnections extend far into the political, social and economic realm. The chapters in this book have shown that trade-offs are a ubiquitous feature of conservation planning and action. In many instances trade-offs are unavoidable, and very often they are unconscious, hidden from view and not subject to technical or policy debate. This tends to have serious implications. It affects the efficiency of conservation, i.e. the ratio of effort

Trade-offs in Conservation: Deciding What to Save, 1st edition.   Edited by N. Leader-Williams,
W.M. Adams and R.J. Smith.   © 2010 Blackwell Publishing Ltd.

Figure 20.1 **The final paragraph of Charles Darwin's *On the Origin of Species* described 'an entangled bank' containing species that were 'different from each other, and dependent on each other in so complex a manner'. Part of Darwin's inspiration came from Downe Bank, shown here, close to his home in Kent. (Photograph by Nigel Leader-Williams.)**

to outcome and the effectiveness of policies and programmes. It also affects the acceptability of conservation actions among conservationists themselves, and more importantly among the wider publics with whom conservationists need to engage. The existence of such trade-offs is obviously important for conservation, but their significance is even more serious when they are not addressed explicitly. Blindness to trade-offs risks alienating both supporters and those groups whose goodwill is critical, such as protected area neighbours, resource users, government authorities or politicians.

So trade-offs in conservation are everywhere, and they matter. But how should conservationists deal with them? Here, we will review some of the issues raised in the preceding chapters, examine the factors that influence these trade-offs and discuss whether conservationists can improve their decision-making processes in the future. We will do this by first arguing that trade-offs in conservation can be particularly severe, both because of where conservation projects take place and because much of this work lacks appropriate support. We will then briefly describe the toolkits that are currently available to make these decisions easier and discuss whether they could be more widely adopted. Finally, we will highlight the importance of widening support for conservation and ensuring that decision making takes place at a less centralized level, as this should help make managing trade-offs easier by better aligning conservation activities with societal goals.

Why are trade-offs in conservation so severe?

Current biodiversity loss is still largely driven by habitat transformation and over-harvesting, and addressing these threats generally involves: (i) restricting access by creating protected areas of various kinds; and (ii) changing exploitation patterns by passing laws and regulations (Leader-Williams *et al.*, this volume, Chapter 1). Many other sectors, such as transport and energy, have to resolve trade-offs resulting from competing demands for space and resources. However, there are two reasons for expecting conservation to be particularly impacted by trade-off decisions. The first relates to where conservationists work, as countries with the highest levels of biodiversity are generally those with the most difficult operating conditions. For example, global priority regions for conservation tend to be found in countries with the highest levels of poverty and corruption (O'Connor *et al.*, 2003; Smith *et al.*, 2003; Roe & Walpole, this volume, Chapter 9) and many of these countries lack appropriate capacity and funding (Balmford & Whitten, 2003). Moreover, most important biodiversity is found in relatively remote or inaccessible areas, which can increase costs, make staff retention difficult and hamper monitoring (Smith & Walpole, 2005). At a more extreme level, important biodiversity is often found in places affected by conflict, which exacerbates many of these problems and creates new ones (Hanson *et al.*, 2009; Aveling *et al.*, this volume, Chapter 14).

There is, however, another fundamental reason for the prevalence of trade-offs in conservation. This stems from the conservation ethic being underpinned by value judgments, which are largely derived from a personal appreciation of nature. Thus, different interest groups find it difficult to agree on a single definition of conservation importance or success because people have diverse positions, priorities and interests. This diversity of opinions means that different conservation objectives are perfectly possible, so trade-offs occur because management choices foreclose other outcomes. This creates tension within the conservation community, with the most obvious impact affecting which species become conservation priorities (Samways, this volume, Chapter 6). More subtly, it can also affect which types of conservation management are proposed and supported. For example, conflict can occur over projects that fund conservation through sustainable use of charismatic species (Rosser & Leader-Williams, this volume, Chapter 8) or have implications for animal welfare (Harrop, this volume, Chapter 7). These tensions are the subject of much debate within the conservation community as they impact how effectively conservation goals can be achieved. However, such issues are largely dwarfed by trade-offs between conservation and other sectors, which tend to influence the extent of these goals. Many people consider investing in biodiversity as a low priority, especially when conservation activities impact their livelihoods. Thus, poor people often prefer projects that provide jobs at the expense of biodiversity, and many governments see natural habitats as untapped resources ripe for exploitation (Pulgar-Vidal *et al.*, this volume, Chapter 13).

Finally, the personal aspect of valuing biodiversity means that people are most influenced by their direct experiences of nature. Some people support the conservation of species and habitats that they have only seen in zoos or on screen, but the majority of funds for conservation are raised through taxation and donation and are spent locally or nationally (Halpern, 2006). This spatial pattern has serious implications. Most of the globally important conservation areas are found in developing countries (Balmford & Whitten, 2003), where people are generally more concerned about economic development (Roe & Walpole, this volume, Chapter 9) and where funding for conservation through taxation is far from adequate (Bruner *et al.*, this volume, Chapter 11). Thus, most funding in these countries comes from international sources, which causes a whole new set of problems and trade-offs. The most obvious occur because conservation policy and practice is driven by outsiders (Brosius, this volume, Chapter 17), which at worst leads to small but vocal groups of foreigners advocating simplistic conservation actions in far-off countries (Smith *et al.*, this volume, Chapter 12). However, even mainstream projects tend to neglect or misrepresent local interests and find it difficult to fully integrate local traditions and local power structures into the decision-making process (Homewood, this volume, Chapter 10; Brosius, this volume, Chapter 17). Moreover, the politics and practicalities of dealing with international donors results in generic projects that fail to account for local conditions (Adams, this volume, Chapter 16) and creates perverse incentives that encourage superficial outcomes (Bruner *et al.*, this volume, Chapter 11).

Improving the decision-making process: toolkits and beyond

Conservation decisions that involve different interested parties or limited resources will always involve trade-offs (Bottrill *et al.*, 2008). However, past research on these issues was largely limited to case studies that described the decision-making process and its impacts on particular social groups. In recent years this focus has changed and there is more interest in developing analytical approaches that aim to improve decision making (Ferraro & Pattanayak, 2006), rather than simply document and criticize. A key part of this is measuring conservation success, as it is only by knowing what works and in what context that conservationists can make choices between different approaches in different conditions (Kapos *et al.*, this volume, Chapter 5). Many of these breakthrough analyses have focused on the efficacy of different management techniques, such as the effectiveness of methods for increasing the abundance of threatened invertebrates (Davies *et al.*, 2008). However, recent work has started to focus on broader conservation projects. For example, a recent study showed that spending US$1.5 million on village development schemes did nothing to reduce illegal deforestation around a national park in Indonesia (Linkie *et al.*, 2008).

Developing such measures of success is also part of a broader process to identify trade-offs between different goals. For example, research on maintaining ecosystem services has moved from sterile debates over whether this approach is good or bad for biodiversity conservation to studies that identify when and where there are synergies (Goldman *et al.*, this volume, Chapter 4). Moreover, producing these data also allows much more sophisticated approaches to improving the decision-making process (Wilson *et al.*, this volume, Chapter 2). Although such approaches may be inappropriate in some situations, they have already been used to help guide decisions on where conservation activities should be focused (Murdoch *et al.*, this volume, Chapter 3), as well as when they should be scheduled (Wilson *et al.*, 2007). They are also increasingly being used to look at trade-offs between the costs of collecting data and the extent to which this improves decision making (Grantham *et al.*, 2008). In fact, one key benefit of all these new techniques is that they force conservationists to think more deeply about what they are trying to achieve (Kapos *et al.*, this volume, Chapter 5). Many authors have noted that conservation science is poor at measuring and understanding success (Sutherland *et al.*, 2004). This arises in part from a lack of funding and the short-term mentality that comes from working in a 'crisis discipline'. However, this is changing, and anecdotal evidence suggests these tools encourage people to think more strategically and be more transparent about their aims.

Widening the discussion

Trade-off optimization approaches aim to improve efficiency, transparency and encourage self-reflection, but their effectiveness depends on several other factors. The first is that they are data hungry and this can be problematic. For example, worryingly little is still known about the state of the planet and this impacts decision making (Walpole *et al.*, 2009). However, research has shown that reasonable decisions can often be made with relatively cheap sources of information (Wilson *et al.*, this volume, Chapter 2), so conservation is more often hampered by a lack of resources for targeting and implementing action than a lack of data *per se* (Cowling *et al.*, 2010). A larger problem arises from data uncertainty, which is exacerbated by issues such as climate change (Willis *et al.*, this volume, Chapter 18) and human population growth (Mace, this volume, Chapter 19), where it is ever more necessary to make predictions based on extrapolations outside the current range of data. However, attempts to understand these issues are becoming increasingly sophisticated and have pioneered the use of scenario testing, horizon scanning and other techniques that could be used more widely in conservation (Mace, this volume, Chapter 19). This research also highlights the need to classify the drivers of change, so that decision makers understand the trade-offs their decisions involve for species and ecosystems.

The second factor that could hamper these approaches is the extent to which conservationists can conceptualize the decision-making process. Conservation researchers have traditionally focused on collecting and analyzing biological data, so their recognition that other types of information should be included in decision making has been relatively recent. This new approach has had a particular focus on using financial data, as this relates directly to conservation budgets and the interests of potential donors. However, recent research has highlighted the need for data on conservation opportunities and stakeholder attitudes (Knight & Cowling, this volume, Chapter 15). Indeed, this book has shown that trade-offs come at many levels and involve many issues. This means that a much more sophisticated approach is needed if conservationists are to understand and model the decision-making process adequately.

The first step in this process is to recognize that all conservation decisions are inherently political. This is most obvious when conservationists choose to work in countries that are criticized by the international community (Graham-Rowe, 2005) or work with extractive industries, such as mining or hydrocarbons (Pulgar-Vidal *et al.*, this volume, Chapter 13). However, even low-profile decisions have political implications, as they influence where money is spent and who is affected. Therefore, any study of trade-offs needs to take a multidisciplinary approach, and involve an awareness of the long-term and complex effects that such decisions can influence. In particular, there is a need to work with implementation agencies, to understand the background to the decision-making process (Smith *et al.*, 2009), and other sectors and groups, to have a clearer understanding of the broader impacts. More generally, conservationists need to acknowledge and spread appreciation of the social underpinnings of conservation action. They need to understand what trade-offs people would support. In turn, this requires conservationists to be open to debates on these different issues, and not to treat those who disagree with their goals as foolish or selfish.

All of this suggests there is much scope for improving how conservationists deal with trade-offs, but there is still debate over the extent to which this can be studied using optimization-based methods. Most current research has used relatively simple scenarios to test these approaches, whereas real world decision making often needs data on a much wider range of issues (McDonald, 2009). This can be resolved in some cases by measuring these new factors, but there are three aspects that need further investigation before this approach receives widespread support. The first is whether including a whole new set of factors, some of which are likely to be measured imprecisely, affects stakeholder support for the quality of the results. There is already anecdotal evidence that stakeholders are sceptical of priority area-setting exercises that identify even a small number of sites that are known to be unsuitable (Smith *et al.*, 2006), although this can be overcome by explaining data limitations more clearly and using stakeholder feedback to improve results. The second is whether the financial and bureaucratic costs of collecting and analyzing these data justify the resultant improvements in the decision-making process.

The third problem is more profound and arises when the decision-making process involves weighting different factors. For example, it is now common for spatial conservation planning assessments to include opportunity cost data, as this helps minimize the impacts of establishing new conservation areas on other sectors (Naidoo *et al.*, 2006). Moreover, work has shown that these economic costs tend to vary more than biodiversity value and so have a large influence on the results (Bode *et al.*, 2008). However, the financial value of lost opportunities can be a very crude measure of impact, especially when dealing with different stakeholders. Part of the problem arises from the difficulty of measuring economic values beyond direct consumptive use, such as existence or bequest values. There are serious technical issues of measurement here, which economists are addressing (Turner *et al.*, 2003). A bigger problem is that financial value of a resource is often a weak measure of the extent to which people depend on it for their livelihoods. Thus, care is needed when weighting the value of these natural resources if conservation activities are to avoid further impacting already marginalized people (Adams *et al.*, 2010).

The importance of weighting these factors appropriately again highlights the need to involve social scientists in capturing this information and to use scenario-based techniques that explore how different groups think natural resources should be used in the future. Of course, providing such transparency can be uncomfortable for decision makers who want to hide their objectives, either for selfish reasons or because they fear that powerful interest groups will find it easier to lobby for maintaining the *status quo* (Possingham, 2009). Thus, there is also a need for more research on the political implications of using these decision-making tools, especially as their well-tested scientific underpinnings appear to give them greater legitimacy than other techniques (Brosius, this volume, Chapter 17). This will help identify the circumstances in which optimization approaches are likely to be adopted and their results accepted, and when the system is likely to produce perverse outcomes, either because of lack of capacity or political interference.

Widening support

We have argued that conservationists need to have a much broader under-standing of the impacts of their work and that decision-making tools can help them understand these issues and make more informed choices. However, another way to reduce these problems is to minimize how many trade-offs need to be made. But is this possible? Well it is certainly true that some conservationists deal with far fewer trade-offs when making decisions. For example, the New Zealand Department of Conservation has used an optimization approach to help prioritize between funding recovery programmes for their different threatened species, based on the costs, benefits and chances of success of each project (Joseph *et al.*, 2009). They are able to do this because they have a clear government mandate and a budget for implementation, raised through taxation. Obviously, the size of this budget reflects a series of political trade-offs but these decisions were made by democratically elected officials, who respond to the priorities of their electorate. Thus, allowing for the limitations inherent in any political system, it is the people of New Zealand who pay for and ultimately decide which of their species should receive funding. This avoids some of the trade-offs we have encountered in this book and produces results with a greater level of legitimacy.

The obvious problem with this approach is that New Zealand is a rich country and very few other countries could fund similar schemes through taxation. One partial solution is for conservationists in every nation to widen donor support and encourage the mainstreaming of biodiversity projects (Goldman *et al.*, this volume, Chapter 4; Knight & Cowling, this volume, Chapter 15), so that every sector aims to produce benefits for biodiversity conservation. Such an expansion in support will reduce reliance on donors who are less sympathetic to local conditions and increase legitimacy more generally. In addition, conservation projects that earn revenue should be encouraged, as their ability to raise funds makes then much less dependent on appeasing remote donors or pressure groups. Another potential solution is to provide funds and support for conservation agencies in developing countries directly, so that they can develop and implement their own prioritization schemes (Smith *et al.*, 2009). Such an approach would need a large capacity building component (Rodríguez *et al.*, 2006) and would require effective monitoring that avoids the naïve assumptions made by some previous directly funded projects. However, such efforts would help produce results with a similar sense of national legitimacy, as the decision-making process is much more likely to account for the views of local stakeholders than those led by international organizations.

Some conservationists might argue that shifting the focus to local priority setting could neglect globally important biodiversity, but this seems unlikely. Most countries have signed up to the Convention on Biological Diversity and are committed to maintain their globally threatened biodiversity (Rosser &

Leader-Williams, this volume, Chapter 8). Thus, current conservation problems largely result from funding limitations and lack of local support for conservation actions agreed at the global or regional scale. Moreover, even if the priority projects identified by national conservation agencies do not exactly match those of different interest groups, then this could be easily resolved. Any prioritization scheme can be adjusted to account for available funding, so that donors are matched with priority projects that also meet their requirements (Knight *et al.*, 2007; Smith *et al.*, 2009).

Final conclusions

The history of conservation practice is littered with projects and policies that failed to explicitly consider the relevant trade-offs. As a result, conservationists have often alienated potential supporters and made life more difficult for themselves and others. Moreover, this blinkered approach often led to the conservation community missing out on opportunities to improve their effectiveness and efficiency, further marginalizing their impact. This is beginning to change and there is a particular focus on developing new tools to improve effectiveness and to guide decision making, although there are still uncertainties about when and where they are appropriate. However, at the heart of these new developments is the message that conservationists need to define their goals, recognize the constraints and consider the implications of their work. Just as importantly, they also need to be honest and open about success and failure. This change of approach will require changes in how conservationists perceive themselves. The conservation community should no longer aspire to being plucky underdogs armed only with the best of intentions. Instead conservationists should recognize the broad implications of their work, and the importance to humankind of the challenge they face to conserve biodiversity, and plan accordingly. Moreover, conservationists should seek to incorporate their work into recognized state sectors, so that conservation issues become an accepted part of debates over national social policy. Making such changes will not be easy, but overcoming these problems will probably be among the key challenges that conservationists face during the 21st century.

References

Adams, V.M., Pressey, R.L. & Naidoo, R. (2010) Opportunity costs: who really pays for conservation? *Biological Conservation*, 143, 439–448.

Balmford, A. & Whitten, T. (2003) Who should pay for tropical conservation, and how could the costs be met? *Oryx*, 37, 238–250.

Bode, M., Wilson, K., Brooks, T. *et al.* (2008) Cost-effective global conservation spending is robust to taxonomic group. *Proceedings of the National Academy of Sciences of the USA*, 105, 6498–6501.

Bottrill, M.C., Joseph, L.N., Carwardine, J. *et al.* (2008) Is conservation triage just smart decision making? *Trends in Ecology and Evolution*, 23, 649–654.

Cowling, R.M., Knight, A.T., Privett, S.D. J. & Sharma, G.P. (2010) Invest in opportunity, not inventory of hotspots. *Conservation Biology*, 24, 633–635.

Darwin, C. (1859) *On the Origin of Species.* John Murray, London.

Davies, Z.G., Tyler, C., Stewart, G.B. & Pullin, A.S. (2008) Are current management recommendations for saproxylic invertebrates effective? A systematic review. *Biodiversity and Conservation*, 17, 209–234.

Ferraro, P.J. & Pattanayak, S.K. (2006) Money for nothing? A call for empirical evaluation of biodiversity conservation investments. *PLoS Biology*, 4, 482–488.

Graham-Rowe, D. (2005) Conservation in Myanmar: under the gun. *Nature*, 435, 870–872.

Grantham, H.S., Moilanen, A., Wilson, K.A., Pressey, R.L., Rebelo, T.G. & Possingham, H.P. (2008) Diminishing return on investment for biodiversity data in conservation planning. *Conservation Letters*, 1, 190–198.

Halpern, B.S. (2006) Gaps and mismatches between global conservation priorities and spending. *Conservation Biology*, 20, 56–64.

Hanson, T., Brooks, T.M., da Fonseca, G.A. B. *et al.* (2009) Warfare in biodiversity hotspots. *Conservation Biology*, 23, 578–587.

Joseph, L.N., Maloney, R.F. & Possingham, H.P. (2009) Optimal allocation of resources among threatened species: a project prioritization protocol. *Conservation Biology*, 23, 328–338.

Karieva, P. & Levin, S.A. (eds) (2003) *The Importance of Species: perspectives on expendability and triage.* Princeton University Press, Princeton, NJ.

Knight, A.T., Smith, R.J., Cowling, R.M. *et al.* (2007) Improving the Key Biodiversity Areas approach for effective conservation planning. *BioScience*, 57, 256–261.

Linkie, M., Smith, R.J., Zhu, Y. *et al.* (2008) Evaluating biodiversity conservation around a large Sumatran protected area. *Conservation Biology*, 22, 683–690.

McDonald, R.I. (2009) The promise and pitfalls of systematic conservation planning. *Proceedings of the National Academy of Sciences of the USA*, 106, 15101–15102.

Naidoo, R., Balmford, A., Ferraro, P.J., Polasky, S., Ricketts, T.H. & Rouget, M. (2006) Integrating economic costs into conservation planning. *Trends in Ecology and Evolution*, 21, 681–687.

O'Connor, C., Marvier, M. & Kareiva, P. (2003) Biological vs social, economic and political priority-setting in conservation. *Ecology Letters*, 6, 706–711.

Possingham, H.P. (2009) HUGHBRIS of the decision theorists. *Decision Point*, 27, 2–3.

Rodríguez, J.P., Rodríguez-Clark, K.M., Oliveira-Miranda, M.A., Good, T. & Grajal, A. (2006) Professional capacity building: the missing agenda in conservation priority setting. *Conservation Biology*, 20, 1340–1341.

Smith, R.J. & Walpole, M.J. (2005) Should conservationists pay more attention to corruption? *Oryx*, 39, 251–256.

Smith, R.J., Goodman, P.S. & Matthews, W.S. (2006) Systematic conservation planning: a review of perceived limitations and an illustration of the benefits using a case study from Maputaland, South Africa. *Oryx*, 40, 400–410.

Smith, R.J., Muir, R.D. J., Walpole, M.J., Balmford, A. & Leader-Williams, N. (2003) Governance and the loss of biodiversity. *Nature*, 426, 67–70.

Smith, R.J., Veríssimo, D., Leader-Williams, N., Cowling, R.M. & Knight, A.T. (2009) Let the locals lead. *Nature*, 462, 280–281.

Sutherland, W.J., Pullin, A.S., Dolman, P.M. & Knight, T.M. (2004) The need for evidence-based conservation. *Trends in Ecology and Evolution*, 19, 305–308.

Turner, R.K., Paavola, J., Cooper, P., Farber, S., Jessamy, V. & Georgiou, S. (2003) Valuing nature: lessons learned and future research directions. *Ecological Economics*, 46, 493–510.

Walpole, M.J., Almond, R.E. A., Besançon, C. *et al.* (2009) Tracking progress toward the 2010 biodiversity target and beyond. *Science*, 325, 1503–1504.

Wilson, K.A., Underwood, E.C., Morrison, S.A. *et al.* (2007) Conserving biodiversity efficiently: what to do, where, and when. *PLoS Biology*, 5, 1850–1861.

Index

Note: page numbers in *italics* refer to figures, those in **bold** refer to tables and boxes. Page numbers with suffix 'n' refer to notes